THE ECHINODERMS OF SOUTHERN AFRICA

THE ECHINODERMS OF SOUTHERN AFRICA

BY

AILSA M. CLARK

AND

JANE COURTMAN-STOCK

BRITISH MUSEUM (NATURAL HISTORY)
LONDON : 1976

BMNH 66·75 0·5m 7/76

Publication No. 776

ISBN 0 565 00776 9

Printed in Great Britain by John Wright and Sons Ltd. at The Stonebridge Press, Bristol BS4 5NU

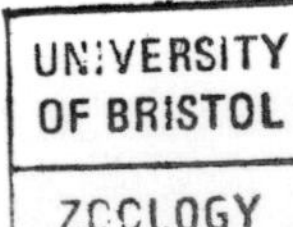

CONTENTS

INTRODUCTION

FOLLOWING the extended surveys of the echinoderm fauna of southern Africa by H. L. Clark (1923) and Mortensen (1933a, of the Asterozoa) only a small number of new taxa have been discovered despite the intensive collections made by the University of Cape Town's Ecological Survey under T. A. Stephenson and J. H. Day over more than thirty years. The time therefore seems right to summarize our knowledge of the echinoderm fauna of this area (except for the holothurians) on the lines used by Professor Day for the polychaetes, although on a smaller scale, the number of species involved being more modest.*

Although Day took as his northern limit the twentieth parallel South, as mentioned in the introductory paper (Clark, 1974) we are using the Tropic of Capricorn (*c.* 23½°S) which excludes the northern half of South-West Africa and the northern three-quarters of Mozambique. Although this eliminates a few West African species considered as not properly belonging to the southern African fauna, it still leaves in a number of tropical Indo-West Pacific species which are carried by the south-west-bound Mozambique–Natal–Agulhas current to southern Mozambique and some also to Natal or even further beyond the limit of reef corals at Delagoa Bay (Lourenço Marques).

The number of species included is *c.* 280, considerably less than the 800-plus polychaetes. Of these, about 48% appear to be endemic to South Africa and nearly 30% are also found in either the whole tropical Indo-West Pacific or the western part of the Indian Ocean, compared with under 10% for various parts of the Atlantic (mainly the north-east), the remainder being known also from the Southern Ocean, deep Indian Ocean, Australasia, or are cosmopolitan (a very small component, less than ten species).

There have been some differences of opinion as to the best way of subdividing the southern African marine fauna into provinces. Day distinguished a Mozambique–Madagascar Province extending south to Delagoa Bay just north of the Mozambique border, a Natal Province extending south to Bashee River (*c.* 32°S) and a Cape and South-west African Province. Ekman (1953, *Zoogeography of the Sea*, London), following Stephenson, recognized a south coast province from Algoa Bay west to Cape Agulhas or Cape Point distinct from that of the west coast to the north of Cape Point and separated from the tropical Mozambique fauna by a transitional zone in Natal. Out of nearly 100 echinoderm species endemic to South Africa, 40 range on both sides of False Bay where the water masses mix to varying extents according to season; 32 others are restricted to the southern coast between Cape Hangklip (on the east side of False Bay) and the vicinity of the Haven, just north of Bashee River, of which six extend into False Bay; while 23 are restricted to the southern end of the area covered by the cold northbound Benguela current limited by Cape Point and (arbitrarily) the Tropic of Capricorn on the west coast, of which

* Unfortunately some additional material collected in 1975 from deep-water off Natal sent by Dr N. A. H. Millard of the South African Museum proves to include an asteroid (*Solaster* sp.) and at least one species of Ophiacanthid (*Ophiotreta matura*) new to the fauna of southern Africa, previously known from the East Indian area.

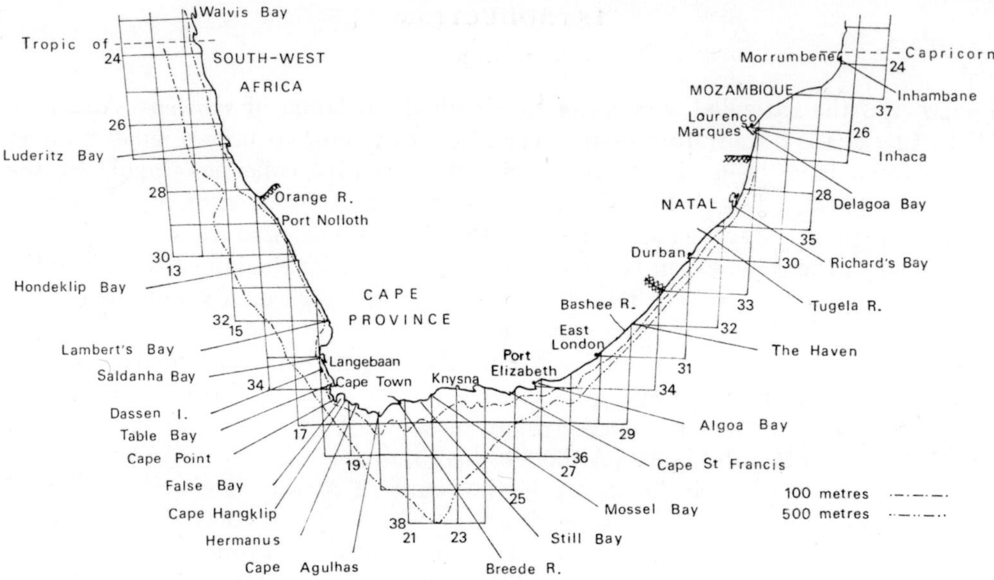

Outline map of southern Africa with latitude/longitude squares adjoining. The even numbers represent latitude south and the odd ones longitude east.

two extend into False Bay ; two species are so far only known from the Bay itself. In order to illustrate the varying distributions of the species involved I propose to break down the region into nine areas limited by points along the coast and treat them in tabular form. In the treatment of the individual species, Day's practice of expressing distribution in latitude/longitude degree squares is followed with the depths given as : i – intertidal ; s – shallow, 0–99 metres ; d – deep, 100–499 metres ; and vd – very deep, 500-plus metres. The squares corresponding to the areas are :

1. *Luderitz Bay area:* Tropic of Capricorn (*c.* 23½°S) to Port Nolloth (*c.* 29°S) ; 23/13, 23/14, 24/13, 24/14, 25/13, 25/14, 26/13, 26/14, 26/15, 27/13, 27/14, 27/15, 28/13, 28/14, 28/15, 28/16, 29/13, 29/14, 29/15, 30/14.
2. *Lambert's Bay area:* Port Nolloth to north of Saldanha Bay (33°S) ; 29/16, 29/17, 30/15, 30/16, 30/17, 31/15, 31/16, 31/17, 31/18, 32/16, 32/17, 32/18.
3. *Cape Town area:* Saldanha Bay to Cape Point (*c.* 34½°S) ; 33/16, 33/17, 33/18, 34/17, 34/18 (except FB).
4. *False Bay:* Cape Point to Cape Hangklip (*c.* 19°E) ; 34/18/FB.
5. *Cape Agulhas area:* Cape Hangklip to Cape Infanta (*c.* 21°E) ; 34/19, 34/20, 35/18, 35/19, 35/20.
6. *Mossel Bay–Knysna area:* Cape Infanta to Cape St Francis (*c.* 25°E) ; 34/21, 34/22, 34/23, 34/24, 35/21, 35/22, 35/23, 35/24.
7. *Port Elizabeth area:* Cape St Francis to The Haven (*c.* 32°S, 29°E) ; 35/25, 35/26, 34/25, 34/26, 34/27, (34/28),* 33/25, 33/26, 33/27, 33/28, 33/29, 32/28.

* Degree squares in brackets are those for which, in practice, no records have so far been made.

8. *Durban area:* The Haven to Kosi Bay on the Natal–Mozambique border (*c.* 27°S) ; 32/29, (32/30), 31/29, 31/30, (31/31), 30/30, 30/31, (30/32), 29/31, 29/32, (29/33), 28/32, 28/33, (28/34), 27/32, 27/33.
9. *Lourenço Marques area:* Mozambique border to the Tropic of Capricorn ; 27/34, 26/32,* 26/33,* 26/34, (26/35), 25/32, 25/33, 25/34, 25/35, (25/36), 24/34, 24/35, 24/36, 23/35, 23/36.

In the text following the distribution table for each class or subclass, diagnoses are only given for genera where more than one species is included. The characters given in the diagnoses are to some extent limited to those found in the species represented. For more comprehensive diagnoses of crinoids and echinoids, the monographs of A. H. Clark and Mortensen, respectively, and for asteroids and ophiuroids the appropriate section in the *Treatise on Invertebrate Paleontology* by Spencer & Wright should be consulted.

The genera and species are treated in alphabetical order under each family.

In the synonymies original references to the species-group taxa and a few other references are given in detail only when they occur in works of a general nature not dealing particularly with the southern African fauna. This avoids both overmuch repetition and the inclusion in the lists of references of too many largely irrelevant works.

Some of the record data are derived from the Survey's lists and are based on material not seen by us but in most cases the records have been checked personally.

Where Mortensen or H. L. Clark have treated a taxon as a variety but it is also distinguished geographically, it is here classed as a subspecies. Non-geographical infraspecific taxa are classed as forms.

In some of the numerical specifications for the species preceding the diagnoses, counts for a single specimen are included in brackets.

The junior author (J. C.-S.) did many of the preliminary identifications of the U.C.T. material, also roughed out diagnoses for the species and prepared the figures for the keys. The final text is the responsibility of A. M. C.

* Strictly speaking, the records for Inhaca Island should be cited as 26/32 but it is so close to the 33° meridian that this combination seems preferable.

Class *CRINOIDEA*

INTRODUCTION. For a detailed study of the recent taxa of this class the monograph of A. H. Clark (1931, 1941, etc.), and Gislén's paper (1938) on the crinoids of South Africa, should be consulted.

The following terms are used in the descriptions of crinoids and may need clarification :

Basals – the ring of interradially aligned plates above the uppermost columnal of a stalked crinoid, or the stalked pentacrinoid larval comatulid (or the centrodorsal of a few adult comatulids) and below the ring of radials, together with which they form the *calyx*, usually numbering five, sometimes fused together but often reduced and completely lost to external view at an early stage of growth in the great majority of comatulids.

Brachials – the calcareous ossicles of the undivided arms, designated Br_1, Br_2, etc.

Centrodorsal – the large plate in the centre of the lower (dorsal or aboral) side, usually bearing jointed *cirri*, except on its apex or dorsal pole, for attachment to the substrate.

Division series – the ossicles between the radials and the first brachials of the undivided arms, rarely absent when the arms arise directly from the radials, usually numbering two or four, the distalmost ossicle of each series an *axillary* and bearing two arms or further division series. The first series is designated IBr with its two ossicles IBr_1 and IBr_2 ; any subsequent series are numbered IIBr, and so on.

Pinnules – the slender jointed appendages arising from the brachials on alternate sides of the arms and bearing extensions of the ambulacral furrow, except for the proximalmost *oral pinnules* which are otherwise modified ; these are followed by *genital pinnules* bearing the gonads and then distal pinnules for food-gathering. The pinnules on the outer (interradial) side of the arm are designated P_1, P_2, etc. and those on the inner side P_a, P_b, etc.

Radials – the ring of usually five plates from which the division series (or arms) arise, only narrowly visible superficially in adults of most comatulid species though conspicuous in many stalked crinoids.

Sacculi – small, rounded, serially arranged sacs along the ambulacra of pinnules, arms and disc, often dark-coloured in preserved specimens, absent in the family Comasteridae.

Synarthry – a ligamentary joint with a large fossa each side of a median vertical ridge which may be extended in some species by median swelling of the dorsal face of both ossicles into a *synarthrial tubercle* ; the two ossicles immediately following the radials and axillaries usually have synarthries between them.

Syzygy – a ligamentary inflexible breaking-joint occurring at intervals along the division series and arms, with radially ridged joint-faces externally appearing discontinuous or undulating, designated by a plus sign.

DISTRIBUTION TABLE FOR CRINOIDEA

	Luderitz Bay area	Lambert's Bay area	Cape Town area	False Bay	Cape Agulhas area	Mossel Bay–Knysna area	Port Elizabeth area	Durban area	Lourenço Marques area	Other localities
BATHYCRINIDAE										
Democrinus chuni (Döderlein)	..	..	vd	..	..	..	..	d vd	..	East Africa, SW. Indian Ocean ridge
Monachocrinus perrieri (Koehler & Vaney)	..	..	vd	..	..	..	..	..	..	Morocco
COMASTERIDAE										
Comanthus wahlbergi wahlbergi (Müller)	..	..	..	i s	i	i s	i s	i s	..	Vema Seamount
Comanthus wahlbergi forma *multibrachia* Gislén	..	..	..	i	..	..	..	..	..	
Comanthus wahlbergi serratus Gislén	..	..	..	..	..	..	s d	d	..	
Comatella africana Gislén	..	..	..	..	..	..	..	d	..	
MARIAMETRIDAE										
Dichrometra afra A. H. Clark	..	..	..	..	..	..	..	d	..	Madagascar, Kenya
HIMEROMETRIDAE										
Heterometra delagoae Gislén	..	..	..	..	..	..	..	..	s	
COLOBOMETRIDAE										
Decametra durbanensis A. M. Clark	..	..	..	..	..	..	..	s	..	
Embryometra mortenseni Gislén	..	..	d	..	..	s	d	..	..	

Distribution Table for Crinoidea (*cont*)

	Luderitz Bay area	Lambert's Bay area	Cape Town area	False Bay	Cape Agulhas area	Mossel Bay–Knysna area	Port Elizabeth area	Durban area	Lourenço Marques area	Other localities
Gislenometra perplexa A. H. Clark	..	..	..	..	..	..	s d vd	d	..	
Oligometra serripinna occidentalis A. H. Clark	..	..	..	..	..	..	..	s	..	Red Sea, South Arabia, Mauritius
TROPIOMETRIDAE										
Tropiometra carinata carinata (Lamarck)	..	..	i ?	s	..	s	s	s	i	Tropical Atlantic, St Helena, western Indian Ocean
Tropiometra magnifica A. H. Clark	..	..	..	..	..	..	..	..	d	East Africa, Gulf of Aden
THALASSOMETRIDAE										
Crotalometra magnicirra (Bell)	..	..	..	..	..	..	– vd	s d	..	SW. Indian Ocean ridge
CHARITOMETRIDAE										
Glyptometra sclateri (Bell)	..	..	..	..	..	..	vd	..	..	SW. Indian Ocean ridge
ANTEDONIDAE										
Annametra occidentalis (A. H. Clark)	..	..	s	i s	..	i	..	..	..	
Cyclometra multicirra A. H. Clark	d	..	..	..	..	..	..	..	..	
PENTAMETROCRINIDAE										
Pentametrocrinus varians (P. H. Carpenter)	..	..	vd	..	..	..	..	..	..	Maldive Is – S. Japan

The number of species included is 17, compared with 12 by Gislén (for the area between Saldanha and Delagoa Bays). The additional ones are: *Decametra durbanensis* A. M. Clark, *Oligometra serripinna occidentalis* A. H. Clark, *Gislenometra perplexa* A. H. Clark, *Tropiometra magnifica* A. H. Clark and *Cyclometra multicirra* A. H. Clark. (*Tropiometra clarki* Gislén has been reduced to subspecific rank but South African records are currently treated as belonging to *T. carinata carinata*.)

Key to the Crinoidea

1 A stalk present in the adult. BATHYCRINIDAE 2

– Adults without stalk, attaching by a cluster of cirri 3

2 Only one or two of the topmost stalk segments or columnals shorter than broad, the other columnals elongate; calyx with the five basals distinct and markedly longer than the radials (Fig. 1); arms simple, arising direct from the radials (often lost) . . . . . . ***Democrinus chuni*** (Döderlein, 1907) (p. 10)

– Fifteen to twenty proximal (upper) columnals shorter than broad; basals fused together, similar in length, or slightly shorter than, the radials (Fig. 2); arms paired, arising from division series based on the radials ***Monachocrinus perrieri*** (Koehler & Vaney, 1910) (p. 10)

3 Five arms. PENTAMETROCRINIDAE ***Pentametrocrinus varians*** (P. H. Carpenter, 1882) (p. 21)

– Ten or more arms 4

4 Distal segments of the oral pinnules each with a large tooth forming a comb-like grasping organ (Fig. 5); mouth usually markedly excentric (Fig. 3). COMASTERIDAE 5

– No terminal comb on the oral pinnules (though in *Annametra* the distal segments are short and the pinnule tip capable of coiling; the presence of sacculi easily distinguishes *Annametra*) (Figs 6, 7); mouth more or less centrally placed on the disc (Fig. 4) 8

5 IIBr (second division) series consisting of only two ossicles; the first syzygy of the arms often between brachials 1 + 2 (Fig. 8) ***Comatella africana*** Gislén, 1938 (p. 12)

– IIBr series usually all of four ossicles, or absent if only ten arms are present; the first arm joint not a syzygy (Fig. 7) 6

6 Arms few, up to 12 but rarely more than ten ***Comanthus wahlbergi serratus*** (Gislén, 1938) (p. 12)

– Rarely less than 15 arms 7

7 Rarely more than 20 arms ***Comanthus wahlbergi wahlbergi*** (Müller, 1843) (p. 11)

– Nearly 40 arms, many IIIBr division series being present (in the only recorded specimen) . . ***Comanthus wahlbergi*** forma ***multibrachia*** Gislén, 1938 (p. 12)

8 Pinnule ambulacra bordered by series of comparatively large rounded side and cover plates, clearly visible with a hand lens (Fig. 10) 9

– No conspicuous plates in the pinnule ambulacra (Fig. 11) 10

9 Cirri very long, approximately half the arm length, with numerous (50–65) segments, the distal ones keeled on the dorsal (incurled) side so that the profile is serrated; first syzygy usually at brachials 3 + 4 on the arms arising from IIBr series. THALASSOMETRIDAE . . ***Crotalometra magnicirra*** (Bell, 1905) (p. 18)

– Cirri stout and curling, equal to only about a quarter of the arm length, not more than 20 segments; first arm syzygy often between brachials 1 + 2. CHARITOMETRIDAE . . . . . . ***Glyptometra sclateri*** (Bell, 1905) (p. 19)

10 More than ten, usually more than 20 arms; prominent single dorsal spines on the distal cirrus segments (Fig. 13) 11

– Not more than ten arms; distal cirrus segments either smooth dorsally or with a transverse ridge or a pair of tubercle-like spines (Figs 14, 15) 12

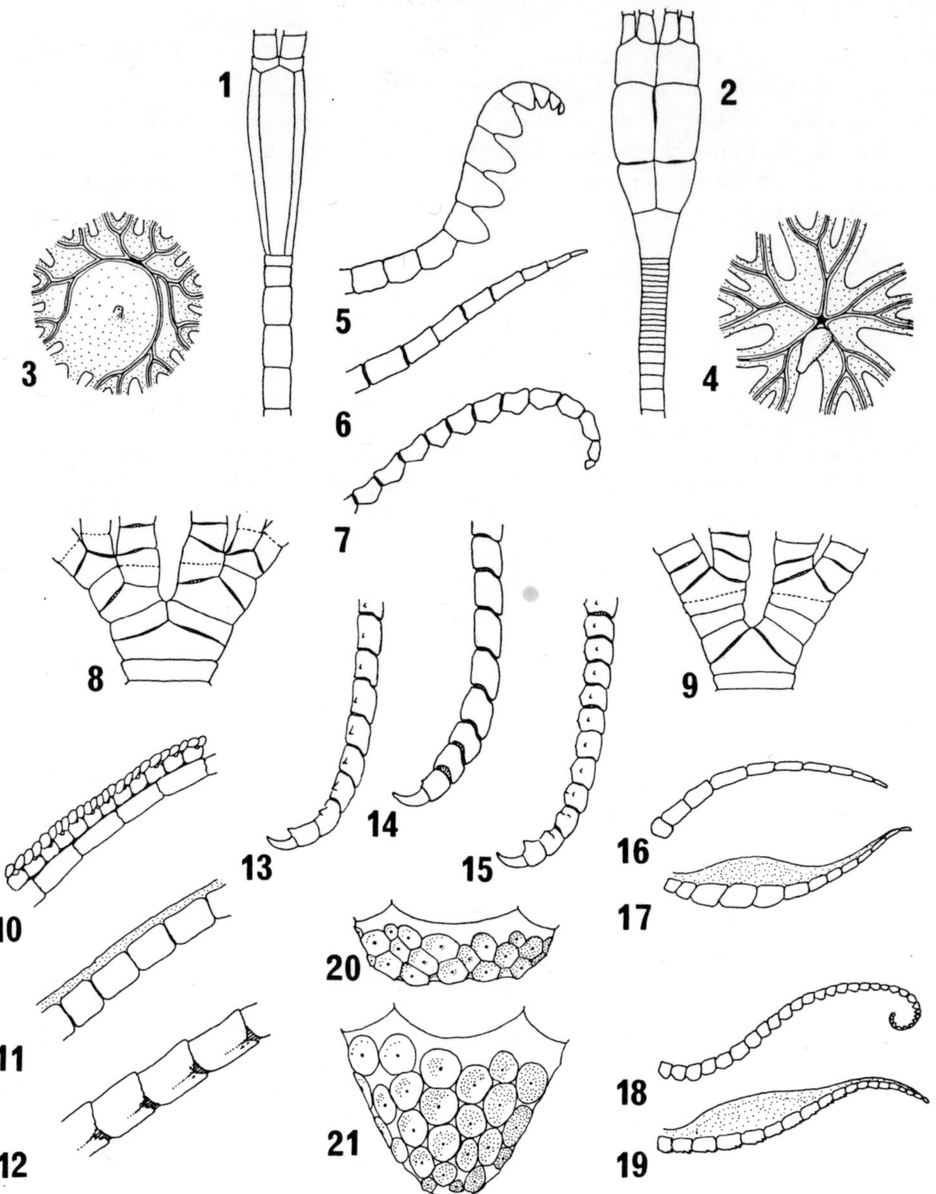

FIGS 1–21 Crinoidea. 1, 2: top of stalk and calyx, showing sutures, of *Democrinus chuni* and *Monachocrinus perrieri*. 3, 4: discs of *Comanthus wahlbergi* and *Heterometra* sp. 5, 6, 7: tips of oral pinnules of *Comanthus wahlbergi*, *Tropiometra carinata* and *Annametra occidentalis*. 8, 9: division series and arm bases of *Comatella* sp. and *Comanthus wahlbergi*. 10, 11, 12: parts of pinnules of *Glyptometra sclateri*, *Heterometra delagoae* (both in side view) and *Tropiometra carinata* (in oblique dorsal view). 13, 14, 15: cirrus tips of *Heterometra delagoae*, *Tropiometra carinata* and *Decametra durbanensis* (all in oblique view). 16, 17, 18, 19: P_1 and a genital pinnule of *Embryometra mortenseni* (16, 17) and *Annametra occidentalis* (18, 19). 20, 21: centrodorsals of *Annametra occidentalis* and *Cyclometra multicirra* (side view).

11 All or most of the IIBr series with four ossicles ; all the cirrus segments distinctly broader than long. HIMEROMETRIDAE ***Heterometra delagoae*** Gislén, 1938 (p. 14)

– IIBr series all with two ossicles ; some proximal cirrus segments longer than broad. MARIAMETRIDAE . . . ***Dichrometra afra*** A. H. Clark, 1912 (p. 13)

12 All the pinnules markedly prismatic (Fig. 12). TROPIOMETRIDAE . . . 13

– Only the proximal pinnules, if any, with crests 14

13 Arms with a prominent dorsal keel, at least in the proximal half ; size moderate, arm length rarely exceeding 150 mm ; dorsal pole of centrodorsal flat ***Tropiometra carinata carinata*** (Lamarck, 1816) (p. 17)

– Arms rounded dorsally ; adult size very large, arm length often exceeding 250 mm ; dorsal pole convex . . . ***Tropiometra magnifica*** A. H. Clark, 1937 (p. 17)

14 Distal cirrus segments each with a transverse dorsal ridge or a pair of tubercles (Fig. 15). COLOBOMETRIDAE (part) 15

– Distal cirrus segments (except for the opposing spine on the penultimate) smooth or with a slight median keel (Fig. 14). 16

15 Cirrus segments numerous (the peripheral ones with over 30 in the only recorded specimen), most of the middle ones with a pair of dorsal tubercles ; P_a (the first pinnule on the inner, adradial, side of Br_4) usually absent ***Decametra durbanensis*** A. M. Clark, 1951 (p. 15)

– Up to 22 cirrus segments, the distal ones with a transverse ridge on the dorsal side ; P_a present . . ***Oligometra serripinna occidentalis*** A. H. Clark, 1911 (p. 16)

16 P_1 and P_2 with not more than 12 slender segments (Fig. 16) ; genital pinnules with two or three of the segments alongside the gonad markedly expanded (Fig. 17). COLOBOMETRIDAE (part) 17

– P_1 and P_2 with about 30 short segments (Fig. 18) ; no expanded segments on the genital pinnules (Fig. 19). ANTEDONIDAE. 18

17 Size very small, arm length not known to exceed *c.* 25 mm ; pinnule series usually complete ; P_1 stiff, though smaller than P_2 ***Gislenometra perplexa*** A. H. Clark, 1947 (p. 15)

– Arm length up to *c.* 65 mm ; P_a and usually also P_4 absent on most arms ; P_1 slender and flexible medially and expanded at the joints ***Embryometra mortenseni*** Gislén, 1938 (p. 15)

18 Cirri numbering up to XL but with less than 20 segments, irregularly placed around the discoidal centrodorsal ; axillaries with proximal edge almost straight ***Annametra occidentalis*** (A. H. Clark, 1915) p. 20)

– Cirri L–LXXX with up to 36 segments, arranged in columns around the conical centrodorsal ; axillaries rhombic with marked proximal angles ***Cyclometra multicirra*** A. H. Clark, 1951 (p. 20)

Order BOURGUETICRINIDA Sieverts-Doreck

Crinoids with the stalk retained throughout life ; lacking true cirri but attaching to the substrate either by branching irregular radicular 'cirri' arising from the more distal (lower) columnals or by an enlarged terminal plate.

Family **BATHYCRINIDAE** Bather

See : A. M. Clark, 1973, *Bull. Br. Mus. nat. Hist.* (Zool.) **25** : 286.

Bourgueticrinids with new columnals only developed singly at the top of the stalk, one or more of the uppermost ones being discoidal, though most are longer than broad, attachment by radicular 'cirri' ; calyx thick-walled, relatively small, inverted conical or more or less cylindrical in shape, the plates of the well-developed basal ring either separate or fused together, longer or shorter than the radials above and alternating with them, the radials either fused or separate or occasionally even fused with the basals; radials bearing either simple arms – usually five in number – or division series of normally two ossicles, the second one axillary and bearing a pair of arms; brachials mostly with alternating muscular and syzygial ligamentous joints, the proximal ones lacking pinnules. Mostly slender and delicate forms, the arms usually lost in preserved specimens.

Democrinus chuni (Döderlein)

Rhizocrinus chuni Döderlein, 1907, *Siboga Exped.* **42a** : 14–15, pl. 1, fig. 5, pl. 6, fig. 6.
Bythocrinus chuni : H. L. Clark, 1923 : 229.
Democrinus chuni : Gislén, 1938a : 21–22, pl. 2, fig. 8; A. M. Clark, 1972 : 146–150, fig. 17, tab. 15.

Stalk : number of columnals 26–60+ (60+); total length 19–95+ (95+) mm; top diameter 0·3–1·2 (0·85) mm.
Calyx : length 1·5–8·5 (7·6) mm; maximum breadth 1·0–2·5 (2·0) mm; basals length 1·0–7·9 (7·2) mm; radials length 0·3–0·8 (0·6) mm.
Arms : number 5; length (7·4) mm; number of brachials (17+1 or 2).

A species of Bathycrinidae with only one or two of the topmost columnals not longer than broad; calyx inverted conical or more or less cylindrical, basals separate, markedly elongated, their height more than four times the diameter of the whole calyx at the lower end, radials much shorter than basals; arms simple, normally five, barely exceeding the calyx in length (usually broken off at the first joint), arising direct from the radials without intervening division series, the first post-radial ossicle (Br_1) broader than long and shorter than the radial.

TYPE LOCALITY. Off Somalia, 1644–1668 metres.

SOUTHERN AFRICAN RECORDS. 34/17/vd; 30/31/vd; 29/31/d; 30/32/vd; 27/32/d; 376–1800 metres. [The minimum depth from a new record at 27°33′S 32°44′E].

NATURE OF BOTTOM. Globigerina ooze; grey mud; soft clay and mud.

Monachocrinus perrieri (Koehler & Vaney)

Bathycrinus perrieri Koehler & Vaney, 1910, *Bull. Mus. natn. Hist. nat. Paris* **16** : 27–29, figs 1, 2.
Monachocrinus coelus H. L. Clark, 1923 : 229–230, pl. 8, fig. 1.
Monachocrinus perrieri : Gislén, 1938b, *Lunds Univ. Arsskr.* **34** (10) : 22.

Stalk : [incompletely known] top diameter 0·25–0·3 mm.
Calyx : length 1·0–1·2 mm; maximum breadth 0·8–1·0 mm; basals length 0·4–0·5 mm; radials length 0·6–0·75 mm.
Arms : number 10; length 8+ mm; number of brachials 20+.

A species of Bathycrinidae with 15–20 of the proximal columnals broader than long; calyx inverted conical, the fused basal ring about as long as its diameter at the top or less than twice its diameter at the bottom, radials slightly longer than basals ; arms normally ten, paired, arising from division series mounted on the radials, the first post-radial ossicle (IBr_1) longer than broad as well as longer than the preceding radial.

TYPE LOCALITY. Off Morocco, 2212 metres.

SOUTHERN AFRICAN RECORDS. 34/17/vd ; 1650–1830 metres.

NATURE OF BOTTOM. Grey mud.

Order COMATULIDA A. H. Clark

Crinoids only attached in the larval stage, adults free-living but attaching themselves temporarily by a cluster of jointed cirri arising from a median ossicle, the centrodorsal, broken off from the top of the larval stalk ; calyx relatively reduced during ontogeny, basals rarely visible in adults.

Family **COMASTERIDAE** A. H. Clark

See : A. H. Clark, 1931 : 76.

A family of Comatulida with the cirri usually moderate to small in size, their distal segments variously armed with a median tubercle or small transverse ridge or else smooth, arising from the periphery of a discoidal, sometimes reduced, centrodorsal ; arms numbering ten or more, with the division series on which they are based arising horizontally from the radials ; oral pinnules modified as grappling organs, long and flexible, especially near the tip where each segment bears a large ventral tooth, forming a comb-like structure ; mouth usually more or less offset towards the edge of the disc, the anal tube taking its place in the centre ; no sacculi.

Genus ***COMANTHUS*** A. H. Clark

See : A. H. Clark, 1931 : 527.

A genus of Comasteridae with usually more than ten arms and some or all of the IIBr division series consisting of four ossicles ; the first brachial syzygy at 3+4 and the first arm pinnule on Br_2.

The tropical Indo-Pacific species *Comanthus parvicirrus* (J. Müller) should, strictly, speaking, be included here since I have recorded it (1972 : 77) from 'Anton Bruun' st. 372Q, 25°57'S 33°02'E, that is off southern Mozambique, in 42 metres. The more or less reduced cirri should distinguish it from *C. wahlbergi*.

Comanthus wahlbergi wahlbergi (J. Müller)

Alecto wahlbergii J. Müller, 1843, *Arch. Naturgesch.* **1843** : 131.

Actinometra parvicirra : P. H. Carpenter, 1888 : 331 ; Bell, 1905c : 141. [Non *Alecto parvicirra* J. Müller, 1841.]

Comanthus wahlbergii : H. L. Clark, 1923 : 231–232 ; A. H. Clark, 1931 : 588–593, pl. 65, fig. 183 ; Stephenson, Stephenson & du Toit, 1937 : 380 ; Gislén, 1938a : 8–10, fig. 7 ; Stephenson, Stephenson & Bright, 1938 : 18 ; Eyre, 1939 : 298 ; Stephenson, 1944 : 348.

Comanthus wahlbergi : A. M. Clark, 1952 : 202 ; Morgans, 1959 : 414, 422, 427 ; 1962 : 303, 312 ; Day, Field & Penrith, 1970 : 82 ; A. M. Clark, 1972 : 77–78.

Arms : number 10–22, usually 16–20 ; length up to 80 mm.
Cirri : number XII–XXV, usually XV–XX ; number of segments 12–17, usually 15 or 16 ; length up to *c.* 10 mm.
Pinnules : P_D segment number 30–40, length 6–12 mm ; P_1 segment number 25–35, length 6·0–9·5 mm ; P_2 segment number 18–25, length 3·5–6·5 mm.

A subspecies of *Comanthus* with the cirri arranged peripherally around the discoidal centrodorsal, their longest segments not more than twice as long as their median breadth, the distal segments each bearing a short transverse ridge across the dorsal (incurled) side, resembling a small tubercle when seen in side view ; IIBr series mostly of four ossicles with a pinnule, P_D, on the second one, almost rectangular in cross-section being somewhat flattened laterally and dorsally, the division series and brachials with rugose transverse ridges near their distal ends. Colour in life black, yellow orally ; yellow arms, dark brown pinnules.

The forma (or variety) *multibrachia* Gislén, 1938a was based on a single anomalous specimen from False Bay with as many as 38 arms. Judging from Gislén's fig. 6, the profile of the distal cirrus segments is flatter than is usual in *C. wahlbergi*.

TYPE LOCALITY. Durban.

SOUTHERN AFRICAN RECORDS. 34/18/FB/i, s ; 34/19/i ; 34/21/i, s ; 34/23/i ; 33/25/i, s ; 33/27/i, s ; 32/28/s ; 30/30/i ; 29/31/s ; intertidal – 47 metres.

NATURE OF BOTTOM. Stones and rock ; sand ; sand and lithothamnion ; shell.

Comanthus wahlbergi serratus (Gislén)

Comissia serrata Gislén, 1938a : 7–8, figs 3, 4, pl. 1, fig. 1.
Comanthus wahlbergi serratus : A. M. Clark, 1972 : 78.

Arms : number 10, rarely 11 or 12, length up to *c.* 40 mm.
Cirri : number XX–XXIII, segment number 13–15, length 7–9 mm.
Pinnules : P_1 segment number *c.* 30, length *c.* 9 mm ; P_2 segment number *c.* 20, length *c.* 4·5 mm.

Differs only from *C. wahlbergi wahlbergi* by the consistently small number of arms. Also found at a greater depth.

TYPE LOCALITY. ? Natal.

SOUTHERN AFRICAN RECORDS. 34/25/s–d ; 33/27/s ; 29/31/d ; 68–130 metres.

NATURE OF BOTTOM. Rock and fine sand ; coarse sand and shell ; rock.

Comatella africana Gislén

Comatella africana Gislén, 1938a : 5–7, figs 1, 2, pl. 2, fig. 5. [Only one specimen known.]

Arms : number *c.* 30, length 55 mm.
Cirri : number XXIV, segment number 14–16, length 9–12 mm.
Pinnules : P_1 segment number *c.* 27, 'longer than P_2'.

A species of Comasteridae with the cirri arranged peripherally around the discoidal centrodorsal, the longest segment about four times as long as its median breath, the distal segments (judging from the figure) with small dorsal projections (described as 'lengthrunning cristae'); IBr, IIBr and external IIIBr division series all with only two ossicles; the first brachial syzygy of the internal (adradial) arms at brachials 1+2 and the first pinnule of each free arm normally on Br_2.

TYPE LOCALITY. NE. of Durban.

SOUTHERN AFRICAN RECORD. 29/31/–.

Family **MARIAMETRIDAE** A. H. Clark

See: A. H. Clark, 1941: 391.

A family of Comatulida with cirri bearing spines or other projections on their more distal segments in the median dorsal plane; adults with more than ten arms, the IIBr division series with only two ossicles; division series articulated either horizontally from the radials or obliquely upwards to some degree; oral pinnules without comb-like tips but with a basal crest along the dorsal surface or along the edge facing the arm tip, the more distal pinnules not markedly prismatic and without conspicuous side and covering plates; mouth more or less centrally placed on the disc; sacculi present.

Dichrometra afra A. H. Clark

Dichrometra flagellata var. *afra* A. H. Clark, 1912, *Smithson. misc. Collns* **60**: 23–24; Gislén, 1938a: 4.
Liparometra multicirra H. L. Clark, 1923: 232–233, pl. 8, fig. 2.
Dichrometra afra: A. H. Clark, 1937: 104; A. M. Clark, 1972: 97–99, fig. 7.

Arms: number 20–50 (36), length up to 100 mm (c. 85 mm).
Cirri: XX–XLV (XLV), segment number 24–42 (25), length 15–32 mm (18 mm).
Pinnules: P_1 segment number 17–25 (18), length 6–10 mm (6 mm); P_2 segment number 22–34 (33), length 9·5–16 mm (10·5 mm); P_3 segment number 22–34 (29), length 10–17 mm (11 mm).

A species of Mariametridae with the cirri arranged around a flattened hemispherical centrodorsal, the longest segments slightly longer than broad, the more distal ones each with a prominent median dorsal spine; arms crowded basally, the division series and arm bases flanged laterally with a straight free edge, dorsally smooth; P_2 and P_3 not markedly dissimilar in length and thickness, not noticeably angular in cross-section. Colour in life (from H. L. Clark) pale fawn, soft parts very dark brown to black on the ambulacral grooves of the disc.

TYPE LOCALITY. Madagascar.

SOUTHERN AFRICAN RECORD. 28/32/d; 164 metres.

Family **HIMEROMETRIDAE** A. H. Clark

See: A. H. Clark, 1941: 180.

A family of Comatulida with cirri bearing spines or other prominences on their more distal segments in the median dorsal plane; ten or more arms in the adult, if more than ten (as in the only species from southern Africa) then the IIBr series are mostly of four ossicles and there is a pinnule (P_D) on the second one ; division series articulated either horizontally from the radials or more often obliquely upwards to some degree ; oral pinnules without comb-like tips but with a basal crest along the dorsal surface or along the edge facing the arm tip, or both, the more distal pinnules not markedly prismatic and without conspicuous side and covering plates ; mouth more or less centrally placed on the disc ; sacculi present.

Heterometra delagoae Gislén

Heterometra africana var. *delagoae* Gislén, 1938a : 10–12, figs 8–11, pl. 1, fig. 3.
Heterometra delagoae : A. H. Clark, 1941 : 334–335 ; A. M. Clark, 1972 : 112–113.

Arms : number 20–40, usually *c.* 30 (30), length up to 85 mm (*c.* 75 mm).
Cirri : number XX–XXXII (XXVIII), segment number 25–35 (35), length 12–24 mm (*c.* 22 mm).
Pinnules : P_D segment number 21–34 (34), length 7·5–12 mm (12 mm) ; P_1 segment number 22–33 (33), length 9·5–13 mm (13 mm) ; P_2 segment number 24–33 (33), length 12–17 mm (17 mm) ; P_3 segment number $<$ 24–28 (28), length 12–14·5 mm (14·5 mm).

A species of Himerometridae with stout cirri arranged around a flattened centrodorsal, the longest peripheral ones with all the segments distinctly broader than long, the more distal segments each with a prominent median dorsal spine ; the division series rounded ; P_2 and P_3 not markedly dissimilar, basally each with two crests or keels, one dorsally, the other on the edge facing the arm tip.

Type locality. Delagoa Bay.

Southern African record. 26/32/– [depth ?].

Family **COLOBOMETRIDAE** A. H. Clark

See : A. H. Clark, 1947 : 1.

A family of Comatulida with some or most of the cirrus segments with a transverse ridge or a pair of tubercles, one each side of the midline dorsally, or slightly keeled (*Gislenometra*) or quite smooth (*Embryometra*) ; often only ten arms (as in all the species from southern Africa), when the second brachial syzygy is usually at 9+10 ; division series aligned obliquely upwards from the radials ; oral pinnules without comb-like tips, basally prismatic or almost rounded, P_2 often enlarged, pinnules without conspicuous side and covering plates ; mouth more or less centrally placed on the disc ; sacculi present.

Note : A. H. Clark (1947) characterized the Colobometridae by the transverse or paired prominences on some of the cirrus segments dorsally, ignoring the total absence of this character in *Gislenometra* and *Embryometra*, both represented in the seas of southern Africa. By making allowance for this in the diagnosis above, the only marked external difference between ten-armed colobometrids and members of the Antedoninae, such as *Annametra occidentalis*, which similarly have the second brachial syzygy at 9+10, is lost. [In this particular instance, *Annametra* can be distinguished by the numerous short segments in P_1, which pinnule approximates in form to the oral pinnules of comasterids, though without a proper comb terminally.] The Colobometridae, as delimited by A. H. Clark, is a somewhat heterogeneous assemblage of genera, as he himself admits.

Decametra durbanensis A. M. Clark

Decametra durbanensis A. M. Clark, 1951 : 1265–1267, fig. 4.
[Only one specimen known.]

Arms : number 10, length 90 mm.
Cirri : number XXVII, segment number *c.* 32, length *c.* 16 mm.
Pinnules : P_1 segment number 13–15, length 6 mm ; P_2 segment number 14–16, length 10 mm ; P_3 segment number 12–14, length 8 mm.

A species of Colobometridae with the stout cirri arranged in an irregular ring around the discoidal centrodorsal, all the segments shorter than broad, those from about the eleventh bearing a pair of nipple-like tubercles on the dorsal side, which approximate distally and coalesce into a single median tubercle before the penultimate segment ; P_a (on the inside or adradial side of Br_4) usually absent ; P_2 distinctly longer and stouter than P_1 or P_3.

Type locality. NE. of Durban, 29/31/68 metres.

Nature of bottom. Coarse sand.

Embryometra mortenseni Gislén

Embryometra mortenseni Gislén, 1938a : 12–16, figs 12–15, pl. 2, fig. 6 ; A. H. Clark, 1947 : 251–254.

Arms : number 10, length up to *c.* 65 mm (65 mm).
Cirri : number XII–XXII (XVI), segment number 14–19 (16–18), length 6–7 mm (6 mm).
Pinnules : P_1 segment number 7–10 (10), length 1·5–3·5 mm (3·5 mm) ; P_2 segment number 10–12 (12), length 3·5–5·0 mm (5·0 mm) ; P_3 segment number 8–10 (10), length 2·5–3·5 mm (3·5 mm).

A species of Colobometridae with the cirri arranged in two partial or complete irregular series around the flattened hemispherical centrodorsal, longest segments of the peripheral cirri slightly longer than broad, proximal segments with a slight collar-like rim overlapping the succeeding segment dorsally, no dorsal tubercles or spines ; P_2 distinctly longer and stouter than P_1 or P_3 ; P_a and usually also P_4 and P_d absent ; pinnule segments only slightly expanded opposite the gonads.

Type locality. SW. of Cape Town, 290 metres.

Southern African records. 34/17/d ; 34/23/s ; 34/25/d ; 49–325 metres.

Nature of bottom. Shell ; rock and khaki sand ; sand, mud, rock.

Gislenometra perplexa A. H. Clark

Pachylometra sclateri (part) : H. L. Clark, 1923 : 234.
Pachylometra sclateri (?) : Gislén, 1938a : 18–20, figs 16, 17, pl. 2, fig. 7.
Gislenometra perplexa A. H. Clark, 1947 : 57–61.

Arms : number 10, length up to *c.* 30 mm (*c.* 27 mm).
Cirri : number X–XVII (XV), segment number 15–21 (17), length 5–9 mm (5 mm).
Pinnules : P_1 segment number 6–9 (6), length 2·0–2·5 mm (2 mm) ; P_2 segment number 8–9 (8), length *c.* 2·2 mm (2·2 mm) ; P_3 segment number *c.* 11, length 'slightly longer than P_2'.

A small species of Colobometridae with the cirri arranged in one or two partial irregular series around a flattened conical or hemispherical centrodorsal, the longest cirrus segments slightly longer than broad and most of them slightly keeled dorsally ; P_3 similar in length to P_2 or slightly longer ; P_1 smaller but stiff and erect ; P_a and P_4 present ; genital pinnules markedly expanded over the gonads.

TYPE LOCALITY. SE. of East London, 146–238 metres.

SOUTHERN AFRICAN RECORDS. 34/25/s, d ; 33/28/d, vd ; 32/28/s ; 29/31/d ; 47–567 metres.

NATURE OF BOTTOM. Rock ; rock and fine sand. Also mud ; coral rock (H. L. Clark, 1923) or sandy mud (Gislén, 1938).

Oligometra serripinna occidentalis A. H. Clark

Oligometra serripinna var. *occidentalis* A. H. Clark, 1911, *Proc. U.S. natn. Mus.* **40** : 3 ; 1947 : 239–240, pl. 26, fig. 136, pl. 28, figs 150, 151, pl. 30, fig. 163.
Oligometra serripinna occidentalis : A. M. Clark, 1974 : 424–426.

Arms : number 10, length up to *c.* 80 mm (*c.* 50 mm).
Cirri : number XI–XX (XX), segment number 15–29 (16–22), length 9–10 mm (10 mm).
Pinnules : P_1 segment number 12–15 (14), length 3·5–5·0 mm (4 mm) ; P_2 segment number 12–19 (16, 17), length 5·0–7·0 mm (5·5 mm) ; P_3 segment number 12–14 (14), length 3·5–5·0 mm (4·0 mm).

A subspecies of Colobometridae with the stout cirri arranged in a more or less staggered ring around a discoidal centrodorsal, none of the segments longer than broad, the distal ones each with a transverse ridge across the somewhat flattened dorsal side ; division series and arm bases well separated and rounded laterally ; proximal pinnules very prismatic ; P_2 normally larger than P_1 or P_3 ; P_a usually present. Colour in life banded yellow and brown.

TYPE LOCALITY. Cargados Carajos, north of Mauritius, 55 metres.

SOUTHERN AFRICAN RECORDS. 30/30/s ; 28/32/s ; 27–44 metres.

NATURE OF BOTTOM. Fine sand and rock ; stony.

Family **TROPIOMETRIDAE** A. H. Clark

See : A. H. Clark, 1947 : 259.

A family of Comatulida with very stout, dorsally smooth cirri ; arms numbering ten ; division series aligned obliquely upwards from the radials ; second syzygy irregular in position, though usually on some arms at 9+10 ; oral pinnules without comb-like tips, all the pinnules markedly prismatic, tapering to attenuated tips, side and covering plates inconspicuous ; mouth more or less centrally placed on the disc ; sacculi present.

Genus ***TROPIOMETRA*** A. H. Clark

See : A. H. Clark, 1947 : 260.

Diagnosis as for the family.

Tropiometra carinata carinata (Lamarck)

Comatula carinata Lamarck, 1816, *Histoire naturelle des animaux sans vertèbres*. Paris. **2** : 534.
Antedon capensis Bell, 1905c : 139–140, pl. 2.
Tropiometra carinata : H. L. Clark, 1923 : 233–234 ; A. M. Clark, 1952 : 203 ; Morgans 1959 : 426.
Tropiometra clarki (part) Gislén, 1938a : 16–17, pl. 1, fig. 4.
Tropiometra carinata carinata : A. H. Clark, 1947 : 291–337, pl. 35, figs 183, 184, pl. 36, figs 187, 188 ; Day, Field & Penrith, 1970 : 82.
Tropriometra [sic] *carinata* : Kalk, 1958 : 206, 218, 237 ; B. I. Balinsky *In* : Macnae & Kalk, 1969 : 97, 99, 130.

Arms : number 10, length up to 180 mm, rarely more than 130 mm.
Cirri : number XIII–XXXVIII, usually XX–XXX, segment number 20–32, usually *c.* 24, length 15–36 mm, usually 18–20 mm.
Pinnules : P_1 segment number 20–25, length 10·0–15·5 mm ; P_2 segment number *c.* 20, length similar to P_1 ; P_3 segment number 20–23, length *c.* 14 mm.

A subspecies of Tropiometridae with the smooth stout cirri arranged in one or two crowded irregular series around the thick discoidal centrodorsal, the segments all broader than long and without any dorsal prominences, even the opposing spine on the penultimate segment being reduced ; arms more or less conspicuously carinate, at least in their proximal halves, the brachials with erect median distal spines or tubercles or else keeled ; the first three pinnules similar in length, or P_1 slightly shorter. Colour in life pale yellow (single record from False Bay) ; proximally purplish-brown, middle and distal parts of arms and pinnules mottled white and purple (Gislén) ; elsewhere usually patterned brown (or purple) and yellow, larger specimens darker.

Type locality. Mauritius.

Southern African records. [? 33/18/i] ; 34/18/FB/s ; 34/22/s ; 33/25/s ; 30/30/s ; 29/31/s ; 28/31/s ; 28/32/s ; 26/33/i ; 24/35/i ; 0–51 metres.

Nature of bottom. Rock ; coarse sand, shell and occasional rock ; coarse sand and shell ; shell ; sand ; black mud and rock.

Note : Although the single specimen actually described by Gislén under the name *Tropiometra clarki* was from False Bay, he made it clear that this was a new name for *T. encrinus* (pt.) : A. H. Clark, 1932 – the specimen from Mandapam, S. India, which Gislén designated as type of *T. clarki*. A. H. Clark maintained that South African specimens, including Gislén's, are closer to *carinata*, though he treated the distinction as a subspecific one.

Tropiometra magnifica A. H. Clark

Tropiometra magnifica A. H. Clark, 1937 : 90–91, pl. 1, fig. 1 ; A. M. Clark, 1972 : 131–132 ; 1974 : 426–427.

Arms : number 10, length probably exceeding 300 mm.
Cirri : number XXV–XXXV (XXXIV), segment number 35–45 (44), length up to 80 mm (80 mm).
Pinnules : P_1 segment number 28–39 (39), length 22–32 mm (32 mm) ; P_2 segment number 26–27 (26), length 18–21 mm (21 mm) ; P_3 segment number 25–27 (27), length 18–22 mm (22 mm).

A species of Tropiometridae of very large size with the smooth stout cirri arranged in two irregular crowded series around the flattened hemispherical centrodorsal, the longest segments slightly longer than broad, the distal ones smooth (or 'faintly carinate') dorsally; arms not carinate, brachials without spines or keels; P_1 distinctly longer than P_2 or P_3.

TYPE LOCALITY. Gulf of Aden, 73–200 metres.

SOUTHERN AFRICAN RECORDS. 24/35/d; 110 metres.

Family **THALASSOMETRIDAE** A. H. Clark

See: A. H. Clark, 1950: 1.

A family of Comatulida with long cirri, arranged in 10 or 15 vertical columns around a hemispherical, conical or truncated columnar centrodorsal, one of the more proximal cirrus segments a transition segment, showing an abrupt change of texture from matt to shiny, the following segments also shiny; arms often more than ten (as is usual in the only species from southern Africa); division series aligned obliquely upwards from the radials; oral pinnules without comb-like tips, all the pinnules prismatic and rather stiff with relatively large side and covering plates visible with a hand lens; mouth more or less centrally placed on the disc; sacculi present.

Crotalometra magnicirra (Bell)

Antedon magnicirra Bell, 1905c: 141, pl. 4.
Crotalometra magnicirra: Gislén, 1938a: 17–18; A. H. Clark, 1950: 97–100; A. M. Clark, 1974: 427–429.
Arms: number 10–20, usually 15–20, length up to at least 110 mm.
Cirri: number XV–XXXIII, segment number 50–64, length 55–70 mm.
Pinnules: P_D segment number 22–27, length *c.* 12 mm; P_1 segment number *c.* 19, length *c.* 9 mm; P_2 segment number *c.* 12, length *c.* 5 mm.

A species of Thalassometridae with the cirri about half as long as the arms and arranged in ten columns of two or three each around the centrodorsal, which ranges from flattened hemispherical to more or less truncated conical but is usually about two-thirds as high as broad, the eighth, ninth or tenth cirrus segment a transition one and about twice as long as broad, the distal segments shorter than long and flared at their distal ends, making a serrated profile, especially dorsally; IIBr series of four ossicles, rarely only two; first brachial syzygy usually at 3+4 on all arms, rarely 1+2; genital pinnules without lateral expansions.

TYPE LOCALITY. Off East London, 548–731 (? 822) metres.

SOUTHERN AFRICAN RECORDS. 33/27/vd; 33/28/vd; 30/30/s; 29/31/d; 91–731 (? 822) metres.

NATURE OF BOTTOM. Coral and mud; sand and broken shells; sand and stones; sand, gravel and sponge fragments; sand.

Family **CHARITOMETRIDAE** A. H. Clark

See : A. H. Clark, 1950 : 191.

A family of Comatulida with short cirri, usually arranged in vertical columns around a centrodorsal of various shapes from discoidal to conical but usually with the dorsal pole wide and almost flat, no transition segment in the cirri, the surface of which is matt throughout ; arms often more than ten (as is usual in the only species from southern Africa) ; division series aligned obliquely upwards from the radials ; oral pinnules without comb-like tips, all the pinnules prismatic with relatively large side and covering plates visible with a hand lens ; mouth more or less centrally placed on the disc ; sacculi present.

Glyptometra sclateri (Bell)

Antedon sclateri Bell, 1905c : 140, pl. 3.
Pachylometra sclateri (part) : H. L. Clark, 1923 : 234–235 [adult specimen only ; non *P. sclateri* (?) : Gislén, 1938a : 18–20, which represents *Gislenometra perplexa* A. H. Clark, like H. L. Clark's small specimens].
Glyptometra sclateri : A. H. Clark, 1950 : 268–270 ; A. M. Clark, 1974 : 429–430.

Arms : number 10–21, length up to 90 mm.
Cirri : number *c.* XXXV, segment number 15–18, length 18–22 mm.
Pinnules : P_D segment number 23–30, length *c.* 10 mm ; P_1 segment number 19–27, length 9·5–10·0 mm ; P_2 segment number 16–20, length *c.* 8 mm.

A species of Charitometridae with the cirri about a quarter as long as the arms and usually arranged irregularly around a thick discoidal to flattened hemispherical centrodorsal, less than half as high as broad, the longer cirrus segments about half again as long as broad, the distal ones shorter and slightly flared at their distal ends ; arms 10–21, IIBr series of four ossicles, rarely only two ; first brachial syzygy on arms arising from IIBr series usually at 1+2 but variable in position, sometimes at 2+3 or, as on arms arising from IBr series, at 3+4 ; genital pinnules with several segments from about the fourth markedly expanded laterally.

TYPE LOCALITY. Off East London, 457–548 metres.

SOUTHERN AFRICAN RECORDS. 33/27/vd ; 33/28/d, vd ; 548 (? 457)—731 (? 822) metres.

NATURE OF BOTTOM. Broken shells : hard ground ; sand and stones. Also mud (H. L. Clark, 1923).

Family **ANTEDONIDAE** Norman

See : A. H. & A. M. Clark, 1967 : 39.

A family of Comatulida with the cirri smooth throughout or the distal segments flared distally or with median spines or keels on the dorsal side ; normally ten arms ; division series aligned obliquely upwards from the radials ; oral pinnules without proper toothed combs at the tips (though in *Annametra* the distal segments are short, convex ventrally and the tips are very flexible), all the pinnules rounded in cross-section, the side and covering plates inconspicuous; mouth more or less centrally placed on the disc; sacculi present.

Annametra occidentalis (A. H. Clark)

Cominia occidentalis A. H. Clark, 1915 : 164–165, pl. 10, figs 1–5; H. L. Clark, 1923 : 231.
Annametra occidentalis: Gislén, 1938a : 20–21, figs 18–26; A. M. Clark, 1952 : 203; Day, 1959 : 544; Morgans, 1959 : 426; 1962 : 310, 311, 312, 322; A. H. & A. M. Clark, 1967 : 94–96, fig. 6; Day, Field & Penrith, 1970 : 82.

Arms: number 10, length up to *c.* 50 mm.
Cirri: number XXXV–XL, segment number 14–19, length up to 15 mm.
Pinnules: P_1 segment number 30–35, length 8·5 mm; P_2 segment number 32–35, length 9 mm; P_3 segment number 25–33, length 8–11 mm.

A species of Antedonidae with relatively short, distally expanded cirri of less than 20 segments, arranged in two, sometimes three irregular series around a thick discoidal or flattened hemispherical centrodorsal, the longest cirrus segments less than twice as long as broad, the distal segments with their dorsal and ventral sides parallel; arms with distal syzygies at intervals of four to six muscular joints; P_3 with a similar large number of segments to the preceding pinnules, though it is the first genital pinnule. Colour in life brownish yellow, sometimes with olive green spots; yellow with light, striped pinnules; yellow; 'dark'; banded arms, grey pinnules.

TYPE LOCALITY. False Bay.

SOUTHERN AFRICAN RECORDS. 33/17/s; 33/18/s; 34/18/FB/i, s; 34/23/i; [? 29/31/–*]; 0–40 metres.

NATURE OF BOTTOM. Sand; shelly khaki sand, rocks; khaki/white coarse sand; sand and rock; rock; fine sand; black mud [!].

Cyclometra multicirra A. H. Clark

Cyclometra multicirra A. H. Clark, 1952 : 189–191, pls 15, 16; A. H. & A. M. Clark, 1967 : 534–536.

Arms: number 10, length 50–at least 100 mm.
Cirri: number LV–LXXX, segment number (peripheral cirri) 30–35, length 35–40 mm.
Pinnules: P_1 segment number 32–34, length 12–15 mm; P_2 segment number 24–27, length 12·0–15·5 mm; P_3 segment number 16–19, length 7·5–10·0 mm.†

A species of Antedonidae with numerous, more or less elongated, cirri arranged in *c.* 15 slightly irregular vertical columns around a conical centrodorsal, longest segments of peripheral cirri more than twice as long as their median breadth, distal segments markedly flared distally on their dorsal sides, apical cirri only about half

* The specimen corresponding to this record is in the British Museum collection and was named by A. H. Clark. Although labelled 'Durban' the collective locality given when it was registered in 1913 with other material from the Natal Museum is 'German East Africa', i.e. Tanzania, which is even more unlikely than Durban. There is therefore no authenticated locality from east of Knysna.

† The additional numerical data supplementing that from the holotype are provided by a second and smaller specimen from the same station sent to the British Museum (Natural History) subsequently.

as long as the peripheral ones ; arms with distal syzygies at intervals of about three muscular joints ; P_3 much shorter and with fewer segments than P_1 and P_2.

TYPE LOCALITY. WSW. of Port Nolloth, west coast, 30/14, 461 metres.

NATURE OF BOTTOM. Green mud.

Family **PENTAMETROCRINIDAE**

See : A. H. & A. M. Clark, 1967 : 766.

A deep-sea family of Comatulida with the cirri consisting of very elongated segments, usually about four times as long as broad ; no division series, there being either five radials and five simple arms (as in the species represented) or ten radials and ten arms ; the joint between the first two post-radial ossicles not a syzygy (this distinguishes *Pentametrocrinus* from the similarly five-armed Eudiocrinidae, unknown so far from South Africa) ; oral pinnules without comb-like tips, all the pinnules rounded in cross-section and without conspicuous side and covering plates ; mouth more or less centrally placed on the disc ; sacculi present.

Pentametrocrinus varians (P. H. Carpenter)

Eudiocrinus varians P. H. Carpenter, 1882, *J. Linn. Soc.* (Zool.) **16** : 496–497 ; 1888 : 81–82, pl. 7, figs 3–7.
Pentametrocrinus varians : H. L. Clark, 1923 : 235 ; A. H. & A. M. Clark, 1967 : 804–810 ; A. M. Clark, 1972 : 146.

Arms : number 5, length up to 110 mm.
Cirri : number XII–XXX, segment number *c.* 22, length up to 50 mm.
Pinnules : P_1 segment number *c.* 25, length *c.* 6·5 mm ; P_2 segment number *c.* 25, length 9–11 mm.

A very fragile species of Pentametrocrinidae with relatively long slender cirri, the segments very elongate after the first two or three, the distal ones tapering gradually to a slender point, arranged in one or two irregular rows around the low hemispherical centrodorsal ; the five arms with the first syzygy at 4+5 and the second usually at 10+11 (or one joint earlier or later) ; the first pinnule, P_1, present on the second brachial, usually on its right side.

Note : H. L. Clark's specimen from west of Cape Point (presumably 34/17) lacked arms, pinnules and cirri. As I noted in 1967, it might be conspecific rather with *P. atlanticus*, known from Morocco to southern Ireland and the Caribbean, which differs from *P. varians* in lacking P_1 (on Br_2) and P_a (on Br_3).

TYPE LOCALITY. Off the Philippines, 1920 metres.

SOUTHERN AFRICAN RECORDS. 24/36/vd ; 1600–1630 metres [? 34/17/1646–1830 metres].

Class *STELLEROIDEA*

Subclass *ASTEROIDEA*

INTRODUCTION. In default of a comprehensive monograph on this subclass, reference can only be made to the summarized treatment by Spencer & Wright (1966) in the *Treatise on Invertebrate Paleontology*, supplemented for certain taxa by Fisher's studies of North Pacific asteroids (primarily 1911) and Philippine asteroids (1919), together with Döderlein's 'Siboga' reports on a few families including the Astropectinidae and Luidiidae.

The following terms are used in the descriptions of asteroids and may need clarification :

Abactinal plates – the superficial plates of the upper side, extending on to the lateral faces of the arms in some species with arms cylindrical in cross-section (especially in the Forcipulatida), bordered laterally by the upper of the two rows of marginal plates, the *superomarginals*, which are prominent in most Paxillosida and Valvatida but usually inconspicuous in the Spinulosida and Forcipulatida.

Actinal plates – the corresponding plates of the lower side, below the *inferomarginals*, separated medially by the *adambulacral* plates bordering the ambulacral furrow.

Autotomous – self-dividing but not necessarily across the disc, for which practice the restricted term *fissiparous* is used.

Cribriform organs – in the family Porcellanasteridae multi-folded ciliated structures supported by vertical series of papillae, numbering one or more in each interradius between the consecutive marginal plates, believed to produce respiratory currents in these mud-burrowing animals.

Furrow spines – those on the inner (adradial) edge of the adambulacral plates, usually projecting obliquely over the furrow, backed by the *subambulacral spines*.

Madreporite – the ridged interradial plate on the upper side of the disc forming the external opening of the water vascular system.

Papulae – the small finger-like transparent respiratory processes which project through pores in the body wall, mainly on the upper side ; their arrangement is sometimes characteristic.

Paxillae – modified abactinal plates with a raised median column, sometimes quite slender, crowned with spinelets. This form of plate is restricted to some families, such as the Astropectinidae ; also found in the Luidiidae where the superomarginal plates as well as the abactinal ones are paxilliform.

Pedicellariae – usually minute pinching organs made up of two or sometimes several modified spinelets of various forms, mounted either directly on a plate (or plates), which may or may not be recessed to accommodate the valves, or with a basal piece between or below the two valves and mounted either on the skin covering the spines or on the body wall between the spines in the Forcipulatida.

The measurements generally used for asteroids are : R, the major radius from the centre of the disc to an arm tip, r, the minor radius from the centre to the edge of

DISTRIBUTION TABLE FOR ASTEROIDEA

	Luderitz Bay area	Lambert's Bay area	Cape Town area	False Bay	Cape Agulhas area	Mossel Bay–Knysna area	Port Elizabeth area	Durban area	Lourenço Marques area	Other localities
LUIDIIDAE										
Luidia africana Sladen	d	d	d	s	s	d	d	s ?	..	Morocco–Ivory Coast
Luidia sp. aff. *L. avicularia* Fisher	..	..	..	..	..	..	s	..	..	
Luidia maculata Müller & Troschel	..	..	..	..	..	..	s	s	i	Tropical Indo-West Pacific
Luidia sagamina aciculata Mortensen	..	..	..	..	..	..	..	s	..	St Helena, Gambia–Congo
Luidia savignyi (Audouin)	..	..	..	..	..	..	..	s	i	Tropical Indo-West Pacific
PORCELLANASTERIDAE										
Porcellanaster ceruleus W. Thomson	..	..	vd	..	..	..	..	..	..	Cosmopolitan except polar
ASTROPECTINIDAE										
Astropecten anacanthus H. L. Clark	..	..	..	..	..	..	..	d	..	
Astropecten antares Döderlein	..	..	..	..	..	i s	s	s	?	
Astropecten exilis Mortensen	..	..	..	..	..	..	..	?	?	
Astropecten granulatus natalensis John	..	..	..	..	..	d ?	s	s d	?	
Astropecten hemprichi Müller & Troschel	..	..	..	..	..	..	..	s	i	Indian Ocean, East Indies

Distribution Table for Asteroidea (*cont.*)

	Luderitz Bay area	Lambert's Bay area	Cape Town area	False Bay	Cape Agulhas area	Mossel Bay–Knysna area	Port Elizabeth area	Durban area	Lourenço Marques area	Other localities
Astropecten irregularis pontoporeus Sladen	s d	d	s d	s	s	s	s d	..	..	
Astropecten leptus H. L. Clark	..	..	..	..	..	..	..	d	..	Gulf of Guinea
Astropecten polyacanthus phragmorus Fisher	..	..	..	..	..	..	..	s	s	Persian Gulf, Philippines
Bathybiaster vexillifer (W. Thomson)	..	..	vd	..	..	..	..	..	..	Arctic
Dipsacaster sladeni capensis A. M. Clark	..	d	d	..	..	d	–	..	..	
Persephonaster roulei euryplax Mortensen	..	..	..	..	..	..	..	d	..	
Plutonaster intermedius Perrier	vd	..	vd	..	..	..	..	..	..	NW. Atlantic
Plutonaster proteus H. L. Clark	..	..	vd	..	..	..	..	..	..	
Psilaster acuminatus Sladen	d	d	d	–	vd	..	..	..	..	New Zealand, Australia
Tethyaster pacei (Mortensen)	..	..	..	..	..	..	..	..	..	[Position uncertain]
BENTHOPECTINIDAE										
Benthopecten pedicifer (Sladen)	..	..	..	..	vd	..	..	..	..	West of the Crozet Is.
Luidiaster hirsutus Studer	..	d	d	..	vd	d	..	..	..	Kerguelen, Bouvet I.
Pectinaster filholi Perrier	..	..	vd	..	..	..	..	..	..	N. Atlantic, Crozet Is

ARCHASTERIDAE										
Archaster angulatus Müller & Troschel	..	..	..	..	..	..	..	..	i	Mascarene Is, Philippines, E. Indies
ODONTASTERIDAE										
Odontaster australis H. L. Clark	d	d	d	..	..	..	..	..	..	
GONIASTERIDAE										
Anthenoides marleyi Mortensen	..	..	..	..	..	..	..	d	..	
Calliaster acanthodes H. L. Clark	..	..	..	..	..	d	d	d	..	
Calliaster baccatus Sladen	..	..	..	s	s	s	s d	..	..	
Ceramaster patagonicus euryplax H. L. Clark	..	d	d	..	d	..	..	..	..	
Ceramaster trispinosus H. L. Clark	..	..	d	..	..	..	..	..	..	
Cladaster macrobrachius H. L. Clark	..	..	d	..	..	..	..	..	..	
Hippasteria phrygiana capensis Mortensen	..	..	d	..	..	..	..	..	..	
Hippasteria strongylactis H. L. Clark	..	d	d vd	..	..	..	..	..	..	
Mediaster capensis H. L. Clark	..	..	d	s	..	..	..	..	..	
Mediaster capensis durbanensis Mortensen	..	..	..	..	..	..	..	d	..	
Pseudarchaster brachyactis H. L. Clark	vd	..	d	..	..	d	..	..	..	
Pseudarchaster tessellatus Sladen	..	d	d	s	d	..	..	..	..	
Sphaeriodiscus bourgeti (Perrier)	..	..	..	..	..	..	..	d	..	Cape Verde Is, Azores
Toraster tuberculatus (Gray)	..	d	s d	..	d	s d	d	d	..	

DISTRIBUTION TABLE FOR ASTEROIDEA (*cont.*)

	Luderitz Bay area	Lambert's Bay area	Cape Town area	False Bay	Cape Agulhas area	Mossel Bay–Knysna area	Port Elizabeth area	Durban area	Lourenço Marques area	Other localities
OREASTERIDAE										
Asterodiscus elegans Gray	..	..	..	..	..	..	..	s	i	Seychelles, India, Philippines, ? China
Culcita schmideliana (Retzius)	..	..	..	..	..	..	..	..	i	East Africa, Ceylon
Pentaceraster mammillatus (Audouin)	..	..	..	..	..	..	..	..	i	East Africa, Red Sea, Persian Gulf
Protoreaster lincki (de Blainville)	..	..	..	..	..	..	..	..	i	East Africa–Bay of Bengal
OPHIDIASTERIDAE										
Austrofromia schultzei (Döderlein)	..	..	..	s	..	..	–	..	..	
Hacelia capensis Mortensen	..	..	..	..	..	..	..	s	..	
Leiaster sp.	..	..	..	..	..	..	..	..	s	
Linckia guildingi Gray	..	..	..	..	..	..	..	..	i	'Tropicopolitan'
Linckia laevigata (Linnaeus)	..	..	..	..	..	..	..	..	i	Tropical Indo-West Pacific
Linckia multifora (Lamarck)	..	..	..	..	..	..	..	..	i	Tropical Indo-West Pacific
PORANIIDAE										
Chondraster elattosis H. L. Clark	..	..	d vd	..	..	vd	..	..	..	

3

Spoladaster brachyactis H. L. Clark	..	d	s	..	..	..	s d	..	..	
Tylaster meridionalis Mortensen	..	d	..	..	..	..	..	..	..	
ASTERINIDAE										
Anseropoda grandis Mortensen	..	d	..	..	..	d	..	..	..	
Anseropoda habracantha H. L. Clark	..	..	..	..	..	..	d	d	..	
Anseropoda novemradiata (Bell)	..	..	..	..	..	..	..	s	..	
Asterina burtoni Gray	..	..	..	..	..	..	..	..	i	Tropical Indo-West Pacific
Asterina gracilispina H. L. Clark	..	..	..	s ?	–	s	s	..	..	
Disasterina leptalacantha africana Mortensen	..	..	..	..	..	..	i	d	..	
Patiria formosa Mortensen	..	..	..	s	..	..	..	..	..	
Patiria granifera Gray	i	i s	i s	s	s	i s	i	i	..	
Patiria stellifera (Möbius)	i	..	..	..	..	..	..	..	..	West Africa, E. South America
Patiriella dyscrita (H. L. Clark)	..	..	..	..	i	..	s	..	..	
Patiriella exigua (Lamarck)	i	i	i s	i s	i	i	i	i	..	St Helena, St Paul I., S. Australia
PTERASTERIDAE										
Diplopteraster multipes Sars	..	d vd	d vd	..	vd	..	..	..	..	Arctic, N. Atlantic, N. Pacific
Hymenaster gennaeus H. L. Clark	..	..	vd	..	..	..	..	..	..	
Hymenaster lamprus H. L. Clark	..	..	vd	..	..	..	..	..	..	
Hymenaster latebrosus Sladen	..	..	vd	..	..	..	..	..	..	Southern Ocean, south from Australia

DISTRIBUTION TABLE FOR ASTEROIDEA (*cont.*)

	Luderitz Bay area	Lambert's Bay area	Cape Town area	False Bay	Cape Agulhas area	Mossel Bay–Knysna area	Port Elizabeth area	Durban area	Lourenço Marques area	Other localities
Pteraster affinis Smith	..	..	d	..	..	..	..	..	..	Kerguelen
Pteraster capensis Gray	d	d	s d	s d	s	..	s	d	..	
Pteraster flabellifer Mortensen	..	d	..	..	..	..	..	..	..	
Pteraster fornicatus Mortensen	..	..	d	..	..	..	..	..	..	
SOLASTERIDAE										
Crossaster penicillatus Sladen	..	d vd	d vd	..	s	s d	–	..	..	St Helena, Tristan, Gough I., Marion I.
Lophaster quadrispinus H. L. Clark	..	..	d	..	..	..	..	d	..	
MITHRODIIDAE										
Mithrodia gigas Mortensen	..	..	..	..	..	..	s	..	..	
ACANTHASTERIDAE										
Acanthaster planci (Linnaeus)	..	..	..	..	..	..	..	..	s	Tropical Indo-West Pacific
ECHINASTERIDAE										
Henricia abyssalis (Perrier)	d	d	s d	..	..	s	d	d	..	Azores–Portugal and Morocco
Henricia ornata (Perrier)	i s	i	i s d	i s	i	i	i	d	..	
Henricia retecta H. L. Clark	..	d	d	s	..	..	s	..	..	

Poraniopsis capensis H. L. Clark	d	..	d	..	..	..	..	..	..	
ASTERIIDAE										
Allostichaster capensis (Perrier)	s	..	s	..	..	..	..	..	..	Tristan, southern South America
Coronaster volsellatus (Sladen)	..	..	d	..	..	..	d	d	..	Philippine Is, Japan
Coscinasterias calamaria (Gray)	..	..	i	i	..	..	..	..	..	Mauritius [? southern Australasia]
Cosmasterias felipes (Sladen)	..	d	s d	s	d	s d	..	..	..	
Marthasterias glacialis forma *africana* (Müller & Troschel)	..	..	i s	i s	..	..	..	..	..	
Marthasterias glacialis forma *rarispina* (Perrier)	..	..	s	? i s	s	i s	i s	..	..	
Perissasterias heptactis H. L. Clark	..	..	d	..	..	..	..	..	..	
Perissasterias obtusispina H. L. Clark	..	..	d	..	..	..	..	..	..	
Perissasterias polyacantha H. L. Clark	..	d	d	..	..	..	..	..	..	
Sclerasterias eustyla (Sladen)	..	d	..	..	..	..	..	..	..	Tristan
Sclerasterias stenactis (H. L. Clark)	..	..	..	..	..	d	..	d	..	
BRISINGIDAE										
Brisinga cricophora Sladen	..	d	d	..	..	d	..	..	..	West Indies
Stegnobrisinga splendens H. L. Clark	..	..	..	..	..	..	..	vd	..	

the disc interradially and sometimes also br, the breadth across the base of the arm. It is advisable to measure from both above and below and take the mean. [In some multiradiate species, notably of the Brisingidae, it is difficult to measure r accurately and then the disc diameter and arm length may be given instead.]

The number of species included is 91, compared with 85 given by Mortensen (1933a), who excluded records from Mozambique.

KEY TO THE ASTEROIDEA

1 Tube feet pointed or ending in a rounded knob 2

– Tube feet with terminal disc 22

2 Lamelliform cribriform organs present between the marginal plates interradially (Fig. 22). PORCELLANASTERIDAE
Porcellanaster ceruleus Thomson, 1877 (p. 47)

– No cribriform organs 3

3 Superomarginal plates small and paxilliform, resembling the abactinal plates, the inferomarginals alone forming the edge of the body (Fig. 24). LUIDIIDAE . 4

– Superomarginals not paxilliform but large and more like the inferomarginals, together with which they form the edge of the body (Fig. 23). ASTROPECTINIDAE 8

4 Seven to nine arms 5

– Five arms 7

5 Some of the lateral paxillae armed with a large central spine (Fig. 25); seven arms
Luidia savignyi (Audouin, 1826) (p. 45)

– No single conspicuous spines on scattered lateral paxillae, though one or more spinelets may be enlarged on *all* the lateral paxillae; usually nine arms . . 6

6 Lateral paxillae squarish, equal in number to the superomarginal paxillae, none of their short central spinelets projecting above the general surface; several actinal plates between each adambulacral and inferomarginal plate, most of them armed with a three-bladed pedicellaria; three adambulacral spines in a simple series at right angles to the furrow . ***Luidia maculata*** Müller & Troschel, 1842 (p. 45)

– Lateral paxillae more or less rounded, more numerous than the superomarginal paxillae, some of them with one, sometimes several, of the central spinelets markedly longer than the rest; only one actinal plate corresponding to each adambulacral and its pedicellaria, if present, is two-bladed; four or five adambulacral spines including a small one deep in the furrow and a pair of outer ones on most plates . . . . ***Luidia*** sp. aff. ***L. avicularia*** Fisher (p. 44)

7 Paxillar spinelets all similar in size,* the single central spinelet of the smaller paxillae no stouter than the peripheral ones (Fig. 26) . ***Luidia africana*** Sladen, 1889 (p. 44)

* In wet specimens there may be an optical illusion that the erect central spinelet is stouter.

FIGS 22–42 Asteroidea. 22: interradius of *Porcellanaster ceruleus* showing cribriform organ (side view). 23, 24: sections of arms of *Astropecten irregularis* and *Luidia* sp. showing superomarginal plates. 25, 26, 27: paxillae of *Luidia savignyi*, *L. africana* and *L. sagamina aciculata*. 28, 29, 30, 31: denuded ventral interradii of *Astropecten irregularis pontoporeus*, *A. granulatus natalensis*, *Bathybiaster vexillifer* and *Plutonaster proteus*. 32, 33, 34, 35: marginal plates of *Astropecten polyacanthus phragmorus*, *A. irregularis pontoporeus*, *A. leptus* and *A. granulatus natalensis* (dorsal view). 36, 37, 38: paxillae of *Astropecten leptus*, *A. granulatus natalensis* and *A. hemprichi*. 39, 40, 41, 42: disc and an arm of *Astropecten anacanthus*, *A. antares*, *A. irregularis pontoporeus* and *A. leptus* (dorsal view).

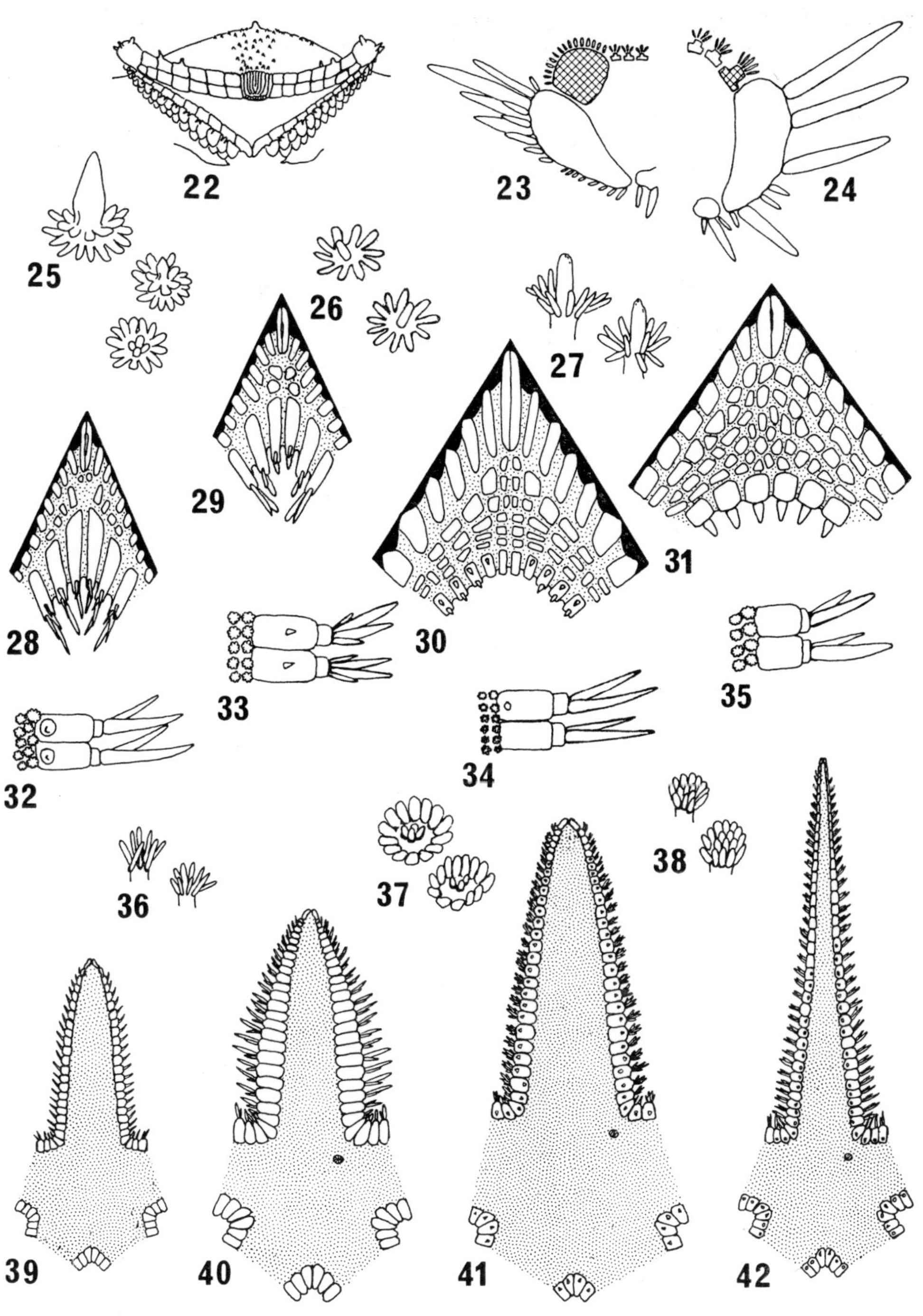
22
23
24
25
26
27
28
29
30
31
32
33
34
35
36
37
38
39
40
41
42

7 A markedly enlarged central spinelet, or small spine, on all the paxillae, at least twice as thick as the peripheral spinelets and about twice as high (0·5–1·0 mm) (Fig. 27) . . . . . ***Luidia sagamina aciculata*** Mortensen, 1933b (p. 45)

8 Actinal plates very few and restricted to the interradii, usually only two or three present each side in a single series (Fig. 29), rarely (*A. irregularis pontoporaeus*) up to six (Fig. 28); main inferomarginal spines projecting abruptly from the upper ends of the plates and forming a conspicuous fringe to the body (Figs 39–42). *Astropecten* 9

– Actinal plates arranged in several series, at least the innermost one extending for about half R (the major radius), if not for the whole arm length (Figs 30, 31); if enlarged inferomarginal spines are present they do not form a fringe to the body but are spaced or appressed 16

9 Superomarginal plates mainly vertical in alignment with a well-developed spine at the upper end of most, if not all of them, the first spine particularly large ***Astropecten polyacanthus phragmorus*** Fisher, 1913 (p. 51)

– Superomarginal plates occupying more of the upper surface of the arms than the lateral, their spines less well developed, often absent altogether or forming a partial series only (Figs 33–35) 10

10 Arms relatively long and narrow, R/r nearly 9/1 (in the only known specimen, R 70 mm) . . . . ***Astropecten exilis*** Mortensen, 1933a (p. 49)

– Arms moderate in length or short, R/r not more than 6/1 11

11 Arms broad petaloid, the superomarginals broadest at about half R and the arms tapering markedly near the tip (Fig. 40), R/r only 2·5–3·0/1 ***Astropecten antares*** Döderlein, 1926 (p. 48)

– Arms not markedly petaloid, superomarginals narrowing fairly evenly (Figs 39, 41, 42), R/r rarely less than 3·5/1 12

12 Usually four actinal plates in each series, though the range is two to six (Fig. 28); upper end of each inferomarginal with a fascicle of elongated spines, though one may be stouter than the others (Fig. 33) (in dorsal view most plates have two first magnitude spines at about the same level, with other spines only a little smaller below (Fig. 41) . ***Astropecten irregularis pontoporaeus*** Sladen, 1883 (p. 50)

– One to three actinal plates only in each series,* inferomarginal plates with only one or two long spines, sometimes an accessory third, at the upper end (Figs 34, 35). 13

13 Arms rather attenuated, tapering more in the proximal than the distal half (Fig. 42); paxillae relatively small, about three corresponding to each superomarginal and with up to only *c.* 10 slender spinelets (Figs 34, 36); two large inferomarginal spines of about equal length on most plates ***Astropecten leptus*** H. L. Clark, 1926 (p. 51)

– Arms stout, tapering evenly from the base (Fig. 39); paxillae stouter, usually four or five bordering two adjacent superomarginals (Fig. 35), the larger ones with 20–40 short spinelets (Figs 37, 38); only one large spine on most inferomarginals with a distinctly smaller accessory one (Fig. 35) 14

14 Three actinal plates in each series; no superomarginal spines ***Astropecten anacanthus*** H. L. Clark, 1926 (p. 48)

– Two, sometimes one, actinal plates each side of the interradius; superomarginal spines may be present on at least part of the series, or occasionally absent altogether 15

15 Peripheral paxillar spinelets forming a raised rim, higher than the central ones (Fig. 37); only *c.* 20 relatively long superomarginals in each series, at least at R. *c.* 40 mm; superomarginal spines rare ***Astropecten granulatus natalensis*** John, 1948 (p. 49)

* Cherbonnier & Nataf (1973, *Bull. Mus. natn. Hist. nat. Paris.* Zool. **120**) find up to five actinal plates in specimens from West Africa which they attribute to *A. leptus*.

15 Paxillar spinelets approximately level at the top (Fig. 38); *c.* 30 superomarginals at R *c.* 40 mm, the proximal ones at least broader than long; an enlarged superomarginal spine usually present on most plates beyond the interradii
Astropecten hemprichi Müller & Troschel, 1842 (p. 50)

16 Madreporite concealed by paxillae; actinal interradial areas large and arcs rounded (Figs 43, 44) 17

– Madreporite not concealed by paxillae, though sometimes inconspicuous; disc more or less compact and arcs usually angular (except in *Tethyaster*) (Fig. 45) . . 19

17 Inferomarginal plates projecting laterally beyond the superomarginals and defining the edge of the body in dorsal view (Fig. 43); no spines or enlarged spinelets on the superomarginals . . . **Dipsacaster sladeni capensis** A. M. Clark, 1952 (p. 52)

– Inferomarginals not projecting (Fig. 44); superomarginals with at least one enlarged spinelet or spine projecting. *Plutonaster* 18

18 A single enlarged subambulacral spinelet on at least the distal adambulacral plates
Plutonaster proteus H. L. Clark, 1923 (p. 54)

– Subambulacral spinelets all diminutive **Plutonaster intermedius** Perrier, 1881 (p. 54)

19 Superomarginal plates convex, broad in dorsal view, the paxillar area making up only about half of the arm breadth proximally
Persephonaster roulei euryplax Mortensen, 1933a (p. 53)

– Superomarginals not convex, mainly lateral in position, the paxillar area making up more than half the arm breadth 20

20 Adambulacral plates with furrow margin only slightly convex or with a very obtuse angle, the furrow spines consequently aligned in gently arched fans (Fig. 46)
Psilaster acuminatus Sladen, 1889 (p. 55)

– Adambulacral plates with markedly V-shaped furrow margins, the median spines projecting between the successive tube feet (at least in preserved specimens) (Fig. 47) 21

21 Interradial arcs angular; most superomarginal plates with a small conical spine, sometimes two, at the upper end (Fig. 45)
Bathybiaster vexillifer (Thomson, 1873) (p. 52)

– Interradial arcs rounded, no spines on the superomarginals
Tethyaster pacei (Mortensen, 1925) (p. 55)

22 Papulae localized in restricted areas (popularia), one at the base of each arm (Figs 48, 49); the two series of marginal plates tending to alternate in position and bearing conspicuous bristling armament. BENTHOPECTINIDAE . . . 23

– Papulae occurring between most of the abactinal plates; marginal plates not alternating, often superficially appearing smooth, or with peg-like spines, or diminutive with only small spines or spinelets 25

23 An odd marginal plate in each series in each interradius, the superomarginal one armed with an exceptionally large spine
Benthopecten pedicifer (Sladen, 1889) (p. 56)

– No odd interradial marginal plates 24

24 Papularia V-shaped, extending further distally on each side than midradially (Fig. 48) . . . . **Luidiaster hirsutus** Studer, 1884 (p. 57)

– Papularia oval (Fig. 49) . . . . . **Pectinaster filholi** Perrier, 1885 (p. 57)

25 Each pair of oral plates bearing a large glassy-tipped spine, pointing distally away from the mouth, in preserved specimens (Fig. 50). ODONTASTERIDAE
Odontaster australis H. L. Clark, 1926 (p. 58)

– No single reflexed spine on each jaw 26

26 Marginal plates more or less enlarged with respect to the adjacent plates, only in some ophidiasterids inconspicuous in dorsal view 27

– Marginal plates not enlarged 50

27 Disc large, form stellate or even pentagonal, at least the lower side flat. . . 28

– Disc small, arms cylindrical, interradii more or less angular. OPHIDIASTERIDAE 45

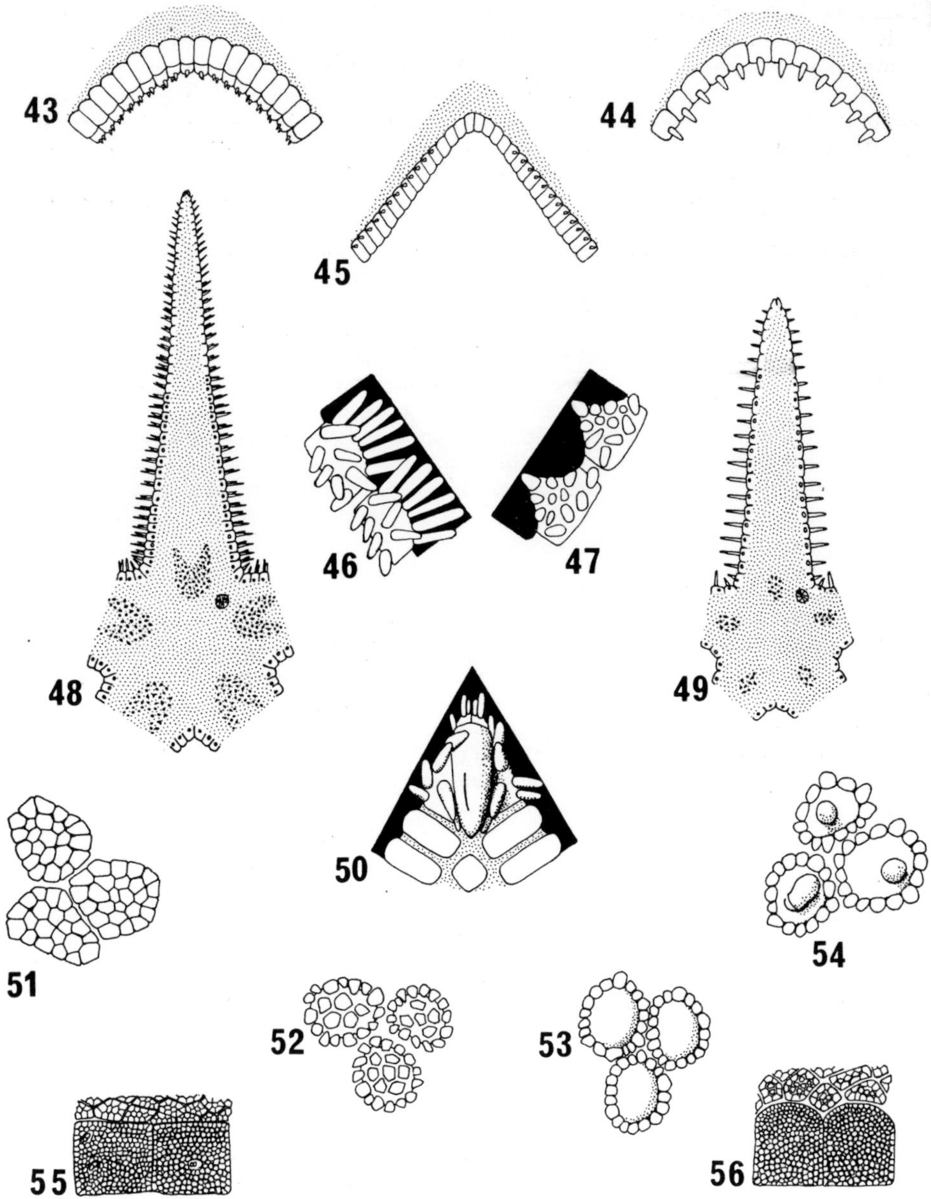

FIGS 43–56 Asteroidea. 43, 44, 45: interradial margin of *Dipsacaster sladeni capensis*, *Plutonaster proteus* and *Bathybiaster vexillifer* (dorsal view). 46, 47: adambulacral plates of *Psilaster acuminatus* and *Bathybiaster vexillifer*. 48, 49: papularia of *Luidiaster hirsutus* and *Pectinaster filholi*. 50: jaw of *Odontaster australis*. 51, 52, 53, 54: abactinal plates of *Ceramaster patagonicus euryplax*, *Pseudarchaster tessellatus*, *Toraster tuberculatus* and *Calliaster baccatus*. 55, 56: superomarginal plates of *Ceramaster patagonicus euryplax* and *C. trispinosus* (dorsal view), the latter reconstructed from the description.

28 Body flattened above as well as below (at least in preserved specimens), superomarginals lateral or dorsolateral in position; papulae single or in small groups . 29
– Body more or less markedly convex above with both series of marginals (at least interradially) ventrolateral in position, or markedly inflated and cushion-like when all but the distalmost superomarginal may be concealed by coarse tubercles (*Asterodiscus*); papulae very numerous, usually forming large patches between the often reticular skeleton. OREASTERIDAE 42
29 Disc small, interradial areas very small and arcs angular, arms narrow basally. ARCHASTERIDAE . . ***Archaster angulatus*** Müller & Troschel, 1842 (p. 59)
– Disc large, interradial areas large and arcs rounded. GONIASTERIDAE . . 30
30 Abactinal plates tabulate and crowned with granules or short spinelets, at least radially (Figs 51, 52) 31
– Abactinal plates of various forms, usually flat but if distinctly convex then the apex is bare (Figs 53, 54) 35
31 Arms short, body form almost pentagonal, R/r less than 2/1, usually less than 1·8/1. *Ceramaster* 32
– Arms more or less prolonged, R/r more than 2/1, usually more than 2·5/1 . . 33
32 Proximal edges of interradial superomarginal plates straight (Fig. 55); four or five furrow spines on each adambulacral plate ***Ceramaster patagonicus euryplax*** H. L. Clark, 1923 (p. 61)
– Proximal edges of superomarginals convex (Fig. 56); three, rarely two, furrow spines . . . . . ***Ceramaster trispinosus*** H. L. Clark, 1923 (p. 62)
33 No enlarged spinelets or inferomarginal plates; superomarginals narrow in dorsal view; some alveolar (tong-shaped) pedicellariae present (at least in the variety *durbanensis*) . . . . . ***Mediaster capensis*** H. L. Clark, 1923 (p. 64)
– One to three enlarged spinelets projecting from the inferomarginal granulation; superomarginals broad in dorsal view, the basal paxillar area breadth approximately half the total arm breadth; alveolar pedicellariae absent. *Pseudarchaster* 34
34 R/r *c.* 3/1 or more . . . ***Pseudarchaster tessellatus*** Sladen, 1889 (p. 65)
– Arms relatively short, R/r 2·0–2·7/1 ***Pseudarchaster brachyactis*** H. L. Clark, 1923 (p. 65)
35 Pedicellariae massive, resembling a bivalved mollusc, with extremely broad though short valves, more or less numerous (Fig. 57). *Hippasteria* 36
– Pedicellariae alveolar (Fig. 58), with spatulate valves or small bivalved (*Anthenoides*, Fig. 59), inconspicuous 37
36 Spines on abactinal and marginal plates heavy and conspicuous ***Hippasteria phrygiana capensis*** Mortensen, 1933a (p. 63)
– Spines inconspicuous . . ***Hippasteria strongylactis*** H. L. Clark, 1926 (p. 63)
37 Many of the larger abactinal and actinal plates, as well as the marginals, armed with a conspicuous blunt tubercle or spine. *Calliaster* 38
– Only small or multiple spinelets or granules present, or else the plates centrally bare 39
38 Adambulacral plates with three or four stout furrow spines; pedicellariae rare; abactinal and marginal spines or tubercles mostly short and blunt ***Calliaster baccatus*** Sladen, 1889 (p. 61)
– Adambulacral plates with six to nine slender furrow spines; pedicellariae more or less numerous; abactinal and marginal spines mostly long and sharp ***Calliaster acanthodes*** H. L. Clark, 1923 (p. 60)
39 Marginal plates very few, only three in each superomarginal series and markedly swollen; body almost pentagonal . ***Sphaeriodiscus bourgeti*** (Perrier, 1894) (p. 65)
– Marginal plates more or less numerous and only the distal superomarginals sometimes convex; arms more or less prolonged (though small specimens of *Toraster* may approach a pentagonal form) 40
40 Actinal armament of uniform short blunt spinelets, almost granuliform, obscuring the limits of the plates; disc large and arms short except in larger specimens (R > 70 mm) . . . . . ***Toraster tuberculatus*** (Gray, 1847) (p. 66)

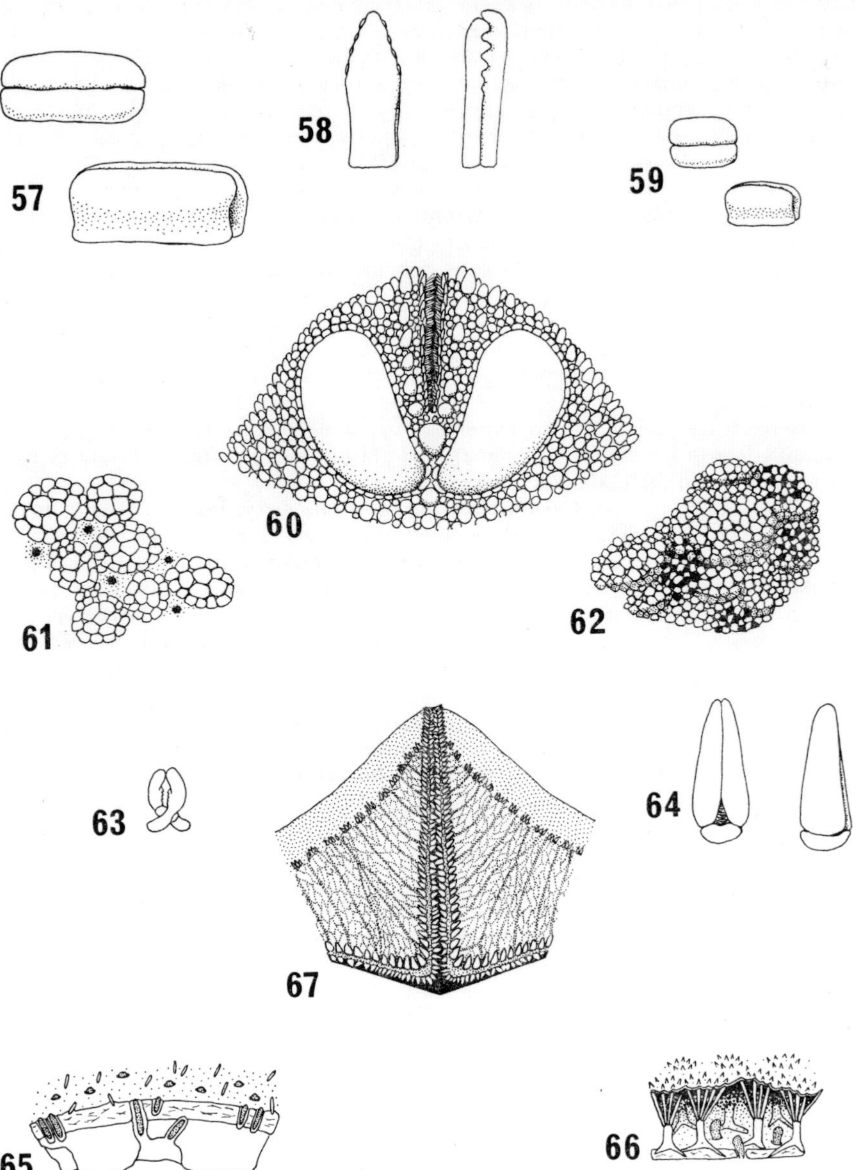

Figs 57–67 Asteroidea. 57, 58, 59: pedicellariae of *Hippasteria phrygiana capensis*, *Calliaster acanthodes* and *Anthenoides marleyi*. 60: arm tip of *Asterodiscides elegans*. 61, 62: abactinal plates with papulae of *Austrofromia schultzei* and *Linckia guildingi*. 63, 64: crossed and straight pedicellariae of *Sclerasterias eustyla*. 65, 66: sections of body wall of *Spoladaster brachyactis* and *Pteraster capensis*. 67: ventral view of *Spoladaster brachyactis*.

40 Actinal plates armed with spaced granules and pedicellariae, their limits distinct . 41

41 Adambulacral plates with fairly numerous furrow spines (seven or eight at R 90 mm); marginal plates not bordered by a continuous ring of granules
Anthenoides marleyi Mortensen, 1925 (p. 60)

– Furrow spines few, only three, sometimes two, at R 40 mm; marginal plates each with a conspicuous border of granules as well as the spaced median granules or tubercles . . . . ***Cladaster macrobrachius*** H. L. Clark, 1923 (p. 62)

42 Arms projecting, triangular, high midradially with a spaced row of tubercles . 43

– Body pentagonal, cushion-like or (in most preserved specimens) flattened on top . 44

43 All or most inferomarginals armed with a tubercle or short spine, the outline of the body serrated interradially; no abruptly projecting distal superomarginals
Pentaceraster mammillatus (Audouin, 1826) (p. 68)

– Inferomarginals unarmed, interradial arcs smooth; several conspicuous conical or spiniform superomarginals projecting at the tip of each arm
Protoreaster lincki (de Blainville, 1834) (p. 68)

44 An isolated pair of large bare marginal plates conspicuous at the tip of each ray (Fig. 60), the whole surface including the remaining marginals covered with granules and tubercles intermixed . . ***Asterodiscides elegans*** Gray, 1847 (p. 67)

– No conspicuous distal pair of superomarginals; tubercles sparse, limited to the meshes of the reticular skeleton above, though continuous below
Culcita schmideliana (Retzius, 1805) (p. 67)

45 Abactinal plates forming regular longitudinal series 46

– Abactinal plates irregular 47

46 Thick smooth skin covering the entire surface . . . . ***Leiaster*** sp. (p. 70)

– Skin inconspicuous, not obscuring the fine granulation; some of the plates bare centrally . . . . ***Hacelia capensis*** Mortensen, 1925 (p. 70)

47 Furrow spine spiniform and projecting; papulae single dorsally (Fig. 61), also present ventrally . . . ***Austrofromia schultzei*** (Döderlein, 1910) (p. 69)

– Furrow spines superficially appearing more or less granuliform, hardly projecting above the surface; papulae in groups dorsally (Fig. 62), absent ventrally. *Linckia* . . 48

48 No granules between the furrow spines on the vertical face of the adambulacral plates; subambulacral spines contiguous obliquely with the furrow spines, giving a herringbone-like appearance to the furrow . ***Linckia guildingi*** Gray, 1840 (p. 71)

– Granules present between the furrow spines; subambulacral spines usually single, tuberculiform, set back from the furrow spines and isolated in the general granulation 49

49 Arms normally five in number and regular in shape, fairly stout and blunt at the tip, generally finger-like in shape; madreporite single; colour in life blue (sometimes rose) when R > *c.* 50 mm . . . ***Linckia laevigata*** (Linnaeus, 1758), (p. 72)

– Arms five or six, slender and often attenuated distally, usually not all equal in length; usually two madreporites; colour drab or patchy
Linckia multifora (Lamarck, 1816) (p. 72)

50 Crossed (forcipiform) pedicellariae absent; tube feet normally in two rows; shape very varied 51

– Numerous small crossed pedicellariae present (Fig. 63), often in wreaths around the spines; tube feet in four rows except in the brisingids; disc small and arms narrow and more or less elongated. 79

51 Surface skin thick and obvious; body form ranging from pentagonal to short-armed stellate; abactinal skeleton either reduced and embedded in the thickened body wall or concealed by a supradorsal membrane supported by very tall paxillae 52

– Skin inconspicuous; arms either well developed and almost cylindrical, sometimes numbering more than five, or short and flattened so that the form is stellate or even pentagonal; skeletal plates well developed and distinguishable on account of their independent armament of spinelets, granules or spines 61

52 Body wall thickened, abactinal skeleton reduced; no supradorsal membrane (Fig. 65). PORANIIDAE 53
– Body wall not thickened but roofed above by a supradorsal membrane supported by the long spinelets of the paxillae (Fig. 66). PTERASTERIDAE . . . 54
53 Papulae restricted to two parallel bands on each arm, one each side of the mid-radial line; marginal plates very reduced and spineless
Chondraster elattosis H. L. Clark, 1923 (p. 73)
– Papulae arranged in patches all over the upper side, as in the meshes of a reticulum; inferomarginal plates bearing some small spines or spinelets, sometimes forming a continuous fringe (Fig. 67) . . **Tylaster meridionalis** Mortensen, 1933* (p. 74)
Spoladaster brachyactis (H. L. Clark, 1923)* (p. 74)
54 Spines of adambulacral plates in series at right angles to the furrow and linked by skin to form transverse fans (Figs 68, 69); general structure fairly strong, not easily damaged 55
– Spines of adambulacral plates few and isolated in individual sheaths of skin, no fans except for the longitudinal ventrolateral ones (Fig. 70); general structure delicate. *Hymenaster* 59
55 Consecutive adambulacral plates and their fans alternating in form, every second plate projecting further over the furrow than the adjacent ones and the tube feet also staggered in alignment, tending to form four longitudinal series (Fig. 68)
Diplopteraster multipes (M. Sars, 1877) (p. 82)
– Consecutive adambulacral plates similar and tube feet in two rows (Fig. 69). *Pteraster* 56
56 Oral spines projecting over furrow margin not linked by skin into fans but isolated (Fig. 71) . . . . . . **Pteraster fornicatus** Mortensen, 1933a (p. 85)
– Oral marginal spines linked by webs of skin (Figs 72, 73) 57
57 A single fan of oral marginal spines on each jaw, continuous across the apex between the spines of the two oral plates (Fig. 72) . **Pteraster capensis** Gray, 1847 (p. 84)
– Each oral plate with its own marginal fan, though the two fans on each jaw may overlap apically (Fig. 73) 58
58 Paxillae with five or six similar spines **Pteraster flabellifer** Mortensen, 1933a† (p. 85)
– Paxillae usually with about four spines, one of them distinctly larger than the others . . . . **Pteraster affinis** E. A. Smith, 1876† (p. 84)
59 Adambulacral spines numbering three . **Hymenaster latebrosus** Sladen, 1882 (p. 83)
– Adambulacral spines numbering one or two (only at R > 100 mm may occasional plates have three) 60
60 Adambulacral spines two . . . **Hymenaster lamprus** H. L. Clark, 1923 (p. 83)
– Adambulacral spines single . . **Hymenaster gennaeus** H. L. Clark, 1923 (p. 83)
61 Abactinal skeleton formed of imbricating plates with relatively small spaces between less than the area of the plates (Figs 74, 75); body form more or less markedly flattened with a ventrolateral angle. ASTERINIDAE 62
– Abactinal skeleton reticular, the area of the spaces about equal to or more than that of the plates (Figs 76, 77); arms rounded laterally 72
62 Body more or less thickened, central height at least half r; papulae occurring all over the abactinal side except perhaps peripherally 63
– Body very thin and leaf-like; papulae restricted to a few longitudinal series mid-radially. *Anseropoda* 70
63 Abactinal armament very reduced, spinelets present only in restricted areas, if at all; skin covering the plates thick and smooth
Disasterina leptalacantha africana Mortensen, 1933a (p. 78)

* See p. 73.

† Possibly H. L. Clark's South African record of *P. affinis* is based on *P. flabellifer* (then undescribed) like the 'Africana' specimen which I misidentified as *P. affinis* in 1952. If this is so, then *P. affinis* can be deleted from the fauna list.

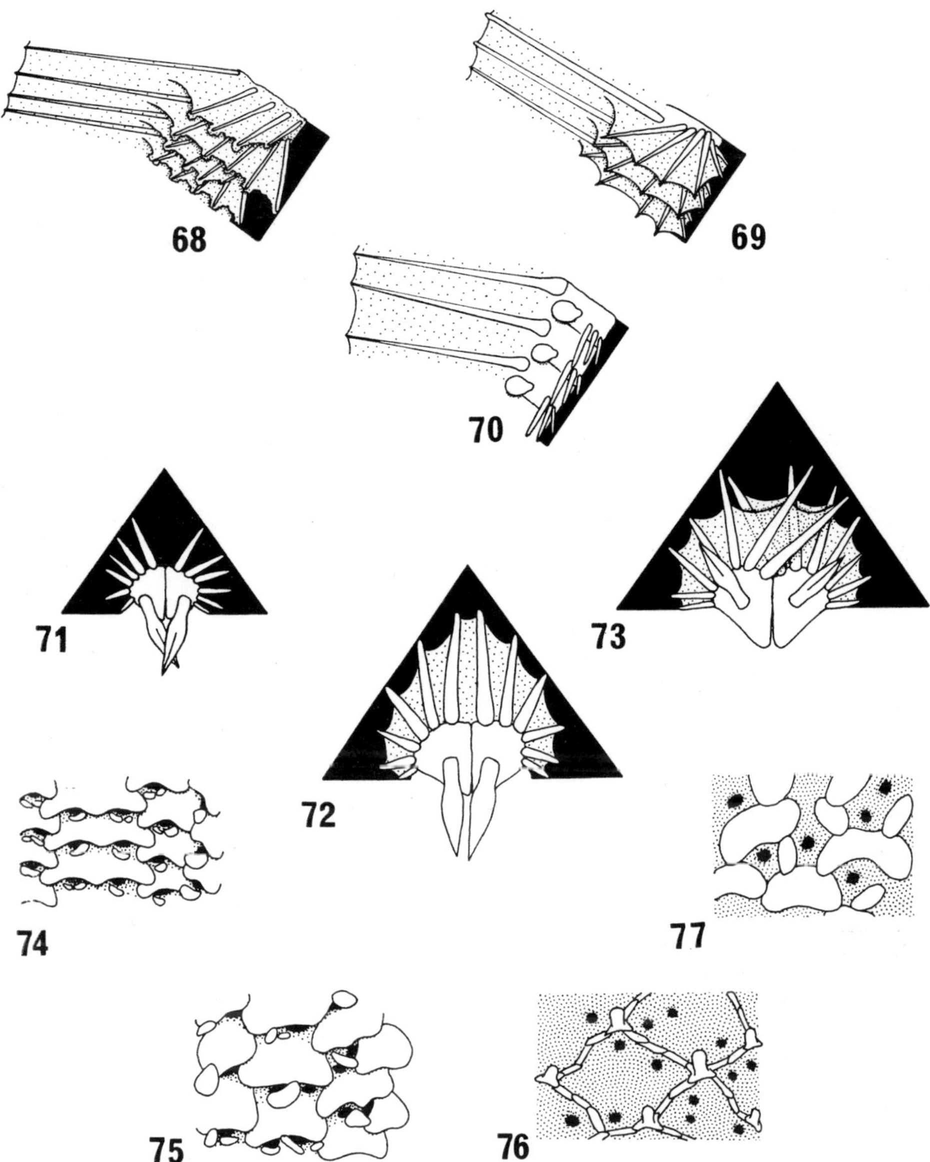

Figs 68–77 Asteroidea. 68, 69, 70: adambulacral and actinal armament of *Diplopteraster multipes*, *Pteraster flabellifer* and *Hymenaster latebrosus*. 71, 72, 73: jaws of *Pteraster fornicatus*, *P. capensis* and *P. flabellifer*. 74, 75, 76, 77: abactinal plates of *Patiriella exigua*, *Patiria granifera*, *Crossaster penicillatus* and *Henricia ornata*.

63 Abactinal spinelets or granules well developed ; skin not noticeable 64

64 Abactinal armament superficially appearing coarse and granuliform ; body form almost or quite pentagonal ; single spines on many actinal plates, also single subambulacral and suboral spines. *Patiriella* 65

– Abactinal armament more or less elongate ; body stellate ; multiple actinal spines, often a fan or a cluster. 66

65 Genital openings on actinal side, two in each interradius
Patiriella exigua (Lamarck, 1816) (p. 81)

– Genital openings abactinal . . **_Patiriella dyscrita_** (H. L. Clark, 1923) (p. 80)

66 Abactinal plates of papular areas somewhat variable in size and shape, the small plates rarely very numerous and the larger ones not markedly elongate in a tangential direction ; size rarely exceeding R 30 mm. *Asterina* 67

– Abactinal plates of papular areas of two magnitudes, large crescentic ones and more or less numerous much smaller ones interstitially or grouped to separate the papular pores but lacking distally and peripherally where the papulae are also lost (Fig. 75) ; size often exceeding R 30 mm. *Patiria* 68

67 Shape stellate, arms of moderate length with broad blunt tips, R/r usually 2·0–2·5/1
Asterina burtoni Gray, 1840 (p. 77)

– Shape almost pentagonal, the short arms petaloid, R/r 1·3–1·5/1
Asterina gracilispina H. L. Clark, 1923 (p. 77)

68 Arms triangular, their tips only slightly blunted
Patiria stellifera (Möbius, 1859)/(p. 80)

– Arms finger-like, broadly blunted at the tip 69

69 Distal primary abactinal plates similar in size to the proximal ones
Patiria granifera Gray, 1847 (p. 79)

– Distal primary plates markedly enlarged and rounded
Patiria formosa Mortensen, 1933a (p. 78)

70 Nine arms . . . . . . **_Anseropoda novemradiata_** (Bell, 1905a) (p. 76)

– Five arms 71

71 Actinal plates with a transverse series of three or four elongated spinelets besides several other smaller spinelets ; known to reach a large size, R 130 mm
Anseropoda grandis Mortensen, 1933a (p. 75)

– Actinal plates with a transverse series of up to ten spinelets ; R only 15 and 16 mm in the two specimens recorded **_Anseropoda habracantha_** H. L. Clark, 1923 (p. 76)

72 Mouth plates large ; furrows broad ; abactinal and marginal plates paxilliform, the spinelets on the inferomarginals at least forming conspicuous tufts (Fig. 78). SOLASTERIDAE 73

– Mouth plates small ; furrows narrow ; abactinal plates not paxilliform and no special marginal armament 74

73 Eight to ten arms* . . . . . **_Crossaster penicillatus_** Sladen, 1889 (p. 86)

– Five arms . . . . . . **_Lophaster quadrispinus_** H. L. Clark, 1923 (p. 86)

74 Nine to twenty-two arms and madreporites more or less numerous ; armament of upper side consisting of huge spike-like spaced spines. ACANTHASTERIDAE
Acanthaster planci (Linnaeus, 1758) (p. 88)

– Usually five arms ; armament of coarse or fine tubercles or spinelets or small spines < 5 mm long 75

75 Abactinal armament consisting of widely spaced, almost spherical knobs, markedly enlarged at the arm tips ; the only specimen known extremely large, R 300–330 mm. MITHRODIIDAE . . **_Mithrodia gigas_** Mortensen, 1935 (p. 87)

* An eight-armed specimen of *Solaster*, possibly *S. paxillatus* Sladen or *S. tropicus* Fisher (known from southern Japan and the East Indies respectively), was collected off Natal (29°40·6′S 32°32·5′E, 722–768 metres) in 1975 and runs down here. It is easily distinguished from *Crossaster pencillatus* by the much finer abactinal and marginal skeleton and spinelets.

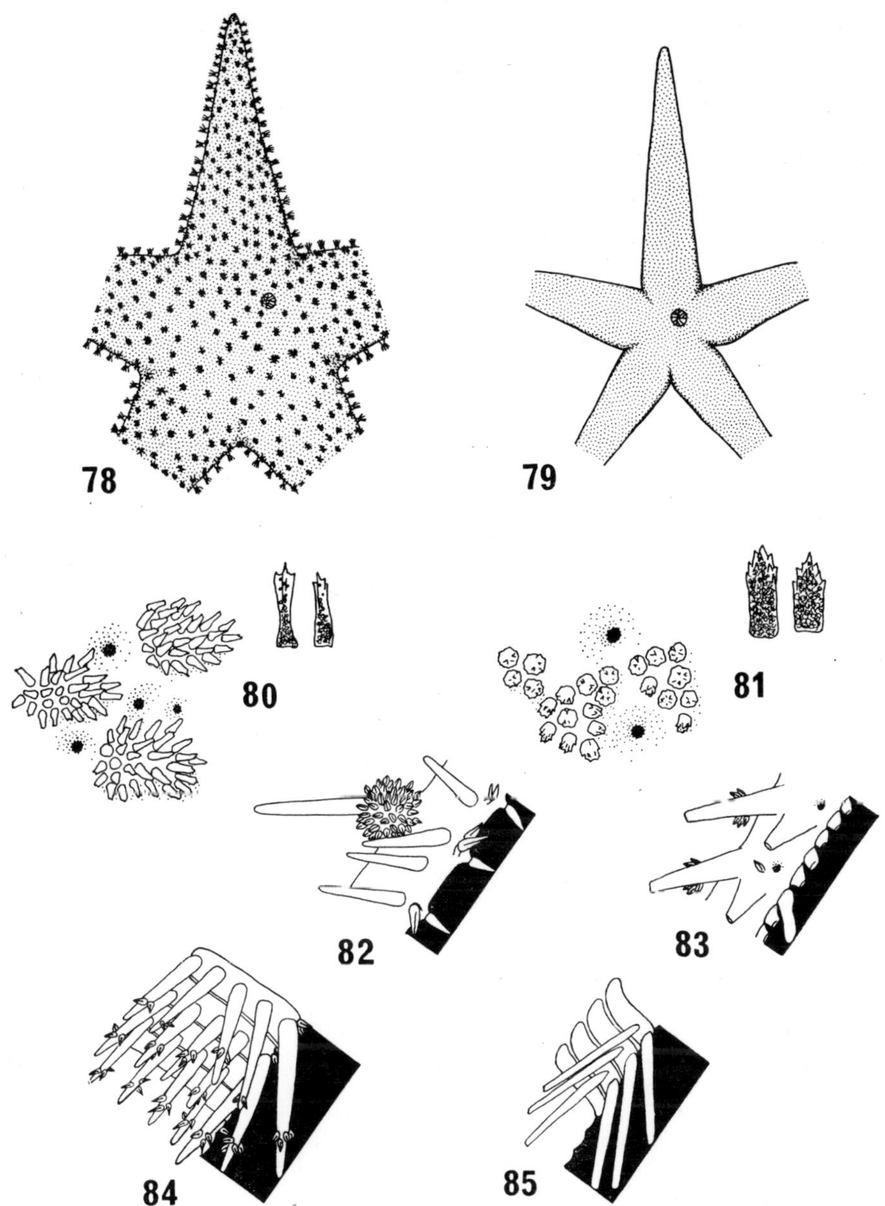

FIGS 78–85 Asteroidea. 78, 79 : outlines of *Lophaster quadrispinus* and *Henricia abyssalis*. 80, 81 : abactinal plates and armament and enlarged spines of *Coronaster volsellatus* and *Coscinasterias* sp. 84, 85 : adambulacral plates and armament of *Perissasterias polyacantha* and *Sclerasterias stenactis*.

75 Abactinal armament of small spinelets or spines, sometimes almost granuliform, not coarser distally ; size unremarkable, R rarely exceeding 70 mm. ECHINASTERIDAE 76

76 Abactinal plates armed with widely spaced spines 1–4 mm long ; disc inflated and arms relatively short, R/r *c.* 2/1 . ***Poraniopsis capensis*** H. L. Clark, 1923 (p. 90)

– Abactinal plates armed with numerous spinelets, rarely > 1 mm long, in clusters or linear series along the reticulations of the skeleton, sometimes spaced ; arms relatively long, R/r usually 4–5/1 77

77 Armament very fine with numerous slender, terminally flanged spinelets on all the plates (Fig. 80), even the adambulacral ones ; actinal plates numerous, extending along the arms ; arms often attenuated (Fig. 79) ***Henricia abyssalis*** (Perrier, 1894) (p. 88)

– Armament of coarser more solid spinelets, superficially often appearing granuliform (Fig. 81), only three or four even coarser ones on each adambulacral, tending to form a linear series at right angles to the furrow ; actinal plates few ; arms usually moderately stout 78

78 Actinal papulae present, sometimes rare ***Henricia retecta*** (H. L. Clark, 1923) (p. 90)

– No actinal papulae . . . . . ***Henricia ornata*** (Perrier, 1869) (p. 89)

79 Tube feet in four series ; abactinal skeleton usually with well-developed longitudinal series for the entire arm, also linked transversely ; found at all depths. ASTERIIDAE 80

– Tube feet in two series ; abactinal skeleton very reduced, consisting of transverse ribs in the proximal part of the arm only ; very delicate in construction, arms nearly always broken off from the disc during collection ;* deep sea only. BRISINGIDAE 90

80 Most or all adambulacral plates with a single spine (Figs 82, 83) 81

– Adambulacrals with predominantly two or more spines (Figs 84, 85) 84

81 Five arms. *Marthasterias* 82

– Six to twelve arms 83

82 Abactinal spines numerous, scattered all over, usually blunt ***Marthasterias glacialis*** forma ***africana*** (Müller & Troschel, 1842) (p. 94)

– Abactinal spines few, usually tapering and restricted to the carinal plates on the arms, though some dorsolateral ones may occur ***Marthasterias glacialis*** forma ***rarispina*** (Perrier, 1875) (p. 94)

83 Nine to eleven very long fragile arms ; only one madreporite ; not recorded as fissiparous ; inferomarginal plates each with a single spine (Fig. 82) ; skeleton delicate . . . . . . ***Coronaster volsellatus*** (Sladen, 1889) (p. 92)

– Usually eight arms of moderate length with stout reticular skeleton ; at least two madreporites ; subject to fission across the disc and regeneration ; two inferomarginal spines (Fig. 83) . . ***Coscinasterias calamaria*** (Gray, 1840) (p. 92)

84 Adambulacral plates with three to seven spines, all bearing several pedicellariae (Fig. 84). *Perissasterias* 85

– Adambulacral plates each with predominantly two spines not bearing pedicellariae, though there may be some straight pedicellariae at the edge of the furrow around the bases of the spines (Fig. 85) 87

85 Seven short stout arms. . . ***Perissasterias heptactis*** H. L. Clark, 1926 (p. 95)

– Five arms† 86

86 Carinal (mid-radial) spines 'bluntly pointed', length : breadth *c.* 4 : 1 ; R often exceeding 250 mm. . . . . ***Perissasterias polyacantha*** H. L. Clark, 1923 (p. 95)

* The multiradiate asteriid *Coronaster volsellatus* is similarly very breakable; it can be distinguished by its longitudinal series of carinal and superomarginal plates, large straight pedicellariae, abactinal papulae and at least partially quadriserial tube feet.

† According to H. L. Clark the arms are much stouter in *P. heptactis* but in fact the ratios derived from the measurements he gives of mean R/maximum arm br are 4·3/1 in *P. heptactis*, 4·75/1 in *P. obtusispina* and only in *P. polyacantha* perhaps significantly different at 5·2 and 5·5/1.

86 Carinal spines 'low and capitate', the holotype (and only known specimen) with R *c.* 160 mm* . . . ***Perissasterias obtusispina*** H. L. Clark, 1926 (p. 95)

87 Abactinal and marginal plates with conspicuous single conical spines, each surrounded by a wreath of crossed pedicellariae. *Sclerasterias* 88

– Abactinal spines numerous, irregular and not themselves bearing pedicellariae, which are mostly scattered between them 89

88 No actinal spines (at least none described) ***Sclerasterias stenactis*** H. L. Clark, 1926 (p. 96)

– An actinal spine present adjacent to the two inferomarginal spines for most of the arm length . . . . . . ***Sclerasterias eustyla*** (Sladen, 1889) (p. 96)

89 Small, fissiparous six-armed species, R not known to exceed 15 mm; inferomarginal spines markedly clavate; arms blunt ***Allostichaster capensis*** (Perrier, 1875) (p. 91)

– Non-fissiparous and five-armed, R up to *c.* 90 mm; inferomarginal spines spatulate or tapering; arms attentuated distally . ***Cosmasterias felipes*** (Sladen, 1889) (p. 93)

90 Ribs or costae on proximal parts of the arms broad and only narrowly separated by areas with numerous thin spineless plates ***Stegnobrisinga splendens*** H. L. Clark, 1926 (p. 98)

– Costae narrow, consecutive ones widely separated, the intercostal areas without plates . . . . ***Brisinga cricophora*** Sladen, 1889 (p. 97)

Order PLATYASTERIDA Spencer

Asteroids with five or more arms; the main ossicles forming transverse series ventrally from the ambulacrals to the single series of (infero-)marginals, with fasciolar grooves on the ventral side at right angles to the long axis of the arm; lateralmost row of abactinal paxillae corresponding in number to the inferomarginals but otherwise barely modified as superomarginals; pedicellariae sessile, more or less obviously modified from spines; tube feet in two rows tapering to end in a rounded knob.

Family **LUIDIIDAE** Verrill

See: Fisher, 1911: 105; Spencer & Wright, 1966: U42.

A family of Platyasterida with massive inferomarginal plates adjoining the adambulacrals, on which are superimposed one or several small actinal plates; arms strap-like, flattened in preserved specimens; lateral abactinal paxillae arranged in two or more regular longitudinal series and also forming short transverse series with the superomarginal paxillae, though the median paxillae are irregularly arranged; pedicellariae incipient (without special sockets or foramina), with two or three valves, ventral ones spiniform with a basal piece, dorsal with short rounded valves, the abactinal ones usually resembling split granules.

Genus ***LUIDIA*** Forbes

See: Fisher, 1911: 105.

Diagnosis as for the family. Most species are delicate and the arms often broken in preservation.

* I doubt whether spine shape is sufficiently constant to be of specific weight.

Luidia africana Sladen

Luidia africana Sladen, 1889 : 256–258, pl. 44, figs 1, 2, pl. 45, figs 1, 2 ; Bell, 1905a : 245–246 ; Mortensen, 1933a : 239–240, figs 3, 4 ; Madsen, 1950, *Atlantide Rep.* **1** : 188–192, fig. 4a–k, pl. 16, figs 3, 4 ; A. M. Clark, 1952 : 195 ; 1953 : 393–394, figs 10, 11 ; Morgans, 1962 : 317, 322, 325 ; Day, Field & Penrith, 1970 : 79.

R up to *c.* 190 mm ; R/r 8–10/1.

A species of *Luidia* with five arms ; abactinal paxillae mostly small and irregularly placed but for two matching longitudinal rows adjacent to the markedly elongated superomarginal paxillae, which they outnumber by about two to one ; one to four middle spinelets on the paxillae, none of them markedly enlarged ; usually two or three large inferomarginal spines, up to six in the largest specimens ; actinal plates in a single series along the arms ; three large adambulacral spines on each plate ; short pedicellariae present on many abactinal paxillae, at least in larger specimens and more slender two-bladed ones on the actinal plates and oral plates ; no patchy coloration dorsally, usually a darker mid-line along the arms.

Type locality. Simon's Bay, False Bay.

Southern African records. 26/14/d ; 30/15/d ; 31/17/d ; 32/16/d ; 32/17/d ; 33/17/d ; 34/17/d ; 34/18/d ; 34/18/FB/s ; 34/19/s ; 32/21/– ; 35/21/d ; 34/22/d ; 34/24/d ; 34/25/d ; 32/28/– ; [? 29/31/s, possibly = *L. sagamina aciculata*] : 66–130 metres.

Nature of bottom. Rocky ; green mud ; coarse sand and shell ; ? sand ; rock and shell ; coarse khaki sand.

Luidia sp. aff. ***L. avicularia*** Fisher

See : Fisher, 1919 : 172–175, pl. 43, fig. 1, pl. 44, fig. 2, pl. 46, fig. 2.

A species of *Luidia* with about nine arms ; abactinal paxillae small and more or less rounded, irregular but for two (possibly three in larger specimens) matching longitudinal rows adjacent to the superomarginal paxillae, of which two correspond usually to three abactinal paxillae ; these lateral paxillae often with one, sometimes two central paxillar spinelets enlarged and projecting above the general surface ; superomarginal paxillae usually with two unequal slightly larger spinelets ; three to five moderately long inferomarginal spines ; actinal plates in a single series along the arms ; usually four large adambulacral spines on each plate, the two lateralmost arranged parallel to the furrow, the proximal one of them sometimes reduced or replaced by a two-bladed pedicellaria ; other two-bladed pedicellariae on the adoral face of each oral plate, deep in the furrow above the curved furrow spine on each adambulacral and on some of the actinal plates.

Type locality. Of *L. avicularia*, Philippines, 215 metres.

Southern African record. 32/28/55 metres.

Note : On the evidence of a single regenerating and poorly preserved specimen, *L. avicularia* cannot be added to the fauna of southern Africa with any confidence.

Luidia maculata Müller & Troschel

Luidia maculata Müller & Troschel, 1842 : 77–78 ; Peters, 1852 : 178 ; Döderlein, 1920, *Siboga Exped.* **46b** : 262–266, figs 4, 13, 16, 23, 24 ; Mortensen, 1933a : 238–239 ; B. I. Balinsky *In* : Macnae & Kalk, 1969 : 99 ; Jangoux, 1973 : 7–9 ; Day, 1974 : 94.

R up to at least 300 mm ; R/br (r is hard to measure with accuracy in these mutiradiate specimens) 9–11/1 when R exceeds *c.* 100 mm.

A species of *Luidia* with usually nine arms, sometimes seven or eight ; abactinal paxillae squarish, forming six to eight regular matching longitudinal rows each side, corresponding in number to the superomarginal paxillae ; adjacent paxillae similar in size, with fine peripheral spinelets encircling 6–20 (usually 15–20 at R > 150 mm) shorter blunt spinelets which may be somewhat inequal but are never abruptly enlarged or projecting, the general surface smooth ; three to five stout but short inferomarginal spines ; several series of actinal plates extending along the arms ; three large adambulacral spines on each plate ; no adoral pedicellariae on the oral plates and sometimes none on the abactinal paxillae but usually numerous three-bladed ones on the actinal plates and often also shorter ones on the inferomarginals ; dorsally beautifully patterned with dark and light colours.

Type locality. Japan.

Southern African records. 29/31/s ; 26/33/i ; 23/35/s ; intertidal–33 metres.

Luidia sagamina aciculata Mortensen

Luidia aciculata Mortensen, 1933b ; *Vidensk. Meddr dansk. naturh. Foren.* **93** : 425–426, fig. 7, pl. 20, figs 7–12 ; Fisher, 1940 : 268–269, fig. M5.
Luidia sagamina var. *aciculata* : Madsen, 1950, *Atlantide Rep.* **1** : 199–203, figs 6, 7.

R up to 140 mm ; R/r 8–10·5/1.

A species of *Luidia* with five arms ; abactinal paxillae mostly small and irregularly placed except for about three matching longitudinal rows adjacent to the slightly elongated superomarginal paxillae, 16–19 abactinal paxillae corresponding to ten superomarginals ; many lateral paxillae with only one central paxillar spinelet or spine present and this is markedly larger than the peripheral spinelets, much stouter and 0·5–1·0 mm long ; no enlarged superomarginal spinelets ; usually about three large inferomarginal spines, tending to alternate in alignment on consecutive plates ; one series of actinal plates extending along the arms ; usually three adambulacral spines on each plate ; short two-bladed pedicellariae sometimes replacing the central spinelets on some of the abactinal paxillae, longer two-bladed ones on the adoral ends of the oral plates and three-bladed ones on some or many of the actinal plates ; no patchy coloration.

Type locality. St Helena.

Southern African record. Probably 30/31/s ; 78–90 metres.

Luidia savignyi (Audouin)

Asterias savignyi Audouin, 1826 *In* : *Description de l'Egypte.* Histoire naturelle. Paris. **1** (4) : 208. [Savigny, Echinodermes pl. 3, fig. 1.]

Luidia savignyi: Lopes, 1939: 78, pl. 1, fig. 1; A. M. Clark, 1953: 385–386, pl. 40, fig. 2; B. I. Balinsky *In*: Macnae & Kalk, 1969: 99.

R up to 200 mm; R/r 6–8/1.

A species of *Luidia* with usually seven arms; abactinal paxillae moderate to relatively large, varying in size, three to five longitudinal rows of squarish lateral ones usually distinguishable and tending to forming transverse series corresponding with the adjacent superomarginal paxillae but the regularity disturbed to some extent by sporadic enlarged paxillae of the third, fourth and sometimes fifth from outermost rows which bear a relatively massive conical spine, similar in magnitude to the inferomarginal spines; other paxillae usually with five to ten short blunt central spinelets distinctly coarser than the peripheral ones; superomarginal paxillae not enlarged; three to five large inferomarginal spines; a single series of actinal plates extending along the arms; three large adambulacral spines on each plate; pedicellariae, mostly three-bladed, on many actinal plates and sometimes shorter ones on the inferomarginals, but not on the oral plates or abactinal paxillae; colour conspicuously patterned dark and light dorsally.

Type locality. Northern Red Sea (or Gulf of Suez).

Southern African records. 29/31/s; 28/32/s; 26/33/i; 23/35/s; intertidal–49 metres.

Order PAXILLOSIDA Perrier

Asteroids normally with five arms (in the taxa represented); main ossicles forming longitudinal series ventrally, with a marginal frame consisting of two well-developed series of plates, inferomarginals and superomarginals, distinct from the furrow frame, which consists of the ambulacrals and adambulacrals, the two frames abutting distally but linked proximally by actinal plates and corresponding only to a limited extent transversely, so that when fasciolar grooves are developed some of them are incomplete or irregular in alignment; pedicellariae simple, sessile, spiniform; tube feet in two rows, usually tapering to points but with sucking discs in the Benthopectinidae.

Family **PORCELLANASTERIDAE** Sladen

See: Fisher, 1911: 22; Spencer & Wright, 1966: U47.

A family of Paxillosida with body form more or less compact, the large disc often almost pentagonal and the arms projecting rather abruptly, the interradial arcs nearly flat proximally; terminal plates of arms relatively large; abactinal plates often reduced and hardly to be termed paxilliform, the spinelets few and sometimes restricted to the proximal plates; papulae extending over most of the upper surface; plates of the two marginal series few in number (often less than ten in each series on each side of the arm), conspicuous, though thin, opposite each other, largely bare and with one or more lamelliform cribriform organs, interradially at least, spanning both series vertically; pedicellariae, when present, simple with two blunt-spiniform blades, apparently absent in adults and absent altogether in some species ; tube feet pointed.

Porcellanaster ceruleus Wyville Thomson

Porcellanaster ceruleus Wyville Thomson, 1877, *The Voyage of the 'Challenger'. The Atlantic.* **1** : 378–380, figs 97, 98.
Porcellanaster cæruleus : H. L. Clark, 1923 : 239 ; Madsen, 1961, *Galathea Rep.* **4** : 126–142, figs 22–24.

R up to 36 mm ; R/r (when R > 12 mm) 2·1–3·4/1, mean 2·7/1.

A species of Porcellanasteridae with fairly broad arms ending in blunt tips, arising from a broad, easily distended disc often surmounted by a central (so-called epiproctal) cone ; general appearance rather naked, abactinal plates inconspicuous, only the proximal ones in adults bearing a few spaced spinelets ; superomarginals relatively few, less than ten, not contiguous across the arms and bare but for a small spine at the upper end, lacking on some or most plates, especially of smaller specimens ; a single broad cribriform organ between the marginals in each interradius ; up to five series of usually naked polygonal actinal plates, forming a pavement in larger specimens and arranged in rows both tangentially and transversely ; adambulacral plates with usually two furrow spines ; small two-bladed pedicellariae only present in some young specimens, occurring on the abactinal, adambulacral and oral plates.

TYPE LOCALITY. Off New York, U.S.A., 2270 metres.

SOUTHERN AFRICAN RECORDS. 33/17/vd ; 34/17/vd ; 34/18/vd ; 1460–2176 metres.

NATURE OF BOTTOM. Green or grey mud.

Family **ASTROPECTINIDAE** Gray

See : Fisher, 1911 : 37 ; Spencer & Wright, 1966 : U45.

A family of Paxillosida with fairly long pointed arms ; interradial areas small (*Astropecten*) to large, the arcs more often angular than rounded ; terminal arm plates inconspicuous ; abactinal plates small, paxilliform, with many spinelets ; papulae widespread over most of the paxillar area ; plates of the two marginal series more or less numerous and conspicuous, opposite each other and armed with an almost continuous covering of spinelets and spines ; no cribriform organs ; pedicellariae, if present, simple fascicular ; tube feet pointed.

Genus ***ASTROPECTEN*** Gray

See : Fisher, 1911 : 55.

A genus of Astropectinidae with a small to medium-sized disc and fairly long triangular, or sometimes petaloid, arms ; superomarginals forming a more or less

broad border to the paxillar area, with or without one or a few enlarged spines; inferomarginals always with at least one very large spine directed horizontally from the upper end of each plate, together forming a conspicuous fringe around the body; actinal plates small and few, one to six each side of each interradius in a single longitudinal series (in the species represented); adambulacrals with an angular furrow margin armed usually with three furrow spines backed by a row of two large subambulacral spines and then several more irregular spines.

Astropecten anacanthus H. L. Clark

Astropecten anacanthus H. L. Clark, 1926 : 4–5, pl. 1, figs 1, 2; Mortensen, 1933a : 233; pl. 8, figs 7, 8.

R up to 44 mm; R/r *c.* 4/1.

A species of *Astropecten* with arms of moderate length, tapering fairly evenly; madreporite hidden by paxillae; paxillar area about two-thirds the arm breadth, paxillae mostly with 20–25 blunt spinelets, four or five paxillae opposite two consecutive superomarginals; superomarginals mainly dorsal in position and mostly broader than long, covered with granules but no enlarged spines; inferomarginals with only one very large spine and a small one proximal to it; three actinal plates in each series.

Type locality. SE. of Durban, 208–280 metres.

Southern African records. 30/31/d; 29/31/d; 125–280 metres.

Nature of bottom. Mud, sand and shells.

Astropecten antares Döderlein

Astropecten antares Döderlein, 1926 : 6–7, pl. 1, fig. 4, pl. 4, fig. 3; Mortensen, 1933a : 233–234, pl. 8, figs 3–6; A. M. Clark, 1952 : 195; Day, Millard & Harrison, 1952 : 396.

R up to 52 mm; R/r 2·5–3·0/1.

A species of *Astropecten* with relatively short, petaloid arms, tapering abruptly in their distal halves but hardly at all proximally; paxillar area about half the arm breadth, larger paxillae mostly with 25–30 similar blunt spinelets, two paxillae opposite each superomarginal; superomarginals mainly dorsal in position and markedly broader than long, without any enlarged spines; inferomarginals with one large but slender aboral spine and one smaller adoral spine at the upper end, with two spines of intermediate size immediately below and a third row of usually three smaller spines; one to three, usually two, actinal plates in each series. Colour 'pinkish above, creamy-white below'.

Type locality. Mozambique, 9 metres.

Southern African records. 34/21/s; 34/22/s; 34/23/s; 33/25/s; 31/29/s; ? 26/33/s (Döderlein); 6(? 2)–64 metres.

Nature of bottom. Sand; muddy sand; rock.

Astropecten exilis Mortensen

Astropecten exilis Mortensen, 1933a : 234–236, figs 1, 2, pl. 8, figs 1, 2.

R 70 mm (only one specimen recorded) ; R/r *c.* 9/1.

A species of *Astropecten* with very long arms, tapering fairly evenly ; paxillar area about half the arm breadth, larger paxillae with about 15 slender peripheral spinelets and four to six more clavate central ones, about five paxillae opposite two superomarginals ; superomarginals high, especially interradially, but also fairly conspicuous when viewed from above, proximal ones broader than long, distal ones as long as broad, without enlarged spines ; inferomarginals with one large flat spine and a much smaller one less than half as long below and distal to it ; three actinal plates in each series.

TYPE LOCALITY. Unknown. 'Probably off the Natal coast or off Portuguese East Africa'.

Astropecten granulatus natalensis John

Astropecten pontoporeus (part) : Bell, 1905a : 244. [Non *A. pontoporeus* Sladen, 1883.]

Astropecten capensis (part) : Bell, 1905a : 244. [Non *A. capensis* Studer, 1884, a synonym of *A. irregularis pontoporeus*.]

Astropecten granulatus : H. L. Clark, 1923 : 250–251. [? Non *A. granulatus* : Macnae & Kalk 1962 ; nec B. I. Balinsky *In* : Macnae & Kalk, 1969 : nec Jangoux, 1973.]

Astropecten granulatus natalensis John, 1948, *Novit. zool.* **42** : 486–489, pl. 1, figs 1, 2 ; A. M. Clark, 1974 : 431–433, pl. 1, fig. 1. [Non *A. granulatus natalensis* : Day, Field & Penrith, 1970 : 79.]

R up to 75 mm ; R/r 3·1–4·5/1.

A subspecies of *Astropecten* with arms of moderate length, tapering evenly or a little more in the distal half to a somewhat blunted tip ; paxillar area about two-thirds the arm breadth ; larger paxillae mostly with about fifteen peripheral and ten shorter and less clavate central spinelets, tending to resemble a daisy, the petal-like peripheral spinelets forming a raised rim ; five or six paxillae opposite two superomarginals ; superomarginals high interradially but more dorsal than lateral for most of the arm, relatively few in number, only 17–22 at R 30–40 mm, so mostly appearing only as broad as long, rarely with any enlarged spines ; inferomarginals with a second large pointed, somewhat flattened, spine below and distal to the main one, about two-thirds to three-quarters as long ; two actinal plates in each series. Colour slate-grey above, dirty white below.

TYPE LOCALITY. Off Natal, position uncertain.

SOUTHERN AFRICAN RECORDS. 32/28/s ; 29/31/s, d [? 26/33/i] ; 38–130 metres.

NATURE OF BOTTOM. Mud ; clay.

Note : In 1958 Dr Macnae sent for checking of identification a small *Astropecten* (R 25–28 mm) from Inhaca labelled *Astropecten granulatus*. At that time I thought it referable to *A. granulatus natalensis* but after re-examination I think that it is more likely to be *A. hemprichi*. There are as many as 24 marginals on each side of the longest arm, though 20 on one with R 26 mm ; even the distal superomarginals are broader than long and about a quarter of the plates have a blunt superomarginal spine. The accessory inferomarginal spine is very small in comparison with the main one. However, judging from the somewhat diagrammatic drawings of both Lopes (1939) and Macnae & Kalk (1969) there is an *Astropecten* at Inhaca with less than 20 superomarginals in each series even at R 35 mm and these are not appreciably broader than long. Jangoux (1973) followed Macnae & Kalk in naming specimens from Inhaca deficient in superomarginal spines as *A. granulatus* Müller & Troschel (known from the tropical west Pacific) and those with spines as *A. hemprichi*. Unfortunately the British Museum collection includes no good samples of *A. hemprichi* but from experience of other species of the genus I am inclined to distrust the value of occurrence of superomarginal spines as a specific character unsupported by other morphological features.

Astropecten hemprichi Müller & Troschel

Astropecten hemprichii Müller & Troschel, 1842, *System der Asteriden*. Braunschweig. p. 71; Peters, 1852 : 178; Simpson & Brown, 1910 : 48; Döderlein, 1917, *Siboga Exped.* **46a** : 139–140, pl. 6, figs 1, 2, pl. 13, fig. 1; Mortensen, 1933a : 232–233; Day & Morgans, 1956 : 308; Jangoux, 1973 : 12–13; Day, 1974 : 94.

Astropecten monacanthus : Lopes, 1939 : 78, pl. 1, fig. 2. [Non *A. monacanthus* Sladen, 1889.]

? *Astropecten acanthifera* : Macnae & Kalk, 1962 : 115. [Non *A. acanthifer* Sladen, 1883.]

? *Astropecten granulatus* : Macnae & Kalk, 1962 : 98, 114; B. I. Balinsky, *In* : Macnae & Kalk, 1969 : 99, 129, fig. 24c, d; Jangoux, 1973 : 10–12. [Non *A. granulatus* Müller & Troschel, 1842.]

R up to 100 mm; R/r 4·2–4·6/1.

A species of *Astropecten* with arms of moderate length, tapering fairly evenly but often somewhat blunted at the tip; paxillar area about half the arm breadth, larger paxillae mostly with 20–25 short blunt spinelets, the central ones coarser, together making a flat surface, usually five paxillae opposite two superomarginals; superomarginals appearing broader than long in dorsal view, at least on the proximal half of the arm, numbering about 30 when R is 30–40 mm, often armed with single short spines, except on the more interradial plates; inferomarginals with only one large spine, the accessory spine below it rarely more than half as long; two actinal plates in each series.

TYPE LOCALITY. Red Sea.

SOUTHERN AFRICAN RECORDS. 29/31/i or s; 26/33/i; 23/35/i.

Astropecten irregularis pontoporaeus Sladen

Astropecten pontoporæus Sladen, 1883, *J. Linn. Soc.* Zool. **17** : 259; 1889 : 210–212, pl. 35, figs 1, 2, pl. 38, figs 10–12; Bell, 1905a : 243–244 (part); H. L. Clark, 1923 : 249–250.

Astropecten capensis Studer, 1884 : 44; Bell, 1905a : 244 (part).

Astropecten irregularis var. *pontoporeus* : H. L. Clark, 1926 : 6; Mortensen, 1933a : 232; A. M. Clark, 1952; 194–195; Day, Field & Penrith, 1970 : 79.

Astropecten irregularis pontoporeus : Fisher, 1940 : 265; Morgans, 1962 : 307–319, 324, 325 (isolated references).

R up to 55 mm; R/r 3·2–3·8/1.

A subspecies of *Astropecten* with fairly short arms, tapering evenly or a little more in the distal half but not really petaloid, the tips somewhat blunted, paxillar area about half the arm breadth, larger paxillae with 20–25 similar spinelets, about five paxillae opposite two superomarginals ; superomarginals high interradially but more dorsal than lateral for most of the arm, appearing broader than long in dorsal view, usually those beyond the interradius armed with one, sometimes two or three, small spines, none close to the adradial edge ; inferomarginals with at least two large similar spines at the upper end and several others only a little shorter below them ; two to six actinal plates in each series. Colour light reddish violet above, or pale apricot to mauve, deeper midradially.

TYPE LOCALITY. Simon's Bay, False Bay.

SOUTHERN AFRICAN RECORDS. 28/16/s ; 29/14/d ; 30/16/d ; 32/17/d ; 33/18/d ; 34/18/s ; 34/18/FB/s ; 34/19/s ; 34/20/s ; 35/20/s ; 34/21/s ; 34/22/s ; 35/22/s ; 34/23/s ; 33/25/s ; 34/25/d ; 32/28/s ; 7–186 metres.

NATURE OF BOTTOM. Dark green sand ; shell, sand and mud ; rock and shell ; shell ; medium sand ; fine sand ; coarse sand, shell and stones ; khaki sand and mud ; sand ; sand and broken shells ; green mud ; mud and dark sand ; rocks with many sponges ; stones ; rock.

Astropecten leptus H. L. Clark

Astropecten leptus H. L. Clark, 1926 : 6–8, pl. 1, figs 3, 4 ; Mortensen, 1933a : 234, pl. 9, figs 3, 4.

R up to 60 mm ; R/r *c.* 5·3/1.

A species of *Astropecten* with fairly long, somewhat attenuated arms ; paxillar area about two-thirds the arm breadth, paxillae small with only 6–10 spinelets, about three opposite each superomarginal ; superomarginals relatively narrow in dorsal view, about as long as broad proximally and relatively longer distally, the first eight to twelve each with an erect pointed spine ; inferomarginals with two similar long pointed spines and a third somewhat smaller one in a vertical or oblique series ; two actinal plates in each series.

TYPE LOCALITY. Off Durban, 287(? 208)–348 metres.

SOUTHERN AFRICAN RECORDS. 30/31/d ; 29/31/d ; 287(? 208)–375 metres.

NATURE OF BOTTOM. Mud, sand and shells ; mud.

Astropecten polyacanthus phragmorus Fisher

Astropecten acanthifer phragmorus Fisher, 1913, *Proc. US natn. Mus.* **43** : 604.
Astropecten phragmorus Fisher, 1919 : 65–67, pl. 11, fig. 5, pl. 14, fig. 1 ; Clark & Rowe, 1971 : 44.
Astropecten polyacanthus : H. L. Clark, 1923 : 249.
Astropecten polyacanthus phragmorus : A. M. Clark, 1974 : 433–434, pl. 1, fig. 2, pl. 2, fig. 1.

R up to 70 mm ; R/r 5·3–6·0/1.

A subspecies of *Astropecten* with fairly long arms, tapering evenly or a little more in the distal half, to a somewhat blunted tip ; paxillar area fairly broad, about two-thirds the arm breadth, the larger paxillae with *c.* 15–20 peripheral and 10–15 similar central spinelets, about three paxillae opposite each superomarginal ; superomarginals almost entirely lateral in position on the proximal half of the arm, though sloping inwards to form a narrow border to the paxillar area in dorsal view, all of them armed with a sharp spine at (proximally) or near (distally) the upper or adradial edge of the plate, the spine on the first plate the largest and the following ones progressively smaller ; inferomarginals with one very large spine and usually two smaller spines below it, even the middle one only about half the length of the main spine ; two actinal plates in each series.

TYPE LOCALITY. Philippines, 30 metres.

SOUTHERN AFRICAN RECORDS. 29/31/s ; 24/34/s ; 22–40 (? 47)metres.

NATURE OF BOTTOM. Fine sand ; shell, sand and rock.

Bathybiaster vexillifer (Wyville Thomson)

Archaster vexillifer Wyville Thomson, 1873, *The Depths of the Sea*. London. p. 150, fig. 25.
Bathybiaster robustus (Verrill, 1895) H. L. Clark, 1923 : 247–248.
Bathybiaster vexillifer : Mortensen, 1927 : 61–62, figs 34, 35.

R up to 130 mm ; R/r 4–6/1.

A species of Astropectinidae with a small disc and fairly narrow pointed arms, the interradial arcs angular ; paxillar area broad, occupying most of the dorsal surface ; paxillae with short blunt polygonal spinelets, forming a smooth surface, basally the plates lobed and contiguous ; madreporite, exposed ; sides of the arms high and flat superomarginals almost entirely lateral, many with a small conical flattened spine at the upper end, often absent proximally and in young specimens ; inferomarginals with one or more slightly longer flattened pointed spines towards the upper end of the plate directed dorsally ; marginals otherwise closely covered with scale-like spinelets ; actinal areas small, interradially five or more crowded, overlapping, oblique series of plates in larger specimens (R > 70 mm) but only two extending far on to the arms ; adambulacrals with markedly angular furrow margins and a V of about seven furrow spines, the median (apical) one abruptly longer than the rest (but usually broken in preserved specimens).

TYPE LOCALITY. Faeroe Channel, N. of Scotland (*c.* 60°N 4°W.), 638 metres.

SOUTHERN AFRICAN RECORDS. 34/17/vd ; 1380–1830 metres. [Otherwise known from 223–2222 metres.]

NATURE OF BOTTOM. Green mud.

Note : Since Mortensen, Fisher and others have referred *Bathybiaster robustus* (Verrill) to the synonymy of *B. vexillifer* and Fisher (1940) has commented on the similarity of the Arctic species to *B. loripes* Sladen of the Falkland area and Kerguelen, it is desirable that the identity of H. L. Clark's South African specimens be reassessed.

Dipsacaster sladeni capensis A. M. Clark

Leptoptychaster kerguelensis : Bell, 1905a : 242. [Non *L. kerguelensis* Smith, 1879.]
Dipsacaster sladeni : H. L. Clark, 1923 : 246–247 ; Mortensen, 1933a : 237.
Dipsacaster sladeni capensis A. M. Clark, 1952 : 204, pl. 17 ; 1974 : 434.

R up to 120 mm ; R/r (when R > 40 mm) 2·6–3·0/1.

A subspecies of Astropectinidae with stellate form, disc relatively large, arms tapering and pointed, interradial arcs rounded ; paxillar area broad, the paxillae with very numerous fine spinelets and arranged in rows both tangentially and at right angles to the superomarginals except for a narrow midradial belt and on the disc ; madreporite hidden by paxillae ; side of the arms angular, superomarginals aligned mainly dorsally and inferomarginals ventrally but the upper ends of the latter jutting out laterally beyond the superomarginals and visible from above ; no superomarginal spines, the plates covered with very fine uniform pointed papillae and fringed by abruptly longer peripheral spinelets ; inferomarginals with distinctly coarser pointed papillae or short spinelets, becoming enlarged towards the upper end on the proximal plates and towards the upper distal corner on the arm plates so that several may be termed spines in larger specimens ($R > 70$ mm) ; actinal areas broad, plates in seven or more series on the disc in larger specimens, several series extending on to the arms ; adambulacrals with relatively long, slightly convex furrow margins armed with at least seven similar furrow spines in larger specimens and numerous shorter subambulacral spines and spinelets. Colour red or reddish orange above.

TYPE LOCALITY. NW. of Cape Town, (33/17), 329–357 metres.

SOUTHERN AFRICAN RECORDS. 32/17/d ; 33/17/d ; 33/18/d ; 34/17/vd ; 34/18/d ; 34/23/d ; 32/28/– ; 110–630 metres.

NATURE OF BOTTOM. Khaki sand ; green sand and mud ; green sand and black specks.

Persephonaster roulei euryplax Mortensen

Persephonaster Roulei var. *euryplax* Mortensen, 1933a : 238, pl. 9, figs 1, 2.

R 120 mm (only one specimen recorded) ; R/r 4·8/1.

A species of Astropectinidae resembling *Astropecten* in shape, disc relatively small, arms tapering evenly to pointed tips and interradial arcs angular ; paxillar area relatively narrow, occupying not more than half the arm breadth, paxillae small, three or four corresponding to each superomarginal ; madreporite exposed, small ; superomarginals broad and very conspicuous dorsally, covered with granules, no enlarged spines ; inferomarginals covered with spinelets, some upper ones being enlarged into small spines but evidently appressed to the arm (judging from the figure) ; actinal areas small but with two or three short series of plates (judging from *P. roulei*) ; adambulacral plates with furrow margin almost flat, armed with seven or eight furrow spines.

TYPE LOCALITY. Off Durban (29/31), 410 metres.

Genus ***PLUTONASTER*** Sladen

See : Sladen, 1889 : 81–84.

A genus of Astropectinidae with a moderately large disc, rounded interradial arcs and fairly long arms tapering evenly until just short of the slightly abbreviated tips ;

madreporite more or less concealed by paxillae ; superomarginals forming a border to the paxillar area, often, at least the proximal ones, armed with a single conical spine or enlarged spinelet ; inferomarginals armed with single stout spines, not projecting horizontally beyond the superomarginals, the two series forming a vertical wall to the body ; actinal plates numerous, arranged in series both longitudinally and between adambulacrals and inferomarginals ; adambulacral plates with relatively long, slightly convex furrow margin bearing a comb of more than six similar spines.

Plutonaster intermedius (Perrier)

Goniopecten intermedius Perrier, 1881, *Bull. Mus. comp. Zool. Harv.* **9** : 25–26.
Plutonaster intermedius : H. L. Clark, 1923 : 242.

R up to 75 mm ; R/r *c.* 3/1.

A species of *Plutonaster* with a moderately large disc ; paxillae low with about ten spinelets, the peripheral ones longer than the more granuliform central ones, about four paxillae corresponding to each superomarginal ; superomarginal spines short, stout-based and conical ; inferomarginal spines similar but longer ; actinal plates polygonal with some tendency to form rows from the adambulacrals to the inferomarginals, armed with (apparently similar) granuliform spinelets ; adambulacral plates with a furrow comb of about seven spines, backed by a row of about six shorter spines and a row of more granuliform spines.

TYPE LOCALITY. West Indies, 1500–1800 metres.

SOUTHERN AFRICAN RECORDS. (? 27/14/vd) ; 34/17/vd ; 34/18/vd ; 1005(? 914)–1390 metres.

NATURE OF BOTTOM. Green mud.

Plutonaster proteus H. L. Clark

Plutonaster proteus H. L. Clark, 1923 : 242–246, pl. 13, figs 3–7.

R up to 58 mm ; R/r 3·2–4·0/1.

A species of *Plutonaster* with a moderately large disc, the arms apparently tapering a little more in the distal half than proximally ; paxillae low, irregular, about three corresponding to each superomarginal ; superomarginal spines short, stout-based, conical ; inferomarginal spines similar but longer ; actinal plates polygonal, with a tendency only to form longitudinal rows, otherwise irregular, armed with short spaced spinelets, a few of which are enlarged into small spines ; adambulacral plates with about eight furrow spines backed by a single subambulacral spine, similar to the inferomarginal spines, and some spaced spinelets.

TYPE LOCALITY. SW. of Cape Point (34/17), 1463–1646 metres.

NATURE OF BOTTOM. Green or grey mud.

Psilaster acuminatus Sladen

Psilaster acuminatus Sladen, 1889 : 225–228, pl. 40, figs 1, 2, pl. 42, figs 7, 8 ; Bell, 1905a : 245 ; H. L. Clark, 1923 : 248–249 ; 1926 : 3–4 ; Mortensen, 1933a : 236–237 ; A. M. Clark, 1952 : 194 ; Day, Field & Penrith, 1970 : 79 ; 1974 : 434.

R up to 78 mm ; R/r (when R > 60 mm) 4·3–6·0/1 (4·6/1 in the two South African paratypes).

A species of Astropectinidae with a fairly small disc, arms tapering evenly to pointed tips, interradial arcs blunted angular ; paxillar area relatively broad, occupying more than half the arm breadth, paxillae small, about five corresponding to two superomarginals ; madreporite exposed but usually inconspicuous ; superomarginals broad but more lateral than dorsal, the edge of the body high, sometimes with an appressed small spine near the distal edge (except on the interradial plates) ; inferomarginals also mainly lateral, not extending horizontally beyond the superomarginals, with several longer, upwardly appressed small spines, mostly towards the distal edge, both marginal series otherwise with a covering of often squamiform spinelets, sometimes reduced near the lower ends of the upper plates and the upper ends of the lower plates ; actinal plates in three to seven series, those on the disc distinctly smaller than the one or two inner series which extend well out on to the arm, armed with spinelets similar to those covering the marginal plates ; adambulacral plates moderately long, with slightly convex or obtusely angled furrow margins bearing five to seven similar spines backed by two or three series of shorter more compressed and somewhat clavate spines. Colour light pink.

Type locality. West of Cook Strait, New Zealand, 274 metres.

Southern African records. 29/14/d ; 29/15/d ; 30/15/d ; 31/15/d ; 31/16/d ; 32/16/d ; 33/17/d ; 34/17/d ; 34/18/d ; 34/18/FB/–* ; 35/18/vd ; 155–547 metres.

Nature of bottom. Dark green mud ; green mud ; sand and mud ; mud ; black specks ; green sand ; polyzoa and rock.

Tethyaster pacei (Mortensen)

Anthosticle pacei Mortensen, 1925 : 147–149, fig. 1, pl. 8, fig. 3.
Tethyaster pacei : A. M. & A. H. Clark, 1954, *Smithson. misc. Collns* **122** (11) : 23.

R 120 mm (only one specimen recorded) ; R/r 4·3/1.

A species of Astropectinidae with a relatively small disc, arms ending rather abruptly and interradial arcs rounded ; paxillar area very broad, occupying most of the upper surface, about two paxillae corresponding to each superomarginal ; madreporite probably exposed but inconspicuous (not described) ; superomarginals mainly lateral, uniformly covered with grains or short spinelets but without enlarged spinelets or spines ; inferomarginals not extending horizontally beyond the superomarginals, with up to five pointed upwardly appressed spines in a series up the middle

* The False Bay entry depends only on Sladen's imprecise 'Challenger' record from Simon's Bay, depth uncertain. Since the minimum depth otherwise recorded for the species is over 150 metres and the maximum depth in False Bay is barely 100 metres (even less in Simon's Bay itself), I suspect that the 'Challenger' record is erroneus. The same applies to *Pseudarchaster tessellatus* and *Cosmasterias felipes*, though the last-named has been recorded in only 79 metres on the *west* side of Cape Peninsula.

of the plate, besides the general covering of flattened spinelets ; actinal plates in several series, of which the innermost extends for about two-thirds the arm length; adambulacral plates with short, markedly angular furrow margins.

TYPE LOCALITY. Unknown.

Family **BENTHOPECTINIDAE** Verrill

See : Fisher, 1911 : 120 ; Spencer & Wright, 1966 : U48.

A family of Paxillosida with long pointed arms ; interradial areas small to moderate in size, arcs rounded or with a blunt angle ; terminal plates inconspicuous ; abactinal plates small, either flattened with few spaced spinelets or small spines, or low paxilliform with more dense, clustered armament ; papulae restricted on the arms to an oval or distally bifurcate area at the base of each arm ; plates of the two marginal series more or less numerous, tending to alternate in position, the superomarginals inconspicuous from above, being mostly aligned laterally, with a single large spaced, horizontally directed spine on each plate, inferomarginals often with an accessory spine more or less well developed below the main spine ; no cribriform organs ; actinal areas relatively small, even the innermost row of plates only extending on to the base of the arms ; pedicellariae more or less well developed, fasciculate with four or five spinelets in a rounded group or, more often, pectinate (shared between two or sometimes abactinally three plates) ; tube feet with sucking discs.

Benthopecten pedicifer (Sladen)

Pararchaster pedicifer Sladen, 1889 : 15–19, pl. 1, fig 3, 4, pl. 4, figs 3, 4.
Benthopecten pedicifer : Fisher, 1911 : 143.

R up to 168 mm ; R/r 11/1.

A species of Benthopectinidae with a small disc and long attentuated arms ; abactinal plates inconspicuous, flat, embedded in the skin and each bearing up to four small spaced spinelets, a few of which near the centre of the disc may be enlarged into small spines ; popularia bifurcate on the arms, extending further laterally than midradially ; an odd interradial marginal plate present in each series, the superomarginal one with a single large spike-like erect spine, somewhat larger than the other superomarginal spines which project more horizontally since the paired superomarginals are purely lateral in alignment ; actinal areas very small ; adambulacral plates with six to eight small furrow spines and three long subambulacral spines in large specimens ; pedicellariae numerous, on the abactinal, inferomarginal and actinal plates.

TYPE LOCALITY. Near the Crozet Is, Southern Ocean, 2926 metres.

SOUTHERN AFRICAN RECORD. 36/19/vd ; 3474 metres. This is based on a very small specimen, R only 14 mm, the specific identification of which is not positive.

Luidiaster hirsutus Studer

Luidiaster hirsutus Studer, 1884 : 47, pl. 4, fig. 7 ; H. L. Clark, 1923 : 241 ; 1926 : 2 ; Mortensen, 1933a : 231–232 ; A. M. Clark, 1952 : 194.

R up to 92 mm ; R/r *c.* 5–6/1.

A species of Benthopectinidae with moderately long arms ; abactinal plates mostly with 8–12 small spinelets, often encircling a single enlarged spinelet or small spine, at least three times their length and thickness ; papularia V-shaped, bifurcate distally, with numerous pores, probably over 100 in each area in larger specimens, R > 70 mm ; no odd interradial marginal plates ; superomarginals mainly lateral in position but forming a narrow border to the upper side, inferomarginals often with a second, more or less enlarged spine below the main one ; actinal plates in about three rows in larger specimens ; adambulacral plates with about six furrow spines and one to three large subambulacral spines, usually more than one in larger specimens ; pedicellariae present, at least on the actinal plates.

TYPE LOCALITY. Kerguelen, 238 metres.

SOUTHERN AFRICAN RECORDS. 30/15/d ; 30/16/d ; 31/15/d ; 31/16/d ; 31/17/d ; 32/16/d ; 32/17/d ; 33/17/d ; 34/17/d ; 34/18/d ; 35/18/vd ; 36/21/d ; 237–547 metres.

NATURE OF BOTTOM. Mud ; green sand.

Pectinaster filholi Perrier

Pectinaster filholi Perrier, 1885, *Annls Sci. nat.* (6) **19** (8) : 70 ; H. L. Clark, 1923 : 240.
Pontaster forcipatus Sladen, 1889 : 43–47, pl. 8, figs 3, 4, pl. 12, figs 3, 4.
Pontaster forcipatus var. *echinata* Sladen, 1889 : 47–48.
Pectinaster echinatus : A. M. Clark, 1962, *Rep. B.A.N.Z. Antarctic Res. Exped.* **B9** : 13.

R up to *c.* 80 mm ; R/r 5·5–6·5/1.

A species of Benthopectinidae with moderately long arms ; abactinal plates of disc with 6–12 small spinelets around a single enlarged spinelet or small spine, reduced on the arms where the spinelets are fewer ; papularia compact, oval, often slightly convex, usually each with 10–20 pores in larger specimens, R > 70 mm ; no odd marginal plates ; superomarginals lateral, each with a single stout spine ; actinal areas relatively small, the plating variable, from two to five series in larger specimens ; adambulacral plates with about seven furrow spines and one large conical subambulacral ; pedicellariae with only four or five triangular spines, fasciculate rather than pectinate, usually present on some abactinal, inferomarginal and actinal plates.

TYPE LOCALITY. Off Mauritania and the Azores, 1258–2330 metres.

SOUTHERN AFRICAN RECORDS. 34/17/vd ; 34/18/vd ; *c.* 1646 (? 1463–1830) metres. Several small specimens, R < 20 mm, from 33/34/1300 metres may be referable to this species, also known from the vicinity of the Crozet Is.

Note : The availability of more North Atlantic material has shown that the actinal plates may be very variable in number and extent in this species and refutes my suggestion of 1962 that

P. forcipatus (Sladen) from the north-west Atlantic and *P. echinatus* (Sladen) from near the Crozet Is in the Southern Ocean can be regarded after all as specifically distinct. The great depth at which the species occurs makes a wide distribution likely and it is possible that some other nominal species, such as *P. mimicus* (Sladen) from the Indian Ocean, will also prove to be synonyms. Sibuet (1975, *Bull. Mus. natn. Hist. nat. Paris*. Zool. **199** : 283, 289) has recorded *P. filholi* from the Gulf of Guinea and off Angola in 2044–3431 metres, noting that the specimens have a close resemblance to Sladen's description of *P. forcipatus*.

Order VALVATIDA Perrier

Asteroids with usually five arms (in the taxa represented), main ossicles forming longitudinal series ventrally, with a marginal frame consisting of two well-developed series of plates, notably fewer than, and distinct from, the adambulacrals and ambulacrals which form the furrow frame ; the two frames linked by actinal plates, at least interradially ; pedicellariae, if present, of various forms from fascicular to bivalved with two very broad valves, or spatulate or tong-shaped, with or without excavations or hollows in the underlying plates to accommodate the valves when open ; tube feet in two rows, ending in sucking discs.

Family **ODONTASTERIDAE** Verrill

See : Fisher, 1911 : 153 ; Spencer & Wright, 1966 : U55.

A family of Valvatida with five arms ; body form usually stellate, though almost pentagonal in some young specimens ; preserved specimens usually flattened dorsally, interradial areas large, the arcs rounded ; both marginal series of plates conspicuous, block-like, without large spines, including an odd interradial plate in each series ; abactinal plates more or less paxilliform, arranged in longitudinal and oblique transverse series but without a mid-radial row of enlarged plates on each arm, or even an enlarged primary radial plate on the disc opposite each arm ; each pair of oral plates bearing either one large median glassy-tipped spine, sometimes with a smaller one each side, or a pair of such spines, pointing away from the mouth ; pedicellariae simple fasciculate, valves either spiniform or granuliform.

Odontaster australis H. L. Clark

Pentagonaster tuberculatus (part) : Bell, 1905a : 247 (no. 2584).
Odontaster australis H. L. Clark, 1926 : 16–17, pl. 1, figs 5, 6 ; Mortensen, 1933a : 246–247 pl. 10, figs 3, 4 ; A. M. Clark, 1952 : 196.

R up to 65 mm ; R/r 2·0–3·0/1.

A species of Odontasteridae with small paxilliform abactinal plates, two or three opposite each of the much larger, non-paxilliform superomarginals, which appear squarish from above, except for the distalmost ones which are shorter and approximate adradially ; actinal spinelets graduating into the adambulacral spines, which are not abruptly larger ; jaws each with a single massive, glassy-tipped reflexed spine.

TYPE LOCALITY. West from Saldanha Bay, 320 metres.

SOUTHERN AFRICAN RECORDS. 29/14/d; 31/16/d; 33/17/d; 34/17/d; 34/18/d; 243–366 metres.

NATURE OF BOTTOM. Dark mud and rock; dark speckled mud and rock.

Family **ARCHASTERIDAE** Viguier

See: Spencer & Wright, 1966: U56.

A family of Valvatida with normally five (occasionally four or six) long, basally narrow arms, interradial arcs more or less angular; both marginal series of plates large, forming a thick margin, the superomarginals spineless and high laterally and only forming a broad border to the paxillar area in smaller specimens $R < 70$ mm; inferomarginals with one or several flattened spines at the upper end but these are inconspicuous, often being appressed or very short; abactinal plates paxilliform, with spiniform or granuliform armament, a distinct midradial series along each arm enlarged but flush with the other paxillae; actinal plates few and restricted to the disc; oral plates with massive glassy-tipped spines; pedicellariae, if present, two-bladed, alveolar without depressions in the underlying plates for the valves.

Archaster angulatus Müller & Troschel

Archaster angulatus Müller & Troschel, 1842; *System der Asteriden.* Braunschweig. p. 66; Jangoux, 1973: 14–17, figs 4–6.

R up to 190 mm; R/r 5·5–7·5/1.

A species of Archasteridae with an almost smooth outline to the body; midradial row of paxillae conspicuously enlarged and regular, the armament of the paxillae consisting of cylindrical, round-tipped spinelets, so closely placed as to appear granuliform superficially; inferomarginals armed with two to five very short, spatulate spines or enlarged squamiform spinelets at the upper end of the plate, appressed or hardly projecting; usually two actinal plates in each series; adambulacral plates mostly with three spines along the furrow margin backed by two rows of subambulacral spines and sometimes an alveolar pedicellaria.

TYPE LOCALITY. Java and Mauritius.

SOUTHERN AFRICAN RECORDS. 26/33/i.

Family **GONIASTERIDAE** Forbes

See: Fisher, 1911: 158; Spencer & Wright, 1966: U56.

A family of Valvatida with form pentagonal or stellate, disc usually large, preserved specimens more or less flat dorsally though probably somewhat convex in life, interradial arcs rounded; both marginal series conspicuous and block-like, no odd interradial marginal plates; abactinal plates ranging in form from paxilliform to flat, armed with spinelets, granules or a single spine or a ring of granules around a flat or convex bare centre, rarely obscured by thick skin, usually arranged in regular longitudinal and oblique transverse rows but with no markedly enlarged mid-radial row of enlarged plates on each arm, though the primary radial plate on the disc opposite the base of each arm may be enlarged ; actinal plates more or less numerous, in several series ; oral plates without massive glassy-tipped spines ; pedicellariae alveolar, with or without depressions for the valves in the underlying plates, the valves spatulate or widened to a bivalved form.

Anthenoides marleyi Mortensen

Anthenoides marleyi Mortensen, 1925 : 149–151, pl. 8, figs 2–4 ; 1933a : 245.

R up to 90 mm ; R/r *c.* 2·7/1.

A species of Goniasteridae with well-developed arms tapering to rounded tips ; abactinal plates almost flat, bearing one to three rounded tubercles and some spaced fine grains ; superomarginals numerous, broader than long, especially the shorter distal ones, three or four of which meet midradially before the terminal plate, their armament similar to that of the abactinal plates ; actinal plates mostly limited to the disc, the innermost series extending only to about the fifth inferomarginal, similarly armed to the other plates ; adambulacral plates with a fairly long furrow margin bearing a comb of up to eight spines backed by one large flattened subambulacral spine in the middle of the plate and usually an adoral spatulate pedicellaria ; other pedicellariae on the abactinal and actinal plates bivalved and more or less broad, especially on the actinal plates.

TYPE LOCALITY. Off Omvot River, Natal.

SOUTHERN AFRICAN RECORDS. 29/31/d ; 256–274 metres.

Genus *CALLIASTER* Gray

See : Döderlein, 1922, *Bijdr. Dierk.* **1922** : 47.

A genus of Goniasteridae with more or less well-developed arms, blunt or even slightly clavate at the tips owing to the breadth and often also the convexity of the distalmost marginals, abactinal and marginal plates predominantly naked, encircled by coarse irregular granules and usually with a central spine or large tubercle, sometimes more than one on some of the marginal plates ; adambulacral plates with three or more furrow spines and two stout subambulacral spines in a row at right angles to the furrow ; pedicellariae, if present, with valves spatulate or sometimes papilliform, longer than broad and alveolar (corresponding to slight depressions in the plates).

Calliaster acanthodes H. L. Clark

Calliaster acanthodes H. L. Clark, 1923 : 264–266, pl. 12, figs 3, 4 ; 1926 : 11 ; Mortensen, 1933a : 244–245, pl. 10, figs 1, 2.

R up to 96 mm ; R/r 2·2–2·9/1.

A species of *Calliaster* with fairly long and often sharp abactinal and marginal spines, the superomarginals with up to three spines in a vertical series and inferomarginals with up to five spines ; adambulacral plates relatively long with six to nine slender compressed furrow spines ; spatulate alveolar pedicellariae, sometimes denticulate and resembling the jaws of a crocodile, more or less numerous on the various plates. Colour orange.

TYPE LOCALITY. SW. of Cape St Francis, 137 metres.

SOUTHERN AFRICAN RECORDS. 34/23/d ; 34/24/d ; 33/28/d ; 32/28/– ; 29/31/d ; 130–411 metres.

NATURE OF BOTTOM. Rock [? green mud].

Calliaster baccatus Sladen

Calliaster baccatus Sladen, 1889 : 280–282, pl. 56, figs 1–4 ; Bell, 1905a : 247 ; H. L. Clark, 1923 : 264 ; Mortensen, 1933a : 244 ; A. M. Clark, 1952 : 196 ; Day, Field & Penrith, 1970 : 79.

R up to 55 mm ; R/r 2·2–2·5/1.

A species of *Calliaster* with the spines stout-based and tips blunt ; marginal plates with only a single spine or tubercle or none at all centrally ; adambulacral plates relatively short, mostly with only three stout furrow spines ; pedicellariae rare or absent.

TYPE LOCALITY. Simon's Bay, False Bay, 9–23 metres.

SOUTHERN AFRICAN RECORDS. 34/18/FB/s ; 34/19/s ; 34/20/s ; 35/20/s ; 34/22/s ; 34/25/s ; 33/27/s ; 33/28/d ; 32/28/– ; 23(? 9)–146(? 238) metres.

NATURE OF BOTTOM. Rock ; gravel ; coarse sand and shell ; coral and rock ; mud and sand.

Genus ***CERAMASTER*** Verrill

See : Fisher, 1911 : 204.

A genus of Goniasteridae with a pentagonal or short-rayed form, the arms ending in blunt points ; abactinal plates regularly arranged, mostly tabular, covered with flat angular granules, making a polygonal mosaic ; marginal plates completely granule-covered or with a more or less extensive bare patch ; actinal plates granule-covered ; pedicellariae small bivalved or spatulate, usually projecting above the granules.

Ceramaster patagonicus euryplax H. L. Clark

Ceramaster patagonicus var. *euryplax* H. L. Clark, 1923 : 262–264, pl. 14, figs 1, 2 ; 1926 : 9–10 ; A. M. Clark, 1952 : 195, 204–205.
Ceramaster chondriscus H. L. Clark, 1923 : 258–260, pl. 14, figs 5, 6 ; Mortensen, 1933a : 242.
Ceramaster patagonicus euryplax : A. M. Clark, 1974 : 435.

R up to 67 mm ; R/r 1·5–1·9/1.

A subspecies of *Ceramaster* with the interradial superomarginals square or rectangular, their proximal edges almost straight ; adambulacral plates with usually four furrow spines, though odd plates may have three, or even two, stouter spines; small or moderate-sized pedicellariae of bivalved form (i.e. with valves broader than long) but sometimes with three, or even four, valves, rather than the usual two, more or less numerous on all the various kinds of plates. Colour brilliant crimson [brick red above, cream below in a dried specimen].

Type locality. South of Cape Point, 156 metres.

Southern African records. 31/16/d; 31/17/d; 32/17/d; 33/17/d; 33/18/d; 34/17/d; 34/18/d; 35/18/d; 156–462 metres.

Nature of bottom. Mud and sand; green mud; green sand and black specks; dark speckled mud and rock.

Ceramaster trispinosus H. L. Clark

Ceramaster trispinosus H. L. Clark, 1923: 260–262, pl. 14, figs 3, 4.

R 41 mm (in the one recorded specimen); R/r 1·95/1.

A species of *Ceramaster* with relatively prolonged arms; interradial superomarginals with markedly convex proximal edges, almost semicircular; adambulacral plates relatively short (*c.* 50 at R *c.* 40 mm, compared with only 33 at R 32 mm in *C. patagonicus euryplax*) with three, rarely two, furrow spines; pedicellariae absent (at least in the holotype).

Type locality. WSW. of Vasco da Gama Peak (34/18), 420 metres.

Cladaster macrobrachius H. L. Clark

Cladaster macrobrachius H. L. Clark, 1923: 268–270, pl. 13, figs 1, 2; 1926: 12–13.

R up to 57 mm; R/r 2·1–2·5/1.

A species of Goniasteridae with fairly well-developed arms tapering to rounded tips; abactinal plates irregular, polygonal, thick and crowded, with spaced granules peripherally encircling a few coarser ones (the armament easily rubbed off in preservation); superomarginals convex, similarly armed with peripheral and spaced central granules, two or three of which at the upper end are enlarged; inferomarginals similar though less convex and without coarser armament; actinal plates irregular with peripheral granules and a bare central area sometimes with a broad-valved pedicellaria; adambulacral plates with three or occasionally two coarse compressed furrow spines backed by a row of three subambulacral spines, the distalmost stoutest, then another row of three spines, of which the first and third are smaller.

Type locality. SW. of Cape Point, 566–915 metres.

Southern African records. 33/17/d; 34/18/d, vd; 381/566 (? 915) metres.

Nature of bottom. Green mud; stones and mud.

Genus *HIPPASTERIA* Gray

See: Fisher, 1911 : 223.

A genus of Goniasteridae with arms of short or moderate length; abactinal plates more or less irregular, roundish, smaller ones covered with granules and sometimes pedicellariae, small spines or tubercles, larger ones mainly bare, often somewhat convex, with peripheral granules and a central spine, tubercle or pedicellaria; marginal plates also mainly bare, with peripheral granules and one to six tubercular spines; actinal plates covered with granules, tubercles and conspicuous very broad bivalved pedicellariae; adambulacral plates with usually one or two furrow spines, rarely more, and usually only one large subambulacral spine, sometimes replaced by a pedicellaria; bivalved pedicellariae numerous and conspicuous, especially on the actinal plates.

Hippasteria phrygiana capensis Mortensen

Hippasteria phrygiana: H. L. Clark, 1923 : 270–272; 1926 : 13.
Hippasteria phrygiana var. *capensis* Mortensen, 1933a : 245–246, pl. 11, fig. 1; A. M. Clark, 1952 : 196, 207.

R up to 135 mm; R/r 1·6–2·5/1.

A subspecies of *Hippasteria* with fairly short arms, especially short in larger specimens (R > 100 mm); convex abactinal plates more often with a stout blunt spine or tubercle than a pedicellaria and with fairly extensive bare areas; marginal plates large, fairly conspicuous in dorsal view, with one or two stout spines; actinal plates mostly with a central bivalved pedicellaria about half the diameter of the plate or less, encircled by granules, larger plates also with a few coarser central granules.

TYPE LOCALITY. SW. of Table Mountain, 357 metres.

SOUTHERN AFRICAN RECORDS. 33/16/vd; 33/17/d; 34/17/d; 34/18/d, vd; 310–980 metres.

NATURE OF BOTTOM. Green mud; green sand and mud; globigerina ooze; stones.

Hippasteria strongylactis H. L. Clark

Hippasteria strongylactis H. L. Clark, 1926 : 13–16, pl. 3; A. M. Clark, 1952 : 196.

R up to 101 mm; R/r 2·8–2·9/1.

A species of *Hippasteria* with fairly prolonged arms; abactinal plates more often with a pedicellaria than a spine or tubercle and without extensive bare areas around these; superomarginal plates small, not conspicuous dorsally (at least in larger specimens), with up to five short thick spines; actinal plates mostly with a huge pedicellaria about three-quarters the diameter of the plate, otherwise covered with small granules.

TYPE LOCALITY. SW. from Saldanha Bay, 350 metres.

SOUTHERN AFRICAN RECORDS. 32/17/d ; 33/16/vd ; 33/17/d, vd ; 34/17/d ; 320–980 metres.

NATURE OF BOTTOM. Mud ; green mud ; globigerina ooze.

Mediaster capensis capensis H. L. Clark

Mediaster capensis H. L. Clark, 1923 : 256–258, pl. 16, figs 1, 2 ; Mortensen, 1933a : 241, pl. 8, fig. 10 ; A. M. Clark, 1952 : 195 ; Day, Field & Penrith, 1970 : 79.

R up to 53 mm ; R/r 2·4–2·8/1.

A species of Goniasteridae with well-developed, rather attenuated arms ; abactinal plates in regular rows, those of the midradial row slightly enlarged, paxilliform with a cluster of short, thick, blunt spinelets ; superomarginals fairly numerous, squarish, forming a distinct though narrow border to the paxillar area in dorsal view, the distal ones not enlarged, all covered with granules ; actinal plates granule covered ; adambulacral plates with three or four furrow spines backed by two series of subambulacral spines, none of them particularly enlarged ; pedicellariae of small bivalved form (though sometimes with three or even four valves) more or less numerous on the abactinal and superomarginal plates. Colour red.

TYPE LOCALITY. False Bay, 38 metres.

SOUTHERN AFRICAN RECORDS. 34/18/d ; 34/18/FB/s ; 38–170 metres.

NATURE OF BOTTOM. Green mud and sand ; green mud ; fine sand.

Mediaster capensis durbanensis Mortensen

Mediaster capensis var. *durbanensis* Mortensen, 1933a : 242, pl. 8, fig. 9.

Differs from *M. capensis capensis* in having the marginal plates 'not nearly so well marked off from the disk', inferomarginals much broader and encroaching more on the ventral interradii.

TYPE LOCALITY. Off Durban (29/31), 410 metres.

Note : In my view the geographical difference warrants a subspecific distinction.

Genus ***PSEUDARCHASTER*** Sladen

See : Fisher, 1911 : 179.

A genus of Goniasteridae with moderate to well-developed arms ; abactinal plates low paxilliform, regularly arranged, those of the midradial row slightly enlarged ; superomarginals large, more or less broad in dorsal view, the distal ones not meeting midradially, covered with granules or flattened spinelets, no large spines, the profile smooth ; inferomarginal and actinal plates similarly armed but with up to four projecting spinelets on the inferomarginals and scattered ones usually on the actinal plates ; adambulacral plates with convex or angular furrow margin and a fan of about six furrow spines ; pedicellariae rare or absent.

Pseudarchaster brachyactis H. L. Clark

Pseudarchaster tesselatus (part) : Bell, 1905a : 242 (no. 2463).
Pseudarchaster brachyactis H. L. Clark, 1923 : 254–256, pl. 12, figs 1, 2 ; 1926 : 8–9, pl. 2.

R up to 60 mm ; R/r 2·1–2·7/1.

A species of *Pseudarchaster* with arms of moderate length, relatively short for the genus, tapering evenly to rounded tips ; superomarginals very broad in dorsal view, the interradial ones making up about a quarter of the minor radius, r ; actinal armament granuliform.

Type locality. Agulhas Bank (36/21), 366 metres.

Southern African records. 27/14/vd ; 33/17/d ; 34/17/d ; 34/18/d ; 36/21/d ; 366–576 (? 730) metres.

Nature of bottom. Green sand ; sand with black specks ; mud.

Pseudarchaster tessellatus Sladen

Pseudarchaster tessellatus Sladen, 1889 : 112–114, pl. 17, figs 3, 4, pl. 18, figs 9, 10 ; H. L. Clark, 1923 : 253–254 ; 1926 : 9 ; Mortensen, 1933a : 240–241 ; A. M. Clark, 1952 : 195.
Pseudarchaster tesselatus : Bell, 1905a : 242 ; ? Day, Field & Penrith, 1970 : 79.

R up to 105 mm ; R/r 2·9–3·7/1.

A species of *Pseudarchaster* with fairly long, narrow arms ; superomarginals not very conspicuous dorsally, the interradial ones only about a sixth of r ; actinal armament of clavate spinelets. Colour uniform red.

Type locality. Simon's Bay, False Bay, no details. [In view of the absence of subsequent records from False Bay, there must be some doubt about this locality.]

Southern African records. 30/15/d ; 30/16/d ; 33/17/d ; 34/17/d ; 34/18/d [? 34/18/FB/– ; see note on p. 55] ; 155–402 metres.

Nature of bottom. Light green sand ; dark green sand ; green sand and mud ; green mud ; black specks and rocks.

Sphaeriodiscus bourgeti (Perrier)

Stephanaster Bourgeti Perrier, 1885, *Annls Sci. nat.* **19** (8) : 31–34 ; 1894, *Expeditions scientifiques du 'Traveilleur' et du 'Talisman'. Stellerides.* Paris. pp. 403–406, pl. 26, fig. 1.
Sphaeriodiscus bourgeti : H. L. Clark, 1926 : 10–11.

R up to 35 mm ; R/r 1·3–1·4/1.

A species of Goniasteridae of almost pentagonal form ; abactinal plates flat, polygonal, covered with granules and occasional pedicellariae ; superomarginals swollen, only six or eight on each of the five sides, the two interradial ones smaller than the immediately adjacent plate each side which is the largest, the distal one or two plates diminishing again, marginals almost bare but for a scattering of granules and one or two rings of peripheral ones ; actinal plates granule covered ; adambulacrals with about five furrow spines ; alveolar pedicellariae with long, spoon-shaped blades found on all the various plates, which bear depressions for them (except those on the adambulacrals). Colour pale brick red.

TYPE LOCALITY. North of the Cape Verde Is and west of the Azores, 410–760 metres.

SOUTHERN AFRICAN RECORD. 30/31/d ; 420 metres.

NATURE OF BOTTOM. Mud, sand and shells.

Toraster tuberculatus (Gray)

Astrogonium tuberculatum Gray, 1847a, *Proc. zool. Soc. Lond.* **1847** : 79.
Pentagonaster tuberculatus : Bell, 1905a : 246–247.
Tosia tuberculata : H. L. Clark, 1923 : 266–268, pl. 9, figs 1, 2 ; 1926 : 11–12 ; Mortensen, 1933a : 243–244, pl. 10, figs 5–7.
Toraster tuberculatus : A. M. Clark, 1952 : 205–206.

R up to 95 mm ; R/r 1·7–2·1/1.

A species of Goniasteridae with relatively short arms, especially in smaller specimens (R < 40 mm) ; most abactinal plates regularly arranged, larger on the radial areas, especially a midradial row, these enlarged plates becoming more or less convex, bare but for a central hemispherical knob-like tubercle and a ring of peripheral granules, the smaller flatter proximal plates often completely covered with granules ; superomarginals mostly squarish in dorsal view and more or less convex but relatively broader distally and, in some specimens, the last three or four may be markedly swollen and contiguous mid-radially, making the arms somewhat clavate ; marginals at least partially bare ; actinal plates granule covered ; adambulacrals fairly short with usually three furrow spines ; small bivalved pedicellariae housed in depressions in the plates more or less numerous. Colour reddish-orange above, pale below.

TYPE LOCALITY. Port Natal (Durban).

SOUTHERN AFRICAN RECORDS. 30/15/d ; 31/16/d ; 31/17/d ; 32/16/d ; 32/17/d ; 33/17/d ; 33/18/d ; 34/17/d ; 34/18/s, d ; 35/18/d ; 35/19/d ; 35/20/d ; 34/21/s, d ; 35/21/d ; 34/22/d ; 34/23/d ; 34/24/d ; 34/25/d ; 33/28/– ; 29/31/– ; 75–366 metres.

NATURE OF BOTTOM. Rock ; sand and rocks ; rock and shell ; coarse khaki sand ; mud and sand ; green sand and mud ; green mud.

Family **OREASTERIDAE** Fisher

See : Spencer & Wright, 1966 : U63.

A family of Valvatida with normally five arms ; body form either pentagonal, cushion-like or stellate with the disc more or less elevated, the interradial arcs rounded, and the arms carinate, reaching a large size, R usually exceeding 100 mm in adults, young specimens, R < 50 mm, flatter and more goniasterid-like ; superomarginals inconspicuous in dorsal view in large specimens, sometimes reduced, no odd interradial ones, often armed with a median tubercle or short spine ; abactinal plates tending to form a reticulum around the large pore areas in adults but more closely joined in young specimens, the midradial ones often enlarged in a regular longitudinal series in which isolated plates are more convex or bear a tubercle, some interradial and dorsolateral plates similarly projecting in two species represented, or else the whole surface covered with coarse granules and tubercles, or (rarely) with thick smooth skin ; ventral side flat ; actinal plates numerous, in several series, granule covered ; oral plates without massive glassy-tipped spines ; pedicellariae alveolar, their valves either spatulate or broad and short (bivalved).

Asterodiscides elegans Gray

Asterodiscus elegans Gray, *Proc. zool. Soc. Lond.* **1847** : 78 ; B. I. Balinsky *In* : Macnae & Kalk, 1969 : 129.
Asterodiscides elegans : A. M. Clark, 1974 : 435.

R up to 80 mm ; R/r 1·3–1·5/1.

A species of Oreasteridae with almost pentagonal form, the interradii even convex in wet specimens, thick, flat below but more or less convex above in life ; abactinal plates small and almost concealed by the armament of granules and tubercles, the latter hemispherical, mushroom-shaped or truncated conical and placed singly on the plates, those of the primary plates encircled by distinctly smaller granules than the interstitial granules, though this is less distinct in larger specimens, R > 50 mm ; proximal marginal plates few in number, only two or three superomarginals and four or five inferomarginals, all more or less widely spaced from each other by encroaching abactinal or actinal plates and armed like the primary abactinal plates, though the single tubercles they bear may be sufficiently enlarged to reveal the positions of the marginals, in addition one large bare distal superomarginal in each series forming a conspicuous pair with the opposite plate at each arm tip, with four or five progressively smaller, contiguous, granule-covered distal inferomarginals between it and the distal adambulacral plates ; actinal plates with tubercles ringed by coarse granules ; adambulacrals with four or five furrow spines backed by two much coarser blunt spines arranged tangentially ; alveolar pedicellariae with narrow, slightly curved, round-tipped valves, usually present on some abactinal and actinal plates.

Type locality. Unknown.

Southern African records. 29/31/s ; 26/33/i ; 0–49 metres.

Nature of bottom. Sand.

Culcita schmideliana (Retzius)

Asterias schmideliana Retzius, 1805, *Dissertatio sistens species cognitas Asteriarum.* Lundae, p. 11.
Culcita schmideliana : H. L. Clark, 1923 : 274 ; Kalk, 1958 : 206, 218, 237 ; 1959 : 21 ; B. I. Balinsky *In* : Macnae & Kalk, 1969 : 99, 105 ; Jangoux, 1973 : 18–20, pl. 3.
Culcita novaeguineae : H. L. Clark, 1923 : 273–274. [Non *C. novaeguineae* Müller & Troschel, 1842.]

R up to *c.* 140 mm ; R/r 1·2–1·4/1.

A species of Oreasteridae with rounded or pentagonal form, the interradii convex in life and the upper side more or less swollen in wet specimens of larger size, R > 40 mm ; abactinal plates flat and forming a wide-meshed reticulum round the extensive pore-areas in larger specimens, more compact in the young, coated with fine granules and some of them with relatively coarse tubercles or short blunt spines, which may be few and inconspicuous or numerous enough to emphasize the reticular structure; margin granule-coated, obscuring the plates which are laterally aligned in larger specimens but form a broad border to the body in vertical view in young specimens, no single naked distal marginals; actinal plates with groups of coarse tubercles, flatter near the margin, emerging from a fine general granulation; adambulacral plates with usually four or five furrow spines backed by one (distally sometimes two successive) short, very thick spines or tubercles. Colour orange.

TYPE LOCALITY. Unknown ['Magellan Sea'].

SOUTHERN AFRICAN RECORD. 26/33/i.

NATURE OF BOTTOM. Rock.

Pentaceraster mammillatus (Audouin)

Asterias mammillata Audouin, 1826, *In*: *Description de l'Egypte*. Histoire naturelle. Paris. **1** (4) : 209. [Savigny, Echinodermes, pl. 5.]
Oreaster tuberculatus: Peters, 1852 : 178. [Non *O. tuberculatus* Müller & Troschel, 1842.]
Oreaster mammillatus: H. L. Clark, 1923 : 273; Lopes, 1939 : 83, pl. 3, figs 1, 2.
Pentaceraster mammillatus: Macnae & Kalk, 1962 : 108, 112; B. I. Balinsky *In*: Macnae & Kalk, 1969 : 102, 129; Jangoux, 1973 : 20–22.

R up to *c.* 140 mm; R/r 2·2–3·0/1.

A species of Oreasteridae of stellate form with triangular arms of moderate length and rounded at the tips, the disc markedly elevated and crowned by the five large tubercular primary radial plates opposite the base of each arm, each arm sloping laterally from a midradial crest or carina; abactinal plates tending to form a finely granulated reticulum around the pore areas but in larger specimens these areas are more or less confluent, some carinal and, in larger specimens, R > 60 mm, usually also some dorsolateral plates tubercular; some or all the superomarginals markedly convex in larger specimens and crowned with a tubercle, the outline scalloped; interradial inferomarginals each with a longer tubercle or short spine, sometimes also the more distal plates; actinal plates covered with coarse granules, sometimes enlarged in the centres of the plates; adambulacral plates with five to nine, usually about seven, furrow spines backed by one, rarely two, rows of two to four much coarser spines; pedicellariae more or less broad bivalved, usually present, at least dorsolaterally. Colour varied, convexities brown or red against yellow, green, khaki, brown or red ground colour.

TYPE LOCALITY. Egypt, probably Kosseir, Gulf of Suez.

SOUTHERN AFRICAN RECORDS. 26/33/i. [I regard H. L. Clark's record for Mossel Bay (34/22) as extremely unlikely.]

NATURE OF BOTTOM. Sand.

Protoreaster lincki (de Blainville)

Asterias lincki de Blainville, 1830, *Manuel d'Actinologie ou Zoophytologie*. Paris. **60** : 238.
Oreaster linckii: H. L. Clark, 1923 : 273; Lopes, 1939 : 83, pl. 2, fig. 1.

Protoreaster lincki : Kalk, 1958 : 215, 237 ; 1959 : 21 ; Macnae & Kalk, 1962 : 108, 112 ; B. I. Balinsky *In* : Macnae & Kalk, 1969 : 102, 129 ; Jangoux, 1973 : 23–25, pl. 4.

R up to *c.* 150 mm ; R/r 2·3–3·0/1.

A species of Oreasteridae of stellate form with triangular arms of moderate length and rounded at the tips, the disc markedly elevated and crowned by five huge, often spiniform tubercles, each arm sloping laterally from a midradial carina ; abactinal plates coated with a mosaic of polygonal, flat-topped granules, a few proximal plates beside a number of the carinal ones with a large, more or less spiniform tubercle, occasionally all these elevations low hemispherical instead of high conical ; superomarginals almost flat and spineless except for one or more of those near the tip, which bear a spine or high conical tubercle projecting laterally, giving a characteristic, almost smooth, outline except for the abruptly projecting distal armament ; inferomarginals completely spineless ; actinal plates covered by a mosaic of flat granules, like the other plates ; adambulacrals with five to eight furrow spines backed by a row of two or three subambulacral spines ; pedicellariae bivalved, inconspicuous, if present. Colour greyish-brown with convexities red.

Type locality. Unknown.

Southern African record. 26/33/i.

Nature of bottom. Rock or sand.

Family **OPHIDIASTERIDAE** Verrill

See : Spencer & Wright, 1966 : U64.

A family of Valvatida with usually five arms, sometimes six, rarely four or seven ; disc usually small and arms more or less long and cylindrical, or at least convex dorsally, interradial arcs usually angular ; superomarginal plates lateral in position, inconspicuous, not markedly larger than the abactinal plates (in the species represented), rarely armed with spines ; abactinal plates tessellate, regular or irregular in arrangement, armament of most of the plates granuliform, even the furrow spines short and sometimes inset so as to appear granuliform superficially, sometimes the plates partially or completely covered by thick skin ; actinal areas fairly small, though at least one series of plates usually extends well out on the arms, if not for almost the full length ; oral plates without massive, glassy-tipped spines ; pedicellariae, if present, alveolar, with or without sockets for the valves.

Austrofromia schultzei (Döderlein)

Fromia schultzei Döderlein, 1910 : 249–250, pl. 4, fig. 3.
Austrofromia schultzei : H. L. Clark, 1923 : 276 ; Mortensen, 1933a : 247–248, fig. 6, pl. 13, figs 8–10 ; Day, Field & Penrith, 1970 : 79.

R up to *c.* 50 mm ; R/r *c.* 3·5/1.

A species of Ophidiasteridae with fairly small disc, moderately long arms, rounded in cross-section, broadened slightly just beyond the base and then tapering evenly to a blunt tip, interradial arcs rounded angular, abactinal plates irregular with smaller interstitial ones between slightly larger, convex ones, all coated with granules, with pores singly at intervals around the plates ; marginal plates distinctly larger than those above and below them but inconspicuous in vertical view, coated with coarse granules ; actinal plates also with coarse granules, arranged in up to four longitudinal series, only the innermost running the full length of the arm, also in transverse rows of which two correspond to one inferomarginal or to two adambulacrals, single actinal papulae present between the plates ; adambulacral plates with two short thick furrow spines backed by two rows each of two similar but progressively shorter subambulacral spines, transitional to the actinal granulation.

TYPE LOCALITY. False Bay.

SOUTHERN AFRICAN RECORDS. 34/18/FB/s ; 32/28/– ; 15–16 metres.

Note : This species has a superficial resemblance to *Henricia ornata* but the abactinal plates in that species form a close reticulum with several pores in each space of the mesh ; conversely, *Austrofromia schultzei* has conspicuous actinal papulae.

Hacelia capensis Mortensen

Hacelia superba var. *capensis* Mortensen, 1925 : 152.
Hacelia capensis : A. M. Clark, 1974 : 435–436.

[No measurements of holotype given ; U.C.T. specimen with R only 15 mm, R/r 5/1, on which diagnosis is based in default of proper description of holotype.]

A species of Ophidiasteridae with fairly long, almost cylindrical arms, only tapering a little in the distal half to very wide rounded tips ; abactinal plates in three regular longiseries, with the two marginal series each side forming seven regular longitudinal and transverse series with six series of small pore areas between an additional, less regular, pore series below the inferomarginals and the main series of actinal plates, which are twice as numerous (only an incipient second actinal series of plates is present at this small size but, if the generic position is correct, adults should have eleven longitudinal series of plates and ten of pore areas), the plates coated with granules which are coarser centrally except for the more distal plates, many of which have a bare central patch ; adambulacral plates with two furrow spines, no granules on the furrow face and a relatively large blunt subambulacral spine on most plates set back from the furrow and surrounded by the general granulation ; pedicellariae numerous, their narrow valves tapering, with well-defined sockets.

TYPE LOCALITY. 'Off the South African coast.'

SOUTHERN AFRICAN RECORD. 29/32/s ; 70 metres.

NATURE OF BOTTOM. Rock and shell.

Leiaster sp.

Leiaster leachi : B. I. Balinsky *In* : Macnae & Kalk, 1969 : 99.

Despite Balinsky's mention of 'the enormous blood red *Leiaster leachi* found at the foot of the main coral reef' in his account of the echinoderms of Inhaca, the species is not included in either the key or in the list of species on p. 129. Peters (1852) described two other species of *Leiaster*, *L. coriaceus* and *L. glaber* from northern Mozambique, of which the latter was long thought to be a synonym of *L. leachi* until recently (see Clark & Rowe, 1971 : 57, footnote). The identity of the species found at Inhaca, I think, needs further consideration.

The genus *Leiaster* is easily recognized by the very thick skin all over the long narrow cylindrical arms with their regular longitudinal and transverse rows of plates. *L. glaber* has the surface of the plates (revealed by dissolving the skin with bleach) smooth, while *L. leachi* and *L. coriaceus* have crystal bodies embedded in the plates, giving a bumpy, almost granular texture ; they differ in the longitudinal grooves running almost the length of the furrow spines on their inner face in *L. coriaceus* whereas in *L. leachi* grooves are only faintly developed at the bases of these spines, if at all. (I have some doubts about the value of this character. Until better samples of the rather sporadic species of *Leiaster* can be made, the extent of variation cannot properly be appreciated.) Peters gives the colour in life of *L. glaber* as cinnabar red (i.e. vermilion) and of *L. coriaceus* as dirty-green or brownish-green with large irregular brownish-red patches. The colour of Gray's type of *L. leachi* was unknown but de Loriol has recorded dry specimens from Mauritius, the type locality, as dark purplish-red (probably darkened by drying) with patches of lighter red to orange, especially towards the ventral side.

Genus ***LINCKIA*** Nardo

See : Jangoux, 1973 : 25.

A genus of Ophidiasteridae with small disc and five, sometimes four, six, seven or eight cylindrical arms, only tapering close to the tips ; abactinal plates irregular, all coated with granules, with groups of pores between the plates ; pore areas also between the laterally placed marginals ; actinal plates granule-coated, in several series extending along the arms, without papular areas between ; adambulacral plates armed with short or even superficially granuliform spines ; no pedicellariae.

Linckia guildingi Gray

Linckia guildingii Gray, 1840, *Ann. Mag. nat. Hist.* **6** : 285 ; Fisher, 1919 : 401.
Linckia guildingi : A. M. Clark & Rowe, 1971 : 61 ; Jangoux, 1973 : 40.

R sometimes exceeding 200 mm ; R/r 7·5–*c.* 15/1.

A species of *Linckia* with usually five, sometimes six or occasionally four relatively long narrow arms, some, if not all, individuals liable to autotomy of parts of arms and regeneration ; one or two madreporites ; adambulacral plates with subambulacral spines immediately behind the furrow spines and slightly oblique, so that together they make a herringbone pattern, no granules between the furrow spines ; colour usually dull, brownish, orange or sometimes yellow.

Type locality. St Vincent, West Indies.

Southern African record. 26/33/i.

Linckia laevigata (Linnaeus)

Asterias laevigata Linnaeus, 1758, *Systema naturae*. Holmiae. Ed. 10 : 662.
Linckia laevigata : H. L. Clark, 1923 : 276 ; Kalk, 1958 : 206, 218 ; B. I. Balinsky *In* : Macnae & Kalk, 1969 : 99, 105, 129 ; Jangoux, 1973 : 29–32, pl. 5.

R up to *c.* 200 mm, though rarely over 120 mm ; R/r usually 6–7/1.

A species of *Linckia* with five, occasionally more, moderately stout, finger-like arms, not practising autotomy ; madreporite usually single ; adambulacral plates with the furrow spines inset into the furrow so that only their tips emerge on the ventral surface backed by fine granulation in which the usually single very short, almost granuliform subambulacral spines are isolated, some granules also extending into the furrow sandwiched between the furrow spines ; colour bright blue (except in small cryptic specimens, R < 50 mm) sometimes salmon pink (at least in the western Indian Ocean).

TYPE LOCALITY. Unknown ['Mediterranean and Indian Seas'].

SOUTHERN AFRICAN RECORDS. 26/33/i ; 24/35/i.

NATURE OF BOTTOM. Coral.

Linckia multifora (Lamarck)

Asterias multifora Lamarck, 1816 : 565–566.
? *Linckia multiforis* : Meissner, 1892 : 185.
Linckia multifora : H. L. Clark, 1923 : 277 ; Lopes, 1939 : 83 ; Kalk, 1958 : 213, 237 ; Macnae & Kalk, 1962 : 118, 119 ; B. I. Balinsky *In* : Macnae & Kalk, 1969 : 99, 105, 129 ; Jangoux, 1973 : 32–35, pl. 4.

R rarely much exceeding 100 mm, usually much less ; R/r usually 8–9/1.

A species of *Linckia* with five or six fairly narrow arms, occasionally four or seven, liable to autotomy and regeneration, forming comets ; usually two madreporites ; adambulacral plates with the furrow spines inset (as in *L. laevigata*) with granules between them, the very short subambulacral spines being set back from the furrow and isolated in the general granulation. Colour variegated, often khaki with dull reddish or brown patches.

TYPE LOCALITY. Unknown ['European Seas'].

SOUTHERN AFRICAN RECORD. 26/33/i. [Meissner's record from Cape Town of a five-armed specimen with two madreporites, must be an error of either identity or locality ; Sander also collected at Zanzibar and Fiji, where *L. multifora* may also be found.]

NATURE OF BOTTOM. Under boulders and in coral debris.

Order SPINULOSIDA Perrier

Asteroids with five or more arms ; main ossicles forming longitudinal series ventrally, though an enlarged marginal frame is not a feature of adults, both series of marginal plates being usually inconspicuous ; pedicellariae very rare, fasciculate (Mithrodiidae) or with two tapering blades (Acanthasteridae), if present ; tube feet in two rows, except in one genus of Pterasteridae where they are staggered to form four rows, ending in sucking discs.

Family **PORANIIDAE** Perrier

See : H. L. Clark, 1923 : 292 (Cryasteridae) ; Spencer & Wright, 1966 : U69.

A family of Spinulosida with body form pentagonal or short-rayed stellate, flat or slightly convex below, well arched above, often cushion-like with the dorsal body-wall more or less thickened ; abactinal plates obscured or concealed in the body-wall, rounded or oblong, forming a loose reticulum or reduced and dissociated during growth, even lost altogether in large specimens of a few species ; marginals similarly inconspicuous or reduced, especially the superomarginals, usually only one series of skin-covered spinelets or small spines fringing the line of junction between abactinal and actinal areas, if any external armament ; actinal areas large, plating reduced or obscured by the thick skin which is marked with parallel channels at right angles to the furrows, no actinal papulae ; adambulacral plates usually with only one or two furrow spines and one or two subambulacrals, rarely more, all sheathed in thick skin ; no pedicellariae.

Note : Madsen (1959, *Vidensk. Meddr dansk. naturh. Foren.* **121** : 153–160) has reviewed the North Atlantic and South African nominal species of Asteropidae (i.e. Poraniidae) with reduced skeletons, including *Chondraster, Pseudoporania, Sphaeriaster, Spoladaster* and *Tylaster*, in the light of variation in armament of *Porania pulvillus* and concludes that an aborted skeleton may be an individual variation or a normal process of ageing. He notes that *Chondraster elattosis* from South Africa 'probably should not be kept specifically distinct' from *C. grandis* (Verrill, 1879) from northeastern U.S.A. and he is also inclined to treat *Chondraster* as only a subgenus of *Porania*. As I implied in 1952, *Tylaster meridionalis* and *Spoladaster brachyactis* may well prove to be synonymous with each other. Zoogeographically, it would not be surprising if they were also conspecific with North Atlantic specimens, possibly of *Pseudoporania* or *Sphaeriaster*. As Madsen says, much more material is needed for a proper assessment of the specific limits of these asteroids with reduced skeletons.

Chondraster elattosis H. L. Clark

Chondraster elattosis H. L. Clark, 1923 : 274–275, pl. 8, fig. 4 ; 1926 : 17–18 ; Madsen, 1959 (see above) : 156–157.
Chondraster elattiosus : A. M. Clark, 1952 : 196.

R up to 115 mm ; R/r 1·25–1·55/1.

A species of Poraniidae of almost pentagonal form ; abactinal skeleton not apparent externally, probably very reduced or lost in adults, apparently no distinct external armament, the only abactinal projections being the papulae which are arranged in two bands along each ray with a narrow bare mid-radial line between, also leaving bare a deep V-shaped peripheral area across each interradius ; marginal plates present but obscured, unarmed or with spines hidden in the body wall ; actinal areas unarmed ; adambulacral plates with three or four furrow spines backed by two or three, rarely four, stouter subambulacral spines.

TYPE LOCALITY. South from Cape Infanta, 1025 metres.

SOUTHERN AFRICAN RECORDS. 34/17/d, vd ; 36/21/vd ; 402–1025 metres.

NATURE OF BOTTOM. Green sand ; green mud.

Spoladaster brachyactis (H. L. Clark)

Culcita veneris : Bell, 1905a : 248. [Non *C. veneris* Perrier, 1879.]
Cryaster brachyactis H. L. Clark, 1923 : 293–294, pl. 11, figs 1, 2 ; Mortensen, 1933a : 249, pl. 12, fig. 14.
Spoladaster brachyactis : Fisher, 1940 : 135–136 ; A. M. Clark, 1952 : 196, 208.

R up to 80 mm ; R/r *c.* 1·5/1.

A species of Poraniidae of very short-rayed stellate or pentagonal cushion-shaped form ; abactinal plates lost early in development, except near the arm tips, body-wall thick, the numerous small skin-sheathed superficial spinelets scattered over the abactinal surface dissociated, papulae also arising all over the abactinal area but sometimes discernibly in rounded groups ; superomarginals lost except distally but traces of inferomarginals retained in adult specimens though obscured by the body-wall, usually bearing several spinelets or some of them a spine ; actinal plates reduced or obscured by the thick skin in which there may be some scattered spinelets ; adambulacral plates with one, sometimes two, furrow spines backed by a larger subambulacral spine.

TYPE LOCALITY. South of Algoa Bay, 104 metres.

SOUTHERN AFRICAN RECORDS. 30/15/d ; 34/18/s ; 33/25/s ; 34/25/d ; 47(? 42)–205 metres.

NATURE OF BOTTOM. Mud ; sand and mud.

Tylaster meridionalis Mortensen

Tylaster meridionalis Mortensen, 1933a : 249–250, pl. 12, figs 11–13 ; A. M. Clark, 1952 : 196, 207–208.

R up to 28 mm ; R/r *c.* 1·25/1.

A species of Poraniidae of pentagonal cushion-like form ; abactinal plates, if present, concealed, not associated with the superficial spinelets scattered all over the abactinal area (papulae not described but presumably also widely distributed abactinally) ; (infero-)marginals with a conspicuous fringe of short spines, three or four on each plate ; actinal areas smooth but for a few distal interradial superficial spines in some specimens, the plates concealed ; adambulacral plates usually with one furrow spine and one subambulacral spine.

TYPE LOCALITY. SW. from Port Nolloth, 197 metres.

SOUTHERN AFRICAN RECORDS. 30/15/d ; 197–205 metres.

NATURE OF BOTTOM. Sand and mud.

Family **ASTERINIDAE** Gray

See : Fisher, 1911 ; 253 : Spencer & Wright, 1966 : U68.

A family of Spinulosida with body form stellate or sometimes pentagonal, arms occasionally more than five, usually short and tapering to blunt tips, flat below, usually somewhat convex above but sometimes almost leaf-like ; abactinal plates relatively small, often crescentic or multilobed, similar or of two distinct magnitudes, flat or convex above, usually imbricating with the midradial plates overlapping proximally and the lateral ones more adradially, armed with small spinelets or granules, papulae restricted to parts of the upper side ; marginals usually distinct but not enlarged in the species represented, the inferomarginals defining a more or less sharp ventrolateral angle ; actinal plates forming transverse as well as longitudinal series, small, imbricating and armed with one or more spinelets, sometimes in combs ; adambulacral plates with armament not conspicuously coarser than the actinal armament ; pedicellariae, if present, simple alveolar or fasciculate with few spiniform valves.

Genus ***ANSEROPODA*** Nardo

See : Fisher, 1906, *Bull. U.S.Fish Commn* **1903** (3) : 1088.

A genus of Asterinidae with five, sometimes more, short rays, body-form extremely thin and leaf-like, only narrow midradial bands at all thickened, often reaching a relatively large size (for this family) of R $>$ 100 mm ; abactinal plates thin, flat and multilobed on the midradial bands to which the papulae are restricted in a few rows each side of the midline, markedly different from the remaining abactinal plates which are superficially rhombic (though with an internal projection like the handle of a skillet) and closely imbricated together in longitudinal and transverse series, so as to resemble snakeskin, armed with groups of small spinelets ; actinal plates also regularly arranged in two directions, armed with longer spinelets or spines ; adambulacral plates with a fan of furrow spines, usually webbed with skin backed by several subambulacral spines, often in linear straight or curved series.

Note : Examination of an intact specimen of *Anseropoda grandis* from off Cape St Francis (34°30′S 24°50′E) and of the paratypes of *A. novemradiata* from off the Tugela River has allowed amplification of the previously published data regarding these species.

Anseropoda grandis Mortensen

Anseropoda (*Palmipes*) *grandis* Mortensen, 1933a : 251–252, fig. 7, pl. 12, figs 1, 2.

R up to 130 mm ; R/r *c.* 2/1.

A species of *Anseropoda* of maple-leaf-like form, with five rays and the interradial arcs well indented ; midradial abactinal plates without any conspicuous tubercle-like prominences and accommodating two or three longitudinal series of papulae each side of the midline, other abactinal plates each with a group of two to four small spinelets at the proximal edge, otherwise bare ; actinal plates mostly with a comb of three or four rather longer spinelets or fine spines (Mortensen's measurement of 15 mm is clearly a mistake for 1·5 mm) ; adambulacral plates with a fan of seven to nine slender furrow spines backed by six or seven subambulacral spines, the inner four or five in a row almost parallel to the furrow or somewhat oblique.

TYPE LOCALITY. NW. from Lambert's Bay, 315 metres.

SOUTHERN AFRICAN RECORDS. 31/16/d ; 34/24/d ; 275–315 metres.

Anseropoda habracantha H. L. Clark

Anseropoda habracantha H. L. Clark, 1923 : 287–288, pl. 17, figs 4, 5.
Anseropoda (Palmipes) habracantha : Mortensen, 1933a : 251.

R up to 16 mm (only two specimens known) ; R/r 1·45/1.

A species of *Anseropoda* of nearly pentagonal form, the interradial arcs only slightly concave ; midradial abactinal plates accommodating only a single series of papulae each side of the midline (though it is probable that larger specimens – if they exist – would have additional series), other abactinal plates with clusters of 10–12 radiating spinelets ; actinal plates with combs of up to ten longer slender spinelets ; adambulacral plates with a fan of four, sometimes three, slender furrow spines backed by about five subambulacral spines in a curved or oblique comb.

TYPE LOCALITY. Off East London, 155 metres.

SOUTHERN AFRICAN RECORDS. 33/28/d ; 29/31/d ; 125–155 metres.

Anseropoda novemradiata (Bell)

Palmipes novemradiatus Bell, 1905a : 248.
Anseropoda novemradiata : H. L. Clark, 1923 : 287 ; 1926 : 18–19.

R up to 70 mm ; R/r 1·2–1·6/1.

A species of *Anseropoda* with nine (sometimes eight) short rays ; a few to many midradial and some other abactinal plates surmounted by a coarse rounded tubercle-like ossicle itself bearing spinelets, at least in larger specimens, R $>$ 40 mm, papulae in two zig-zag rows each side of the midline of each ray, other abactinal plates armed with a cluster of three to six small radiating spinelets with isolated additional spinelets scattered between the groups ; actinal plates with combs of three to seven longer spinelets, the median ones in each comb longest ; adambulacral plates with a fan of five or six furrow spines backed by three to five subambulacral spines in a linear series either straight and at right angles to the furrow or curving distally at the inner end.

TYPE LOCALITY. Cape Natal (Durban), 88 metres.

SOUTHERN AFRICAN RECORDS. 29/31/s ; 46–88 metres.

NATURE OF BOTTOM. Fine sand ; rocks and stones ; sand and shells ; sand.

Genus *ASTERINA* Nardo

See : Fisher, 1919 : 409.

A genus of small Asterinidae, R rarely exceeding 30 mm, of pentagonal or stellate form with five or sometimes more arms, thick centrally but with the ventrolateral angles acute ; abactinal plates of papular areas often somewhat irregular in size and shape in larger specimens, R > 15 mm, some of the larger plates concave on the proximal or adradial side but then compact crescentic, not markedly elongate in a tangential direction and not subtending groups of much smaller plates, these being relatively few and scattered ; all plates armed with groups of spinelets ; papulae extending over much of the abactinal side, except distally and peripherally ; both series of marginal plates usually distinct though not enlarged, the inferomarginals defining the angle ; actinal plates with combs of spinelets, sometimes clusters ; adambulacral plates usually with the furrow and subambulacral spines in two fans.

Asterina burtoni Gray

Asterina burtonii Gray, 1840, *Ann. Mag. nat. Hist.* **6** : 289 ; H. L. Clark, 1923 : 283.
Asterina burtoni : Kalk, 1959 : 21 ; B. I. Balinsky *In* : Macnae & Kalk, 1969 : 105, 129 ; A. M. Clark & Rowe, 1971 : 68–69 ; Jangoux, 1973 : 35–38 ; A. M. Clark, 1974 : 437–438 ; Day, 1974 : 55, 59, 94.
? *Asterina coronata* : Kalk, 1958 : 215 ; Jangoux, 1973 : 38–39, pl. 8. [Non *A. coronata* von Martens, 1866.]

R up to 32 mm, rarely > 25 mm, R/r up to 2·8/1, usually 2·0–2·4/1.

A species of *Asterina* with relatively long blunt-tipped arms numbering usually five, sometimes six and in certain parts of the range, notably the Red Sea, sympatric and probably conspecific with fissiparous asterinids with up to eight arms and also with more than one madreporite ; abactinal plates mostly armed with five to ten short stumpy spinelets, length : breadth *c.* 2 : 1 ; actinal plates with two to six spinelets, usually in a comb but sometimes this is double and they appear clustered ; adambulacral plates with a fan of about five furrow spines and another of about four subambulacrals. Colour orange (at least in Mozambique, elsewhere usually mottled khaki and brown).

Type locality. Red Sea.

Southern African records. 26/33/i ; 25/32/i ; 24/35/– ; 23/35/i ; LW–LWS.

Nature of bottom. Under stones.

Asterina gracilispina H. L. Clark

Asterina gracilispina H. L. Clark, 1923 : 286–287, pl. 16, figs 3, 4 ; Mortensen, 1933a : 255–256 ; A. M. Clark, 1974 : 437.
? *Asterina (Patiriella) gracilispina* : Day, Field & Penrith, 1970 : 79.

R up to 18 mm ; R/r 1·3–1·5/1.

A species of *Asterina* with five very short arms, often distinctly petaloid ; abactinal plates with 10–15 very small spinelets ; actinal plates with combs of two to five spinelets ; adambulacral plates with a fan of three to five furrow spines backed by two or three subambulacral spines.

Type locality. SW. of Cove Rock, East London, 40 metres. [H. L. Clark's 'Coke' Rock is a misprint.]

Southern African records. ? 34/18/FB/s ; 34/20/– ; 34/22/s ; ? 34/23/s ; 34/25/s ; 33/27/s ; 25–86 metres.

Nature of bottom. Rock and broken shell ; rock ; rock and khaki sand.

Disasterina leptalacantha africana Mortensen

Disasterina leptalacantha var. *africana* Mortensen, 1933a : 259–260, pl. 12, fig. 3.

R 18 mm ; R/r *c.* 1·8/1 (judging from Mortensen's photograph).

A subspecies of Asterinidae with five fairly short, thick, tapering, blunt-tipped arms ; abactinal plates obscured by thick gelatinous skin, almost naked, spinelets only present around the anus and sometimes on some interradial plates ; actinal plates also covered with thick skin, armed with single spines or spinelets ; adambulacral plates not described (presumably with about five furrow spines backed by a single subambulacral spine, as in *D. leptalacantha* from Australia).

Type locality. Type not designated (owing to varietal ranking), presumably Port Elizabeth, rock pool, since this specimen is the better preserved and more likely for selection as holotype.

Southern African records. 33/25/i ; 29/31/d ; 0–366 metres.

Genus ***PATIRIA*** Gray

See : Fisher, 1940 : 269.

A genus of Asterinidae of thick stellate form, with usually five arms of moderate length, evenly tapering or, more often, finger-like with blunted tips, though more or less broadened basally, adult size usually between 30 and 60 mm R, ventrolateral angles often only about 90° for much of their extent, though somewhat acute in one species represented ; abactinal plates of papular areas of two distinct magnitudes, the larger primary ones often crescentic or with proximal or adradial concavities, the smaller ones dividing up the papular areas, which extend over most of the abactinal side, except peripherally, the plates armed with spinelets at least twice as long as broad ; marginal plates distinct but not enlarged (in the species represented) ; actinal plates with spinelets in linear series proximally, but often grouped towards the margin ; adambulacral plates with subambulacral spines in fans either oblique or parallel to the furrow behind the fan of furrow spines, barely larger than the adjacent actinal armament.

Patiria formosa (Mortensen)

Parasterina formosa Mortensen, 1933a : 261–262, pl. 12, figs 9, 10.
Patiria formosa : Morgans, 1962 : 308 ; Day, Field & Penrith, 1970 : 79.

R up to *c.* 50 mm ; R/r 2·2–3·0/1.

A species of *Patiria* with thick finger-like blunt-tipped arms, almost semicircular in cross-section, the sides more or less vertical for much of their length and the ventrolateral angle about 90° ; primary abactinal plates compact crescent-shaped or squat V-shaped proximally but distally becoming larger and superficially almost circular, though separated by secondary plates except at the very tip of the arm ; actinal plates with two to seven spinelets, sometimes only one on a few proximal plates, in groups on the more peripheral plates but otherwise in combs ; adambulacral plates with a fan of three or four furrow spines backed by an oblique series of three subambulacral spines ; pedicellariae not observed. Colour : blue-grey ; reddish with purple papulae ; pale purple centrally grading to orange distally, underside white.

TYPE LOCALITY. False Bay, 55 metres.

SOUTHERN AFRICAN RECORDS. 34/18/FB/s ; 12–55 metres.

NATURE OF BOTTOM. Sand ; sand, shells and *Phyllochaetopterus* ; shelly sand ; rock ; coarse sand and shell.

Patiria granifera Gray

Patiria granifera Gray, 1847a, *Proc. zool. Soc. Lond.* **1847** : 82 ; Fisher, 1940 : 271–272 ; A. M. Clark, 1956 : 380–382, fig. 4 ; Day, 1959 : 544 ; Morgans, 1959 : 424, 426, 427 ; 1962 : 303, 308 ; Grindley & Kensley, 1966 : 12 ; Penrith & Kensley, 1970 : 201, 233.

Patiria bellula Sladen, 1889 : 385–386, pl. 63, figs 1, 2, pl. 64, figs 5, 6 ; Koehler, 1908 : 296 ; 1914 : 171 ; Fisher, 1940 : 269–270 ; A. M. Clark, 1952 : 196 ; Day, Millard & Harrison, 1952 : 396.

Asterina penicillaris : Meissner, 1892 : 187. [Non *Asterias penicillaris* Lamarck, 1816.]

Asterina coccinea (part) : Bell, 1905a : 248. [Non *A. coccinea* Gray, 1840.]

Asterina granifera : H. L. Clark, 1923 : 281–282, pl. 17, figs 1, 2 ; Mortensen, 1933a : 256.

Parasterina bellula : H. L. Clark, 1923 : 280 ; Mortensen, 1933a : 260–261 ; Stephenson, Stephenson & du Toit, 1937 : 363, 380 ; Bright, 1937 : 59 ; Stephenson, Stephenson & Bright, 1938 : 10, 18 ; Eyre, 1939 : 298 ; Stephenson, Stephenson & Day, 1939 : 368 ; Stephenson, 1944 : 317, 334, 347.

Asterina granifera var. *sporacantha* H. L. Clark, 1923 : 282–283, pl. 17, fig. 3.

Callopatiria bellula : Döderlein, 1928 : 296.

R up to *c.* 60 mm ; R/r 2·2–2·9/1.

A species of *Patiria* with thick, finger-like, blunt-tipped arms almost semicircular in cross-section, the sides more or less vertical for much of their length and the ventrolateral angle about 90° ; primary abactinal plates compact crescent-shaped or squat V-shaped, similar in size or smaller distally ; actinal plates with one to eight spinelets, in linear combs proximally but grouped peripherally ; adambulacral plates with a fan of about four furrow spines backed by an oblique series of about three subambulacral spines ; pedicellariae not observed. Colour deep orange above, pale orange below.

TYPE LOCALITY. Unknown.

SOUTHERN AFRICAN RECORDS. 26/15/i ; 32/18/i, s ; 33/18/i, s ; 34/18/FB/i, s ; 34/19/s ; 34/21/i ; 34/22/i, s ; 34/23/i ; 33/25/i ; 33/27/i ; 29/31/i ; LW–82 metres.

NATURE OF BOTTOM. Rock ; sand and rock ; shell ; fine green sand and shell.

Patiria stellifera (Möbius)

Asteriscus stellifer Möbius, 1859, *Abh. Geb. Naturw. Hamburg* **4** (2) : 4–5.
Asterina luderitziana Döderlein, 1908, *Jb. nassau. Ver. Naturk.* **61** : 296–298, pl. 2, figs 1–4; 1910 : 250–252, pl. 4.
Asterina marginata [Hupé, 1857, nom. nud.] : Koehler, 1914 : 171–173, pl. 6, figs 9–13.
Asterina stellifer : Madsen, 1950, Atlantide Rep. **1** : 213–214.
Patiria stellifera : Fisher, 1919 : 410 [by inference].

R up to 65 mm; R/r 1·6–2·4/1.

A species of *Patiria* with moderately short arms, triangular in outline, tapering fairly evenly to almost pointed tips and also almost triangular in cross-section, the ventrolateral angles less than 90°; crescentic primary abactinal plates sufficiently elongated to appear banana-shaped, especially when denuded, not enlarged distally; actinal plates with two to six spinelets, usually three, in tangential combs, sometimes single spinelets on a few proximal plates; adambulacral plates with a fan of usually three furrow spines backed by two subambulacral spines aligned parallel to the furrow, all sheathed in thick skin; some simple fasciculate pedicellariae, with usually two to four valves barely modified from spinelets, often present abactinally. Colour light orange.

Type locality. Rio de Janeiro.

Southern African record. 26/15/i.

Note : Fisher's synonymy in 1919 of *Enoplopatiria* Verrill, 1913 (of which this is the type-species) with *Patiria* Gray, 1840 implied that the proper combination is *Patiria stellifera*. Although subsequent students of West African asteroids such as Madsen (1950) and myself (1955, *Jl W. Afr. Sci. Ass.* **1**) followed Koehler's earlier (1914) usage of the generic name *Asterina*, Tortonese (1956, *Ann. Mus. Stor. nat. Genova* **68**) and some others working on the South American fauna have reverted to the use of *Patiria*, which I now think is preferable, owing to the secondary abactinal plates and large size.

Genus *PATIRIELLA* Verrill

See : Dartnall, 1971, *Proc. Linn. Soc. N.S.W.* **96** (1) : 39–40.

A genus of Asterinidae with up to 11 short rays, normally five in the species represented which have the body form almost pentagonal, flattened without being leaf-like, the upper side slightly but evenly convex, the marginal angle acute and the adult size small, R up to only *c.* 20 mm; abactinal plates of the fairly extensive papular areas of two magnitudes, the large, often crescentic, ones with their concave sides subtending small groups of smaller plates separating the papulae, armed with coarse granules or superficially granuliform short spinelets, length < twice breadth; marginal plates distinct but not enlarged; actinal plates armed with one or two spinelets or small spines; adambulacral plates with usually only two furrow spines and one subambulacral spine. Pedicellariae not observed.

Patiriella dyscrita (H. L. Clark)

Asterina dyscrita H. L. Clark, 1923 : 284–285, pl. 16, figs 5, 6.
Asterina (*Patiriella*) *exigua* (part); Mortensen, 1933a : 252–255.

Patiriella dyscrita : A. M. Clark, 1974 : 438.

R up to *c.* 17 mm ; R/r 1·3–1·5/1.

A small species of *Patiriella* with normally five arms ; gonopores on the upper surface. Colour (above) various shades of green, mottled with specks of red, blue, yellow, etc., below not consistently blue-green.

TYPE LOCALITY. Between [Breede] River and Sebastian Bluff (St. Sebastian Point), LW.

SOUTHERN AFRICAN RECORDS. 34/20/i ; 33/25/s ; intertidal to 'below LWS'.

Patiriella exigua (Lamarck)

Asterias exigua Lamarck, 1816 : 554.

Asterina kraussi Gray, 1840, *Ann. Mag. nat. Hist.* **6** : 289.

Asterina calcarata : Koehler, 1908 : 632 ; H. L. Clark, 1923 : 285–286. [Non *A. calcarata* Gay, 1854.]

Asterina exigua : Sladen, 1889 : 392 ; Döderlein, 1910 : 250 ; Koehler, 1914 : 171 ; H. L. Clark, 1923 : 285 ; Stephenson, Stephenson & du Toit, 1937 : 358, 359, 363, 380 ; Bright, 1937 : 76, 86, 87 ; Stephenson, Stephenson & Bright, 1938 : 10, 18 ; Eyre & Stephenson, 1938 : 38, 43 ; Eyre, Broekhuysen & Crichton, 1938 : 96, 110 ; Eyre, 1939 : 298, 304, pl. 21 ; Stephenson, Stephenson & Day, 1939 : 357, 368 ; Stephenson, 1944 : 286, 305, 317, 334, 347 ; Day, Millard & Harrison, 1952 : 389, 393, 396 ; Day & Morgans, 1956 : 308 ; Macnae, 1957 : 362, 363, 367 ; Kalk, 1958 : 200, 223, 237 ; Day, 1959 : 544 ; Morgans, 1959 : 398, 424 ; 1962 : 327 ; B. I. Balinsky *In* : Macnae & Kalk, 1969 : 105 ; Penrith & Kensley, 1970 : 201, 206, 233.

Asterina (*Patiriella*) *exigua* : Mortensen, 1933a : 252–255, pl. 12, figs 4–8 ; Macnae & Kalk, 1969 : 129 ; Day, Field & Penrith, 1970 : 79.

Patiriella exigua : Fisher, 1940 : 272 ; Dartnall, 1971 (see above) : 40–43, pl. 4, fig. c.

R up to *c.* 20 mm ; R/r 1·2–1·5/1.

A small species of *Patiriella* with usually five arms, sometimes six or four ; gonopores on the undersurface, two in each interradius separated from the oral plates by only about three actinal plates. Colour above dull green or brown ground colour with armament red, orange, purple, brown or pastel greens or browns, often making geometrical patterns, below consistently blue-green.

TYPE LOCALITY. 'Les mers d'Amérique'. Neotype selected by Dartnall from False Bay.

SOUTHERN AFRICAN RECORDS. 26/15/i ; 29/16/i ; 32/18/i ; 33/18/i, s ; 34/18/i ; 34/18/FB/i, s ; 34/19/i ; 34/21/i ; 34/22/s ; 34/23/i ; 34/24/i ; 33/25/i ; 34/25/i ; 26/33/i ; intertidal–3 metres.

NATURE OF BOTTOM. In rock pools on algae or worm tubes or under stones.

Family **PTERASTERIDAE** Perrier

See : Fisher, 1911 : 343 ; Spencer & Wright, 1966 : U68.

A family of Spinulosida with body form ranging from pentagonal to short-rayed stellate with blunt-tipped arms, usually cushion-like but sometimes flattened above (though this may be exaggerated by poor preservation of these relatively soft deep-water animals), arms five or occasionally six in the species represented ; upper surface made up of an additional supradorsal membrane supported by the long slender spines of the tall paxilliform abactinal plates, the cavity so formed ventilated by numerous small spiracular openings and a central osculum, acting also as a brood chamber ; marginal plates undistinguished and actinal plates absent, the supradorsal membrane continuing downwards laterally to meet a longitudinal fringe-like actinolateral membrane supported by the elongated outermost adambulacral spines, the remaining adambulacral spines aligned in series at right angles to the furrow and often linked together by skin into fans ; oral plates with several spines along the furrow margin, which may be linked into longitudinal fans by skin, and one, sometimes two, more or less enlarged suboral spines on the ventral face, often with glassy tips ; no pedicellariae.

Diplopteraster multipes (Sars)

Pteraster multipes M. Sars, 1865, *Forh. VidenskSelsk. Krist.* **1865** : 200.
Diplopteraster multipes : Fisher, 1911 : 371–373, pl. 107, figs 1, 2 ; H. L. Clark, 1923 : 300 ; 1926 : 22 ; Mortensen, 1933a : 271–272 ; A. M. Clark, 1952 : 198.

R up to *c.* 110 mm ; R/r 1·3–1·6/1.

A species of Pterasteridae with body form almost pentagonal, thick and cushion-like ; abactinal plates with an enlarged central spine ringed by *c.* 8 smaller peripheral spines ; actinolateral membrane well developed but hardly, if at all, projecting laterally ; adambulacral plates with alternating armament, webbed together so that fans of three or four spines alternate with others having an extra spine and projecting further across the furrow in which the tube feet are staggered to form four rows ; oral plates with the spines along their furrow margins webbed together in a continuous fan across the apex of each jaw.

Type locality. Drøbak, southern Norway.

Southern African records. 31/16/d ; 32/17/d ; 33/17/vd ; 34/17/d ; 34/18/d ; 35/18/vd ; 293–547 metres.

Nature of bottom. Green sand, black specks ; green mud ; mud ; green mud and sand.

Genus ***HYMENASTER*** Wyville Thomson

See : Sladen, 1889 : 491 ; Fisher, 1911 : 373.

A genus of deep-sea Pterasteridae with the body flattening when preserved and usually almost pentagonal ; actinolateral membrane large and projecting laterally like a web, partially supported by very long spines not linked transversely by skin to the few free adambulacral spines ; an aperture present in the membrane between the outer (abradial) ends of each two adambulacral plates guarded by an operculum-like aperture papilla ; oral furrow spines not webbed.

Hymenaster gennaeus H. L. Clark

Hymenaster gennaeus H. L. Clark, 1923 : 302–303, pl. 10 ; 1926 : 22.
R up to 75 mm ; R/r 1·25/1·4/1.

A species of *Hymenaster* of pentagonal form ; actinolateral spines relatively short, up to *c.* 11 mm long at R 75 mm ; adambulacral armament likened to that of *H. nobilis* and therefore presumably consisting of only a single free spine on each plate ; oral plates with four (rarely three) spines along the furrow margin.

TYPE LOCALITY. SW. of Cape Point, 1370–1463 metres.

SOUTHERN AFRICAN RECORDS. 33/16/vd ; 34/17/vd ; 1463 (? 1370)–2330 metres.

NATURE OF BOTTOM. Green mud ; globigerina ooze.

Hymenaster lamprus H. L. Clark

Hymenaster lamprus H. L. Clark, 1923 : 301–302, pl. 11, figs 3, 4 ; Mortensen, 1933a : 267, pl. 14, figs 8, 9.
R up to 110 mm ; R/r *c.* 1·3/1.

A species of *Hymenaster* of almost pentagonal form ; actinolateral spines rather short, up to *c.* 6 mm long at R 42 mm ; adambulacral plates with two free spines, occasionally three in larger specimens ; oral plates with only two spines along the furrow margin.

TYPE LOCALITY. SW. of Cape Point, 1463–1646 metres.

SOUTHERN AFRICAN RECORDS. 34/17/vd ; 1463–1646 metres.

NATURE OF BOTTOM. Green mud.

Hymenaster latebrosus Sladen

Hymenaster latebrosus Sladen, 1882, *J. Linn. Soc.* Zool. **16** : 230 ; 1889 : 514–515, pl. 92, figs 4, 5, pl. 93, figs 7–9 ; H. L. Clark, 1923 : 300–301.
R up to 30 mm ; R/r 1·6–1·8/1.

A species of *Hymenaster* of short-rayed stellate form ; actinolateral spines well developed, up to *c.* 5 mm long at R 22 mm ; adambulacral plates with three free spines ; oral plates with two or three marginal spines.

TYPE LOCALITY. Southern Ocean, south from Australia (*c.* 54°S), 3560 metres.

SOUTHERN AFRICAN RECORD. 34/17/vd ; 1463–1646 metres.

NATURE OF BOTTOM. Green mud.

Note : H. L. Clark (1923 : 301) has 'with much hesitation' referred 13 relatively small poorly preserved specimens to *Hymenaster membranaceus* Sladen, 1882, otherwise known from west of Portugal, although the oral armament (which I think provides the only obvious distinction between *H. membranaceus* with four or five furrow spines and *H. latebrosus* with three) is not distinct in any of them. The basis for inclusion of *H. latebrosus* itself in the South African fauna is the single specimen so named by H. L. Clark, which in fact came from the same station as some of those which he called *H. membranaceus*. More and better preserved material is clearly needed to settle the true identity of these specimens, which is admittedly difficult in view of their fragility and the great depths from which they have to be dragged. With such deep-sea species, wide geographical ranges are to be expected and the tally of 50 nominal species of *Hymenaster* currently recognized, most of them known only from the type material, is ludicrous.

Genus *PTERASTER* Müller & Troschel

See : Fisher, 1911 : 344 ; 1940 : 191.

A genus of Pterasteridae with body form thick and cushion-like, pentagonal to short-rayed stellate, the species represented with five or sometimes six arms ; actino-lateral membrane narrow or better developed but hardly, if at all, projecting laterally and rarely visible from above, linked by skin to the transverse fans formed by the (other) adambulacral spines, all of which fans are similar, not alternating ; oral plates with their marginal furrow spines either webbed together or separate ; tube feet in two rows (in the species represented).

Pteraster affinis E. A. Smith

Pteraster affinis E. A. Smith, 1876, *Ann. Mag. nat. Hist.* (4) : **17** : 108 ; H. L. Clark, 1923 : 300 ; Mortensen, 1933a : 397 ; A. M. Clark, 1962, *Rep. B.A.N.Z. Antarctic Res. Exped.* 1929–31. **B9** : 59–63. [Non *P. affinis* : A. M. Clark, 1952 : 197, 211.]

R up to 35 mm ; R/r 1·7–2·8/1.

A species of *Pteraster* with normally five short arms ; abactinal paxillae with fairly low columns (pedicels) less than three times as high as broad, crowned by usually only three or four spines, one of them distinctly larger than the rest ; actinolateral membrane fairly well developed ; oral plates each with four to six spines along the furrow margin, webbed together with skin but not continuous with the corresponding fan of the adjacent plate at the apex of the jaw.

TYPE LOCALITY. Observatory Bay, Kerguelen, on roots of *Macrocystis*.

SOUTHERN AFRICAN RECORD. 34/18/d ; 420 metres.

Note : As I remarked in 1962 (p. 63) further consideration prompted me to refer the two '*Africana*' specimens, which I had named *P. affinis* in 1952, to *P. flabellifer* Mortensen. I am only retaining *P. affinis* in the South African fauna list because Mortensen (in a postscript) apparently did not consider the Cape specimen named *affinis* by H. L. Clark to be conspecific with his own *P. flabellifer* but thought it 'not all improbable' that it *is* conspecific with the species from Kerguelen. Possibly he was put off by the difference in arm number ; the Cape specimen presumably has five arms since it is not stated to the contrary, whereas the holotype of *P. flabellifer* has six. However, Mortensen himself was inclined to think that two five-armed specimens of *Pteraster* are referable to *P. flabellifer*, being mainly deterred by the fact that their fans of oral marginal spines do not overlap at the apex of the jaw – something I would attribute to differences in preservation.

Pteraster capensis Gray

Pteraster capensis Gray, 1847a, *Proc. zool. Soc. Lond.* **1847** : 83 ; H. L. Clark, 1923 : 298–300, pl. 9, figs 3, 4 ; 1926 (part) : 21–22 ; Mortensen, 1933a : 267–269, fig. 9, pl. 14, figs 1–3 ; A. M. Clark, 1952 : 197, 210–211 ; Grindley & Kensley, 1966 : 12 ; Day, Field & Penrith, 1970 : 80.

Retaster capensis : Bell, 1905a : 250.

Pteraster gibber : A. M. Clark, 1952 : 197 [lapsus].

R up to 53 mm ; R/r 1·3–1·8/1.

A species of *Pteraster* with form ranging from pentagonal to short-rayed stellate; abactinal paxillae with tall pedicels crowned by usually six to eight spines of similar thickness; actinolateral membrane narrow and inconspicuous; oral marginal spines five or six on each plate and webbed to form a single continuous fan across the apex of each jaw.

Type locality. 'Cape of Good Hope.'

Southern African records. 28/15/–; 29/15/d; 31/16/d; 32/16/d; 32/17/d; 33/17/s, d; 33/18/—; 34/17/d; 34/18/s, d; 34/18/FB/s; 34/22/s; 34/25/d; 32/28/d; 34(? 19)–370 metres.

Nature of bottom. Black specks; green sand and black specks; green mud; fine sand; sand and mud; sand, broken shell and patches of *Phyllochaetopterus*; polyzoa and rock; rock.

Pteraster flabellifer Mortensen

Pteraster flabellifer Mortensen, 1933a: 269–270, fig. 10, pl. 14, figs 6,7; A. M. Clark, 1962, *Rep. B.A.N.Z. Antarctic Res. Exped.* 1929–31, **B9**: 62, 63.
Pteraster affinis: A. M. Clark, 1952: 197, 211. [? *P. affinis*: H. L. Clark, 1923: 300; Mortensen, 1933a: 397.] [Non *P. affinis* Smith, 1876.]

R up to 40 mm; R/r 1·9–2·5/1.

A species of *Pteraster* with five or six short arms; abactinal paxillae with tall slender columns about six times as high as broad, crowned with five or six similar spines; actinolateral membrane fairly well developed; oral plates each with about seven spines along the furrow margin, webbed together with skin but not continuous with the fan of the corresponding plate across the apex of the jaw.

Type locality. NW. of Cape Town, 366 metres.

Southern African records. 31/15/d; 31/16/d; 272–366 metres.

[Mortensen also thinks a specimen from off Durban may be referable to this species, as well as one from near the type locality off the west coast.]

Nature of bottom. Polyzoa and rock.

Pteraster fornicatus Mortensen

Pteraster capensis (part): H. L. Clark, 1926: 21–22. [Non *P. capensis* Gray, 1847.]
Pteraster fornicatus Mortensen, 1933a: 270–271, fig. 11, pl. 14, figs 4, 5.

R up to 8 mm; R/r *c.* 1·3/1. (Only two specimens recorded.)

A species of *Pteraster* of almost pentagonal form; abactinal paxillae with five or six spines; actinolateral membrane of moderate size (from the photograph); oral plates each with six (sometimes five) independent spines along the furrow margin, not forming fans.

Type locality. SW. of Cape St Martin (33/17), 330 metres.

Nature of bottom. Green mud.

Family **SOLASTERIDAE** Perrier*

See : Fisher, 1911 : 305 ; Spencer & Wright, 1966 : U66.

A family of Spinulosida with a large disc and five to fifteen arms, rounded in cross-section and tapering to narrow tips ; abactinal plates more or less paxilliform, with or without a central column, their lobed bases imbricating to form an open irregular reticulum with groups of papulae in the meshes and the columns crowned with clusters of slender spinelets ; one or two marginal series of tuft-like enlarged paxillae representing the marginal plates ; actinal plates mostly restricted to the disc, rarely extending very far along the arms, no actinal papulae ; adambulacral plates with a longitudinal comb of furrow spines backed by a transverse comb of subambulacral spines ; no pedicellariae.

Crossaster penicillatus Sladen

Crossaster penicillatus Sladen, 1889 : 446–447, pl. 70, fig. 5, pl. 72, figs 9, 10 ; Koehler, 1908 : 560 ; H. L. Clark, 1923 : 295 ; 1926 : 21 ; Mortensen, 1933a : 272–273 ; A. M. Clark, 1952 : 197.
Solaster penicillatus : Bell, 1905a : 249.

R up to 60 mm ; R/r *c.* 3·0/1.

A species of Solasteridae with eight to eleven, usually ten arms ; abactinal plates with a low column, rarely higher than broad, crowned with about ten spinelets ; only one series of (infero-)marginal plates forming a regular longitudinal row and distinctly larger than the abactinal paxillae ; actinal plates mainly restricted to the disc, armed with a few (one to four) short spinelets ; adambulacral plates with four or five furrow spines and up to eight spines in the subambulacral row. Colour uniform deep red.

Type locality. Tristan da Cunha, 200 metres.

Southern African records. ? 30/16/d ; 31/16/d ; 32/16/vd ; 32/17/d ; 33/16/d ; 33/17/d, vd ; 34/17/d ; 34/18/s, d ; 34/20/s ; 36/21/d ; 34/22/d ; 34/23/s ; 32/28/– ; 56–819 metres.

Nature of bottom. Rock ; polyzoa and rock ; sand and rocks ; rock and khaki sand ; sand ; mud.

Lophaster quadrispinus H. L. Clark

Lophaster quadrispinus H. L. Clark, 1923 : 295–297, pl. 18, figs 1, 2 ; 1926 : 21 ; Mortensen, 1933a : 272 ; A. M. Clark, 1952 : 197 ; 1974 : 438.

R up to 70 mm ; R/r *c.* 3·5/1.

A species of Solasteridae with five arms ; abactinal plates with a high column, two to four times as high as broad, crowned with about ten spinelets ; two marginal series of plates distinct and markedly larger than the abactinal paxillae, especially the lower series ; some actinal plates extending on to the arms but only occasional ones beyond the disc armed with a few spinelets like the plates in the interradial

* See footnote on p. 40.

angles ; adambulacral plates with three to six furrow spines and two to four subambulacral spines.

Type locality. NW. of Lion's Head, Cape Town, 420 metres.

Southern African records. 31/16/d ; 33/17/d ; 34/18/d ; 33/27/– ; 29/31/d ; 330–459 metres.

Nature of bottom. Stones ; yellow clay, sand, rock ; mud, sand and shells ; gr-(-een or grey) sand ; sand ; mud.

Family **MITHRODIIDAE** Viguier

See : Spencer & Wright, 1966 : U71.

A family of Spinulosida with a small disc and normally five long cylindrical arms ; superficial skeletal plates, except for the adambulacrals, forming a reticulum and bearing numerous rounded tubercles or ventrally coarse short spines in the species represented, all invested with nodular skin, no marginal or actinal longitudinal series distinguishable ; adambulacral plates each with a fan of relatively small slender spines along the furrow margin backed by one or two very coarse blunt subambulacral spines ; pedicellariae, if present (not observed in the single specimen of the species represented) fasciculate, with a cluster of small spinelets mounted on a tubular ossicle.

Mithrodia gigas Mortensen

Mithrodia gigas Mortensen, 1935a : 1–4, fig. 1, pl. 1.

R up to 330 mm (only one specimen recorded) ; R/r *c.* 7/1.

A species of Mithrodiidae with armament consisting of numerous rounded tubercles, mostly on the nodes of the reticulum, inconspicuous except at the very tips of the arms where they are abruptly larger and fewer (at least in the huge holotype) ; adambulacral plates with about four relatively slender furrow spines backed and obscured by a much larger club-shaped subambulacral spine, those of adjacent plates close enough to form an almost continuous longitudinal series ; pedicellariae not observed. Colour above purplish-pink, tips of the arms more cinnamon red, below pale yellowish.

Type locality. Off Point Morgan, NE. of East London (32/28), 45–55 metres.

Nature of bottom. Stony.

Family **ACANTHASTERIDAE** Sladen

See : Spencer & Wright, 1966 : U70.

A family of Spinulosida with a large disc and tapering pointed arms numbering more than five, except immediately after metamorphosis, usually 9–18 in adult specimens ; more than one madreporite (though reproduction by fission and regeneration is not known) ; abactinal plates with basal lobes linked to form a reticulum and a high central column crowned by and continuous with a single, often massive, spike-like spine ; marginals undistinguished but at least one series of actinal plates extends well on to the arms ; adambulacral plates with a longitudinal series of furrow spines backed by one much larger subambulacral spine ; pedicellariae two-bladed spiniform, usually present.

Acanthaster planci (Linnaeus)

Asterias planci Linnaeus, 1758, *Systema naturae*. Holmiae. Ed. 10 : 823.
Acanthaster planci : Grindley, 1963 : 265–268, fig.

R up to *c.* 250 mm ; R/r 1·7–2·3/1.

A species of Acanthasteridae with huge, spike-like abactinal spines ; adambulacral plates with two to five, usually three or four furrow spines ; pedicellariae more or less numerous on the disc, sides of the arms and on the furrow face of the adambulacral plates. Colour greenish or khaki with spines orange.

Type locality. Goa, Western India.

Southern African record. 23/35/s ; 6 metres.

Nature of bottom. Living coral.

Family **ECHINASTERIDAE** Verrill

See : Fisher, 1911 : 259 ; Spencer & Wright, 1966 : U70.

A family of Spinulosida with the disc usually small and normally five cylindrical arms, rarely short-armed stellate ; abactinal plates oblong or rounded, imbricating to form an irregular reticulum, often giving a honeycomb-like appearance, armed with multiple spinelets or single, usually small spines, not mounted on a central column, papulae grouped in the meshes of the reticulum on the lower side, usually present but sometimes absent ; marginal plates rarely much distinguished ; actinal plates often forming more distinct transverse than longitudinal series, though this may be emphasized by furrowing of thickened skin ; adambulacral plates with few spines along the furrow margin, usually one to three, and often a small compressed curved spine deep in the furrow ; no pedicellariae.

Genus ***HENRICIA*** Gray

See : Fisher, 1911 : 266.

A genus of Echinasteridae with the disc small and arms more or less elongated, almost circular in cross-section ; abactinal reticulum often compact, with diminutive armament, spinelets rarely more than 1 mm long, usually numerous and clustered but sometimes few and spaced in a single line along the plates ; one marginal series more or less distinct, sometimes both ; adambulacral plates each with a deep furrow spine.

Henricia abyssalis (Perrier)

Cribrella abyssalis Perrier, 1894, *Expeditions scientifiques du 'Travailleur' et du 'Talisman'*. Paris. pp. 144–146, pl. 11, fig. 1.
Henricia ornata : Bell, 1905a : 250–251 (pt) ; H. L. Clark, 1926 : 19–20.

Henricia abyssalis: Mortensen, 1933a: 265–266, fig. 8a, pl. 13, figs 11, 12; A. M. Clark, 1952: 197.
Henricia studeri: A. M. Clark, 1952: 197, 209–210. (This record was based on an unusually stout specimen since redetermined as *H. abyssalis*.)

R up to 70 mm; R/r 4·3–6·4/1.

A species of *Henricia* with slender attentuated arms, narrow at the tips; abactinal plates armed with numerous compactly placed fine spinelets terminating in several glassy points; actinal plates forming fairly regular longitudinal series, as well as transverse, of which two extend well out along the arms in larger specimens, R > 40 mm, actinal papulae more or less numerous in the interstices; adambulacral plates with about three spines along the furrow margin backed by several other spines and then a group of smaller spinelets. Colour orangey-yellow; pinkish-white; white; black [? coated].

TYPE LOCALITY. Off Morocco, 1105–1635 metres.

SOUTHERN AFRICAN RECORDS. 29/14/d; 31/16/d; 31/17/d; 32/17/d; 33/17/d; 33/18/s; 34/17/d; 34/18/s, d; 34/23/s; 34/25/d; 29/31/d; 56–408 metres.

NATURE OF BOTTOM. Mud, sand, and rock; rock; rock and shell; khaki and black sand, gravel; hard with black speckled sand and mud; black rock with black speckled sand; green sand and black specks; rock, khaki sand; sand and mud; sand; green mud; mud.

Henricia ornata (Perrier)

Echinaster (*Cribella*) *ornatus* Perrier, 1869, *Annls Sci. nat.* (5) **12**: 251.
Cribrella ornata: Sladen, 1889: 543; Koehler, 1908: 629–630, pl. 12, figs 105, 106; 1914: 173; Döderlein, 1928: 295.
Henricia ornata: Bell, 1905a: 250–251 (pt); Döderlein, 1910: 252, pl. 4, fig. 2; H. L. Clark, 1923: 289; Fisher, 1940: 272; A. M. Clark, 1952: 197, Day, Millard & Harrison, 1952: 396; Day, 1959: 544; Morgans, 1959: 398, 404, 424, 426, 427; 1962: 303; Grindley & Kensley, 1966: 12; Penrith & Kensley, 1970: 233; Day, Field & Penrith, 1970: 80. [Non *H. ornata*: H. L. Clark, 1926 = *H. abyssalis*.]
Echinaster ornatus: Mortensen, 1933a: 262–264, fig. 8b, pl. 13, figs 1–3; Stephenson, Stephenson & du Toit, 1937: 380; Bright, 1937: 59, 87; Stephenson, Stephenson & Bright, 1938: 10, 18; Eyre, Broekhuysen & Crichton, 1938: 96, 110; Eyre, 1939: 298; Stephenson, Stephenson & Day, 1939: 357; Stephenson, 1944: 317, 334, 347.

R up to 90 mm; R/r 3·4–5·0/1.

A species of *Henricia* of variable form, the arms usually tapering slightly to broad blunt tips but larger specimens may have them more attenuated and smaller ones more cylindrical; abactinal plates armed with short, coarse, often granuliform spinelets in double or sometimes single rows along the plates, occasionally more or less spaced, invested in thick skin; usually only one row of actinal plates extending to the distal part of the arm, their arrangement not very distinct, armed with rather sparse spinelets, actinal papulae lacking (the two gonopores in each interradius should not be mistaken for papulae); adambulacral plates armed with only a few (two to four) coarse blunt spines on their ventral surface, tending to form a line at right angles to the furrow. Colour: reddish-orange; brown-orange.

TYPE LOCALITY. Cape of Good Hope.

SOUTHERN AFRICAN RECORDS. 26/15/i, s ; 29/16/i ; 30/17/i ; 31/17/i ; 33/17/s ; 33/18/i, s ; 34/18/i ; 34/18/FB/i, s ; 34/19/i ; 34/20/i ; 34/21/i ; 34/24/i ; 33/25/i ; 33/26/i ; 33/27/i, s ; 32/28/i ; intertidal–90 metres.

NATURE OF BOTTOM. Rock ; sand and rock ; fine green sand and shell ; coarse sand and shell.

*Henricia retecta nom. nov.**

Echinaster reticulatus H. L. Clark, 1923 : 290–292, pl. 15, figs 1, 2 ; Mortensen, 1933a : 264–265 ; Day, Field & Penrith, 1970 : 80.
Henricia reticulata : A. M. Clark, 1974 : 438–439.

R up to 75 mm ; R/r *c*. 5/1.

A species of *Henricia* with arms tapering slightly to blunt tips ; abactinal plates armed with coarse spinelets in single, slightly spaced rows, all covered with thick skin ; actinal plates only forming indistinct longitudinal series, armed with single spinelets tending to form lines transverse to the axis of the arm, actinal papulae more or less numerous ; adambulacral plates with three, often two, coarse blunt spines on their ventral surface forming a line at right angles to the furrow.

Note : Judging from Mortensen's comments, both the occurrence of actinal papulae and the density of the abactinal armament are so variable that it may not prove possible to maintain a specific distinction between *H. ornata* and *H. retecta*.

TYPE LOCALITY. False Bay, 33–46 metres.

SOUTHERN AFRICAN RECORDS. 31/16/d ; 33/18/– ; 34/17/d ; 34/18/FB/s ; 32/28/s.

NATURE OF BOTTOM. Rocks ; sand.

Poraniopsis capensis H. L. Clark†

Poraniopsis capensis H. L. Clark, 1923 : 289–290, pl. 15, figs 3, 4 ; 1926 : 20–21 ; A. M. Clark, 1952 : 196, 208–209.

R up to 68 mm ; R/r 2·0–2·3/1.

A species of Echinasteridae with short-rayed stellate form, the disc very swollen ; abactinal plates forming a loose reticulum, armed with widely spaced sacculate spines 1–4 mm long ; one or both marginal series of plates more or less distinct, each armed with one, sometimes two, spines ; actinal plates numerous interradially but sparsely armed, not extending far on to the arms ; adambulacral plates with a compressed spine on the edge of the furrow (but no deep furrow spine) and one larger subambulacral spine.

TYPE LOCALITY. WSW. of Vasco da Gama Peak, 420 metres.

SOUTHERN AFRICAN RECORDS. 29/14/d ; 34/18/d ; 238–420 metres.

NATURE OF BOTTOM. Clay, sand and rock.

* New name for *Henricia reticulata* (H. L. Clark) A. M. Clark, 1974, non *H. reticulata* Hayashi, 1940 (*J. Fac. Sci. Hokkaido Imp. Univ.* Zool. **7** : 162). [Retecta = uncovered]

† Madsen (1956, *Acta Univ. lund.* N.S. **52**(2) : 29) has synonymized *P. capensis* with *P. echinaster* Perrier, 1891 from southern Patagonia. This was unfortunately overlooked in the present context in time for full correction.

Order FORCIPULATIDA Perrier

Asteroids with five or more long arms, cylindrical or sometimes pentagonal in cross-section but usually slightly tapering ; main ossicles forming longitudinal series ventrally at least, no enlarged marginal frame ; the superomarginals (if distinguished) hardly different from the abactinal plates and the inferomarginals aligned ventrolaterally ; pedicellariae highly specialized, with only two valves and always an extra basal piece, either below the valves in straight pedicellariae or sandwiched between their proximal lobes in the scissor-like crossed pedicellariae ; tube feet in either two rows (Brisingidae) or in four (Asteriidae), ending in sucking discs.

Family **ASTERIIDAE** Gray

See : Fisher, 1928 : 56 ; Spencer & Wright, 1966 : U75.

A family of Forcipulatida from various depths with five or more arms usually merging into the disc but sometimes more or less sharply marked off from it ; abactinal skeleton reticulate, with longitudinal as well as transverse links, usually well developed but sometimes weak, extending all along the arms, the plates armed with spaced spines or spinelets ; inferomarginal spines inconspicuous from above being aligned ventrolaterally and relatively short and stout ; adambulacral plates very short, not hourglass-shaped, with usually one or two (rarely up to seven) spines in a line at right angles to the furrow ; both straight and crossed pedicellariae usually present ; tube feet in four rows.

Allostichaster capensis (Perrier)

Asterias capensis Perrier, 1875, *Archs Zool. exp. gén.* **4** : 337 (p. 73 of separate) ; H. L. Clark, 1923 : 307. [? *A. capensis* : Bell, 1905a : 252.]
Cosmasterias capensis : Mortensen, 1933a : 276–277, figs 13a, 14a, pl. 15, figs 11, 12.
Allostichaster capensis : Mortensen, 1941, *Results Norweg. Exped. T. da Cunha*, no. 7 : 4.

R up to *c.* 15 mm ; R/r 3·0–3·2/1, in six-armed specimens.

A species of Asteriidae with usually six, sometimes five, fairly short tapering arms, fissiparous, with at least two madreporites (even in the five-armed syntype) ; carinal (midradial) dorsolateral and superomarginal plates forming fairly regular longitudinal and transverse series, armed with short capitate spines, usually two or three to a plate ; inferomarginal plates with two, rarely three, much larger more clavate spines ; a partial row of actinal plates present, some armed with a spine ; adambulacral plates each with two spines (diplacanthid) not bearing crossed pedicellariae ; crossed pedicellariae scattered over the surface, not clustered around the spines, no large felipedal (or unguiculate, cat's paw-like) straight pedicellariae.

TYPE LOCALITY. South Africa.

SOUTHERN AFRICAN RECORDS. 27/15/s ; 33/17/s ; 19–34 metres.

Note : The occurrence of this species in South African waters, doubted by Mortensen (1941), is confirmed by two new records (SWD 84, 27°30′S 15°25′E, 24 metres and WCD 40, 33°6·5′S 17°56·7′E, 19–34 metres) based on three fissiparous six-armed specimens similar in size to the type material. Even so, in the absence of any intertidal records, one cannot help speculating whether the collector, Dr Andrew Smith, whose activities were evidently concentrated on the interior of South Africa, might not have collected the type specimens from the shore at Tristan da Cunha *en route* for the Cape.

NATURE OF BOTTOM. Rock ; rock and gravel.

Coronaster volsellatus (Sladen)

Asterias (*Stolasterias*) *volsellata* Sladen, 1889 : 584–585, pl. 107, figs 1–4.
Asterias volsellata : Bell, 1905a : 252.
Coronaster volsellatus : Fisher, 1919 : 496–497, pl. 135, fig. 4, pl. 151, fig. 2 ; H. L. Clark, 1923 : 306 ; 1926 : 25.

R up to *c.* 200 mm.

A species of Asteriidae with 10 or 11 arms, form superficially brisingid-like, the numerous long arms somewhat swollen proximally and easily broken off from the small round disc ; skeleton of upper side delicate, the carinal and superomarginal series continuous longitudinally and linked by sometimes irregular transverse series to form a weak lattice, distally the transverse plates may be more prominent, the carinal and most of the superomarginal plates armed with single needle-like spines encircled by large ruffs of crossed pedicellariae ; inferomarginals closer to the adambulacral plates than to the superomarginals, sporadic ones also armed with needle-like ruffed spines ; no actinal plates ; adambulacrals armed with single slender tapering spines (monacanthid) ; scattered straight pedicellariae of various sizes, the largest markedly felipedal in form with curved digits, more or less numerous. Colour bright salmon red.

TYPE LOCALITY. Zebu, Philippines, 174 metres.

SOUTHERN AFRICAN RECORDS. 33/17/d ; 34/26/s ; 33/27/d ; 30/30/d ; 29/31/d ; 72–413 metres.

NATURE OF BOTTOM. Sand, shells and stones ; green sand and mud ; mud.

Coscinasterias calamaria (Gray)

Asterias calamaria Gray, 1840, *Ann. Mag. nat. Hist.* **6** : 179 ; de Loriol, 1885, *Mem. Soc. Phys. Hist. nat. Genève* **29** (4) : 4–6, pl. 7, figs 1, 2 ; Bell, 1905a : 251.
Coscinasterias calamaria : H. L. Clark, 1923 : 306. [? *Coscinasterias calamaria* : H. L. Clark, 1938 and other Australasian records.]

R up to 65 mm ; R/r *c.* 6·5/1. [Australasian specimens may reach over 100 mm R.]

A species of Asteriidae with six to nine arms [up to 11 in Australasian specimens], fissiparous, with at least two madreporites (as many as six and four in the only two South African specimens) ; carinal and superomarginal plates all four-lobed and imbricating directly longitudinally,* connected transversely by dorsolateral

* In both the South African specimens and in several from Mauritius which I have examined the successive carinal plates are all similarly four-lobed. Fisher (1928) distinguished *Coscinasterias calamaria* (apparently on the basis of Australasian specimens) from *C. tenuispina* (from the Mediterranean and adjacent North Atlantic) by the presence in the former of alternately oblong carinal plates. Since this does not hold good in specimens from Mauritius, I am doubtful whether they are, after all, conspecific with those from Australasian waters, which also differ in having up to 11 arms (though this may be correlated with the fact that they reach a larger size) and sometimes a single madreporite. Mascarene and South African specimens also have only about two pairs of interradially contiguous adambulacral plates distal to the oral plates whereas the number is up to six in Australasian specimens. If these

plates some of which may form irregular longitudinal series, most of the carinal and alternate superomarginals as well as one (sometimes a partial second) row of dorsolateral plates armed with single, spaced, conical but usually somewhat blunted large spines wreathed in crossed pedicellariae ; inferomarginal plates each with two large spines, the outer one with a clump of pedicellariae on its outer side ; actinal plates inconspicuous, sometimes armed with single spines ; adambulacral plates mostly with a single, usually spatulate, spine but some proximal plates may bear two spines, one behind the other ; straight pedicellariae of various sizes scattered over the surface and at the sides of the furrows, the largest incipiently felipedal but usually with poorly developed digits.

Type locality. 'Isle of France (Mauritius) ; New Holland (Australia)'. In the absence now in the British Museum (Natural History) collections of any material labelled New Holland, an old seven-armed specimen (Leach collection) labelled Isle of France has been presumed to be Gray's type ; the type locality therefore may be restricted to Mauritius.

Southern African records. 33/18/i ; 34/18/FB/i.

Cosmasterias felipes (Sladen)

Stichaster felipes Sladen, 1889 : 433–434, pl. 101, figs 1, 2, pl. 103, figs 7, 8 ; H. L. Clark, 1923 : 304–305 ; 1926 : 22.
Cosmasterias felipes : Mortensen, 1933a : 274–276, figs 12c, 13b, 14b, pl. 15, figs 1–4 ; A. M. Clark, 1952 : 198 ; ? Day, Field & Penrith, 1970 : 80.

R up to 90 mm ; R/r 5·0–6·5/1.

A species of Asteriidae with normally five moderately long arms, almost circular in cross-section, tapering evenly, or more in the distal half, to fairly narrow tips ; plates of upper side in distinct longitudinal series, two dorsolateral ones each side of the broader carinal series, armed with two to nine short blunt similar spinelets ; inferomarginal plates and the two partial series of actinal plates armed with one or two spinelets ; adambulacral plates diplacanthid with two short blunt spines ; numerous crossed pedicellariae scattered all over and some large felipedal straight pedicellariae with big claws, especially interradially and on the jaws. Colour dull reddish.

Type locality. South from False Bay, 275 metres.

Southern African records. 30/15/d ; 32/17/d ; 34/17/s, d ; 34/18/s, d ; 35/18/d ; 36/21/d ; 34/22/s ; 34/23/d [? 34/18/FB/– ; see note on p. 55] ; 79–353 (? 373) metres.

Nature of bottom. Sand ; green sand ; fine green sand ; khaki sand ; green mud ; black sandy mud ; stones and mud ; grey sand and rock ; rock.

---

differences are consistent, then the name *Coscinasterias muricata* (Verrill), type locality New Zealand, should be revived for Australasian material hitherto treated as *C. calamaria*.

Zoogeographically, specimens from the west side of South Africa are more likely to be conspecific with North Atlantic species than with those from the Indian Ocean. However, the two Cape specimens of *Coscinasterias*, though agreeing with *C. tenuispina* in the form of the carinal plates and the jaws, disagree in having the spines much shorter and blunter.

Marthasterias glacialis (Linnaeus)

Asterias glacialis Linnaeus, 1758, *Systema naturae*. Holmiae. Ed. 10 : 661.
Forma *africana* :
Asteracanthion africanus Müller & Troschel, 1842 : 15.
Marthasterias africana : H. L. Clark, 1923 : 306.
Marthasterias glacialis africana : Döderlein, 1928 : 294.
Marthasterias glacialis var. *africana* : Mortensen, 1933a : 273–274, pl. 16, fig. 1 ; Eyre, 1939 : 298 ; A. M. Clark, 1952 : 198 ; Day, Field & Penrith, 1970 : 80.
Marthasterias glacialis forma *africana* : A. M. Clark, 1974 : 439.
Forma *rarispina* :
Asterias rarispina Perrier, 1875, *Archs Zool. exp. gén.* **4** : 327 (62–63 of separate).
Asterias (*Stolasterias*) *africana* : Sladen, 1889 : 589.
Asterias glacialis : Bell, 1905a : 252.
Marthasterias glacialis : H. L. Clark, 1923 : 305 ; Stephenson, Stephenson & Bright, 1938 : 10, 18 ; Stephenson, Stephenson & Day, 1939 : 368 ; Stephenson, 1944 : 287, 317, 334, 347 ; Day, Millard & Harrison, 1952 : 396 ; Scott, Harrison & Macnae, 1952 : 321 ; Morgans, 1959 : 421, 425, 426, 427, pl. 19 (part).
Marthasterias rarispina : H. L. Clark, 1923 : 305–306.
Marthasterias glacialis var. *rarispina* : Mortensen, 1933a : 273, pl. 16, figs 2, 3 ; Stephenson, Stephenson & du Toit, 1937 : 380 ; A. M. Clark, 1952 : 198, 211–212 ; Day, Field & Penrith, 1970 : 80.
Marthasterias glacialis forma *rarispina* : Fisher, 1940 : 273 ; Morgans, 1962 : 303, 308, 311–315, 319, 322, 324, 325.

R up to 110 mm (may be > 400 mm in European specimens) ; R/r 4·5–6·0/1.

A species of Asteriidae with normally five moderately long arms, pentagonal in cross-section, tapering evenly or more in the distal half to slightly blunted tips ; carinal and superomarginal plates at least in regular longitudinal series though the carinal plates often zig-zag, especially distally in forma *rarispina*, armament consisting of either numerous slightly spaced blunt spines of moderate length all over, some plates with several spines (forma *africana*) or widely spaced stout blunt or conical single spines on the carinal and superomarginal plates only, with sometimes a few on some of the dorsolateral plates (forma *rarispina*), the spines wreathed with crossed pedicellariae ; inferomarginal plates armed with two large spatulate spines, of which the outer is more or less encircled by a wreath of crossed pedicellariae ; actinal plates inconspicuous, only the largest specimens (R > 100 mm) sometimes with single spines on the proximal ones ; adambulacral plates monacanthid, the spines somewhat flattened, not tapering, blunt at the tip ; straight pedicellariae scattered, tapering and usually rounded terminally but sometimes approaching a felipedal form with short terminal digits. Colour (MS. note by Mr Croil Morgans) forma *africana* : typically orange with light and dark mottlings ; forma *rarispina* : mottled lilac and bluish, rarely brownish.

Type locality. 'Cape of Good Hope' for both forms.

Southern African records. Forma *africana* : 33/18/i, s ; 34/18/FB/i, s. [? 33/27/i ; ? 29/31/–]. Intertidal to *c.* 12·5 ? 55 metres. Forma *rarispina* : 33/18/s ; 34/18/FB/s ; 34/19/s ; 34/20/s ; 34/21/i, s ; 34/22/s ; 34/23/i, s ; 33/25/i ; 33/26/s ; 33/27/s ; 32/28/s. Intertidal–91 metres. It appears that *africana* ranges primarily only from the shore and shallow depths of Table Bay to the western side of Cape Peninsula, with outliers in False Bay where *rarispina* is only found offshore ; eastwards from False Bay *rarispina* is the predominant (if not the only form) found.

Nature of bottom. Rock ; stones and shell. Rock ; rocks and sand ; sand ; fine khaki sand ; coarse and fine shell ; sand, mud and rock.

Genus *PERISSASTERIAS* H. L. Clark

See : H. L. Clark, 1923 : 307 ; Fisher, 1930 : 238.

A genus of Asteriidae with five to seven arms, stout basally but attenuated distally ; plates of upper side small, irregular, forming numerous longitudinal series which are indistinct but for a carinal one with more or less enlarged armament, all the plates with one to several relatively small spines ; both series of marginal plates lateral, distinct but inconspicuous, armed with one to five spines ; several series of spiniferous actinal plates, at least one extending for most of the arm ; adambulacral plates broad, mostly with three to seven spines in a line at right angles to the furrow ; all the spines, even including the actinal and adambulacral ones, bearing clusters or wreaths of small pedicellariae, mostly crossed ones on the upper spines but straight ones may be numerous on some adambulacral spines, other small pedicellariae, mainly straight, scattered between the spines and along the furrows.

Perissasterias heptactis H. L. Clark

Perissasterias heptactis H. L. Clark, 1926 : 26–27, pl. 5.

R 245–275 mm (only one specimen described) ; mean R/r 5·2/1, R/maximum br 4·3/1 (but the holotype is in an abnormally humped posture which may well have affected the measurements).

A species of *Perissasterias* with seven moderately long arms ; spines of the upper side 'fairly stout, bluntly pointed' ; plates of both marginal series each with two or three spines and most adambulacrals with four or five spines (at least at the size of the holotype).

Type locality. West from Lambert's Bay (32/17), 385 metres.

Perissasterias obtusispina H. L. Clark

Perissasterias obtusispina H. L. Clark, 1926 : 28–29, pl. 7, figs 1, 2.

R 155–160 mm (only one specimen recorded), mean R/r 5·5/1, R/maximum br 4·75/1.

A species of *Perissasterias* with five moderately long arms tapering evenly to pointed tips ; spines 'low capitate' ; plates of both marginal series each with one or two spines and most adambulacrals with four (three to five) spines (at least at the size of the holotype.

Type locality. WNW. from Dassen I. (33/17), 392 metres.

Perissasterias polyacantha H. L. Clark

Perissasterias polyacantha H. L. Clark, 1923 : 307–309, pl. 18, fig. 3 ; 1926 : 29–30, pl. 6 ; Mortensen, 1933a : 278 ; A. M. Clark, 1974 : 439–440.

R up to at least 310 mm ; R/r 8·5–10·0/1 (R/br 5·2–7·2/1).

A species of *Perissasterias* with five relatively long arms (easily lost in preservation); spines 'bluntly pointed', sometimes sharp; plates of both marginal series with two to four or five spines and the adambulacrals with three to seven spines. Colour dull reddish.

TYPE LOCALITY. NW. of Lion's Head, Cape Town, 285 metres.

SOUTHERN AFRICAN RECORDS. 30/15/d; 33/17/d; 34/17/d; 34/18/d; 170–285 metres.

NATURE OF BOTTOM. Dark green sand; light green sand; green sand and mud.

Genus *SCLERASTERIAS* Perrier

See: Fisher, 1928: 105.

A genus of Asteriidae resembling *Marthasterias* from above, with fairly long arms almost pentagonal in cross-section owing to the prominent carinal and superomarginal plates which form regular longitudinal series, many of them being armed with conspicuous single conical spines bearing large wreaths of crossed pedicellariae, a partial and rather irregular longitudinal series of dorsolateral plates between them each side, armed with smaller single spines; inferomarginal plates with two large spines, only the outer with a cluster of pedicellariae on its upper side; single papulae present on the furrow side of the inferomarginals; one series of actinal plates of varying extent, with or without spines; adambulacral plates predominantly diplacanthid, without pedicellariae on the spines or along the furrows; straight pedicellariae not felipedal, the tips of the larger ones rounded.

Sclerasterias eustyla (Sladen)

Asterias (Stolasterias) eustyla Sladen, 1889: 587–588, pl. 106, figs 5–8.
Sclerasterias eustyla: Mortensen, 1933a: 278, pl. 15, figs 7, 8.

R up to 60 mm; R/r *c.* 8·5/1.

A species of *Sclerasterias* with moderately long arms; both the carinal and superomarginal series of plates, at least proximally, with stout-based conical spines only on alternate plates (in the holotype), though the wreaths of pedicellariae around them are so large as to fill the spaces between; inferomarginal plates with the two spines distinctly spatulate, the third spine described by Sladen arising in fact from a separate actinal plate in oblique series with the inferomarginals, the actinal plates and spines extending for much of the arm.

TYPE LOCALITY. Nightingale I., Tristan da Cunha, 183–275 metres.

SOUTHERN AFRICAN RECORD. 30/15/d; 198 metres.

Sclerasterias stenactis (H. L. Clark)

Eustolasterias stenactis H. L. Clark, 1926: 23–24, pl. 4.
Sclerasterias stenactis: Mortensen, 1933a: 277, pl. 15, figs 5, 6; A. M. Clark, 1952: 199.

R up to 77 mm; R/r *c.* 11/1.

A species of *Sclerasterias* with relatively long narrow arms; spines of the upper side slender and sharp, occurring on most carinal plates in the holotype (but only on alternate ones in the Survey specimen, which is notably different from *S. eustyla* in having all the plates of the longitudinal series very short so that the transverse bars of the linking oblong plates are very close together) ; inferomarginal spines only slightly flattened and blunted at the tips ; actinal plates of limited extent and (evidently) lacking spines.

TYPE LOCALITY. NE. of Durban, 355 metres.

SOUTHERN AFRICAN RECORDS. 36/21/d ; 34/23/d ; 29/31/d ; 146–410 metres.

NATURE OF BOTTOM. Mud, sand and shells ; mud ; grey sand.

Family **BRISINGIDAE** G. O. Sars

See : Fisher, 1928 : 4 ; Spencer & Wright, 1966 : U77.

A family of deep-water Forcipulatida with more than five very long narrow arms, sharply distinct and easily broken off from the well-defined circular disc ; abactinal skeleton weak, on the arms usually restricted to the proximal third or half where the arm is somewhat swollen and consisting mainly of transverse ribs or costae, not a reticulum ; a series of spaced long slender needle-like lateral spines present ; adambulacral plates relatively elongate, hourglass-shaped, forming the entire ventral face of the arm outside the furrow ; pedicellariae all crossed, numerous ; tube feet in two rows.

Brisinga cricophora Sladen

Brisinga cricophora Sladen, 1889 : 606–608, pl. 109, figs 6–8 ; H. L. Clark, 1923 : 309–310 ; 1926 : 30–31 ; A. M. Clark, 1952 : 199.

Disc diameter up to *c.* 26 mm ; arm length up to *c.* 350 mm.

A species of Brisingidae with 11–13 arms ; costae very narrow, ribbon-like, often sinuous, widely separated by semitransparent skin, the intercostal areas naked but for occasional spicules and transverse belts of fine pedicellariae not mounted on plates, the costal plates armed with small, isolated spinelets except for the lowest one each side which bears a conspicuous needle-like lateral spine ; adambulacral plates usually with four spines of various sizes, three in an oblique row near the distal end of the plate, the smallest and distalmost of them inset into the furrow and the largest subambulacral one on the ventral face of the plate, the fourth spine small and near the proximal end of the plate, usually on the edge of the furrow or slightly inset, at the base of the arm before the costae develop the large subambulacral spines are stouter and more or less capitate terminally, expanding into a flared crown ; small crossed pedicellariae scattered, in intercostal belts and on the sheaths of the spines. Colour orange.

TYPE LOCALITY. NW. of St Thomas, Virgin Is, West Indies, 715 metres.

SOUTHERN AFRICAN RECORDS. 31/16/d ; 33/17/d ; 34/18/d ; 36/21/d ; 360–457 metres.

NATURE OF BOTTOM. Green sand and stones ; green sand and mud ; dark mud and rock.

Stegnobrisinga splendens H. L. Clark

Stegnobrisinga splendens H. L. Clark, 1926 : 31–33, pl. 7, figs 3, 4.

Disc diameter 38 mm (only one specimen known), arm length > 400 mm.

A species of Brisingidae (arm number not stated) with the proximal costae at least not at all ribbon-like, the intercostal areas faced with numerous thin, fenestrated plates, alternate costae more or less enlarged and only these bear the large lateral spines ; adambulacral plates each with four small furrow spines, two distal and two near the proximal end and one subambulacral spine, no capitate large spines at the base of the arms ; pedicellariae scattered, in bands across the arms and on the spine sheaths.

TYPE LOCALITY. SE. from Durban (30/31), 860 metres.

Subclass *OPHIUROIDEA*

INTRODUCTION. As with the asteroids, in default of a comprehensive monograph on this subclass, reference can only be made to the summarized treatment by Spencer & Wright (1966), supplemented for certain taxa by Koehler's 'Siboga' report (1905) and study of Philippine ophiuroids (1922) and Matsumoto's monograph, primarily of Japanese species (1917).

The following terms are used in the descriptions of ophiuroids and may need clarification :

Adoral shields – the pair of superficial plates in each ventral interradius proximal to the unpaired *oral shield*, superimposed on the distal parts of the thick *oral plates* which make up the jaw together with the bar-like vertical *dental plate* at its apex.

Arm comb – the series of papillae or spinelets arising from a plate distal to each radial shield opposite the base of the arm and continuous below with the *genital papillae* fringing the *genital slits* lateral to each arm base, found in many Ophiuridae.

Fissiparous – self-dividing across the disc, followed by regeneration.

Oral papillae – superficial papillae fringing the oral plates, one or a few of which at the apex of the jaw may be *infradental*, (placed below the lowest tooth) or this position may be occupied by a cluster of *tooth papillae*, as in the families Ophiotrichidae and Ophiocomidae, the former lacking oral papillae altogether.

Oral tentacle scales – single or sometimes paired papillae adjacent to the first one or two tube feet or tentacles, more or less inset into the oral slit, sometimes in series with or indistinguishable from the oral papillae. Ordinary *tentacle scales* border the tentacle pores along the arms lateral to the ventral arm plates.

Radial shields – the pair of bar-like or triangular plates on the upper side of the disc opposite the base of each arm, rib-like in many euryalids, sometimes reduced or concealed by the disc armament.

Rosette – the group of five primary radial plates and the central plate on the dorsal side of the disc of the newly metamorphosed brittle-star, sometimes still distinct in the adult.

The measurements generally used for ophiuroids are : d.d., the disc diameter and a.l., the arm length from the edge of the disc.

The number of species included is 115, compared with 82 given by Mortensen (1933a), who excludes records from Mozambique.

DISTRIBUTION TABLE FOR OPHIUROIDEA

	Luderitz Bay area	Lambert's Bay area	Cape Town area	False Bay	Cape Agulhas area	Mossel Bay–Knysna area	Port Elizabeth area	Durban area	Lourenço Marques area	Other localities
EURYALIDAE										
Asterostegus tuberculatus Mortensen	..	..	..	..	..	..	..	d	..	
Astroceras spinigerum Mortensen	..	..	..	..	..	..	..	d	d	
ASTERONYCHIDAE										
Asteronyx loveni Müller & Troschel	d	d	vd	..	..	..	..	..	..	Cosmopolitan
ASTEROSCHEMATIDAE										
Asteroschema capensis Mortensen	..	..	..	..	..	..	..	s	d	
GORGONOCEPHALIDAE										
Astroboa nuda forma *nigra* Döderlein	..	..	..	..	..	..	..	..	s	East Africa, Red Sea
Astrocladus africanus Mortensen						'South Africa'				
Astrocladus euryale (Retzius)	..	..	..	s	s	s d	s	..	..	
Astrocladus hirtus Mortensen	..	..	..	..	..	..	..	'Natal (or perhaps P.E.A.)'		
Astrocladus hirtus forma *reticulatus* Mortensen	..	..	..	..	..	..	..	s	..	
Astroconus capensis Mortensen	..	..	..	..	..	..	..	d	..	
Astrothorax papillata (H. L. Clark)	..	..	d	..	..	..	s d	d	..	
Gorgonocephalus pectinatus Mortensen	..	..	s d	..	..	..	d	..	..	

OPHIOMYXIDAE										
Ophiomyxa bengalensis Koehler	..	..	..	..	..	..	..	d vd	..	Andaman Is, Philippines, East Indies
Ophiomyxa tenuispina Mortensen	..	..	..	..	..	..	d	..	..	
Ophiomyxa vivipara capensis Mortensen	..	d	s d	..	d	..	d	..	..	SW. Indian Ocean ridge
Ophioscolex (*Ophiolycus*) *dentatus* Lyman	..	d	d	..	d	..	..	..	..	
Ophioscolex dentatus forma *spiniger* Mortensen	..	d	d	..	..	..	..	d	..	
Ophioscolex (*Ophioscolex*) *inermis* Mortensen	..	..	..	..	..	..	..	d	..	
OPHIOTRICHIDAE										
Macrophiothrix aspidota : Balinsky	..	..	..	..	..	..	..	..	i	[Pakistan–East Indies]
Macrophiothrix demessa (Lyman)	..	..	..	..	..	..	..	..	i	Tropical Indo-West Pacific
Macrophiothrix hirsuta cheneyi (Lyman)	..	..	..	..	..	?	..	s	i	E. Africa, S. Arabia
Macrophiothrix longipeda (Lamarck)	..	..	..	..	..	..	..	..	i	Tropical Indo-West Pacific
Ophiocnemis marmorata (Lamarck)	..	..	..	..	..	..	..	s	s	E. Africa–S. Japan and N. Australia
Ophiogymna capensis (Lütken)			'C a p e'							
Ophiogymna fulgens (Koehler)	..	..	..	..	..	..	..	d	..	Gulf of Aden–S. Japan
Ophiothela danae Verrill	..	..	..	..	..	..	?	s d	i s	Tropical Indo-West Pacific
Ophiothela nuda (H. L. Clark)	..	..	..	..	..	..	..	s	s	
Ophiothrix (*Acanthophiothrix*) *proteus* Koehler	..	..	..	..	..	..	..	s	s d	Red Sea–S. Pacific Is
Ophiothrix (*Ophiothrix*) *aristulata* Lyman	..	d	d	..	d	d	d	..	..	Indian Ocean–New Zealand
Ophiothrix echinotecta Balinsky	..	..	..	..	..	..	..	..	i	Zanzibar
Ophiothrix foveolata Marktanner-Turneretscher	..	..	..	..	..	..	..	..	i s	Maldives–Philippines

DISTRIBUTION TABLE FOR OPHIUROIDEA (*cont.*)

	Luderitz Bay area	Lambert's Bay area	Cape Town area	False Bay	Cape Agulhas area	Mossel Bay–Knysna area	Port Elizabeth area	Durban area	Lourenço Marques area	Other localities
Ophiothrix fragilis forma *pentaphyllum* (Pennant)	s	s	i s d	s	..	s	s	..	..	Europe–W. Africa
Ophiothrix fragilis forma *triglochis* Müller & Troschel	..	i	i s d	i s	i s	i s	i s d	s d	..	
Ophiothrix trilineata Lütken	..	..	..	..	..	..	..	..	i	Tropical Indo-West Pacific
Ophiothrix (*Keystonea*) *propinqua* Lyman	..	..	..	..	..	..	..	..	i	Tropical Indo-West Pacific
AMPHIURIDAE : AMPHIURINAE										
Amphilycus scripta (Koehler)	..	..	..	..	..	..	..	..	i	Red Sea–Bay of Bengal
Amphiodia sp. aff. *A. microplax* Burfield	..	..	..	..	..	..	..	s	..	
Amphioplus (*Amphioplus*) *pectinatus* Mortensen	..	..	..	..	..	..	..	d	..	
Amphioplus (*Lymanella*) *furcatus* Mortensen	..	..	..	..	..	..	..	s	..	
Amphioplus (*Lymanella*) *integer* (Ljungman)	..	s	i	i s	–	i s	i s	s	i	Red Sea
Amphioplus (*Unioplus*) *falcatus* Mortensen	..	..	..	..	..	..	..	d	..	
Amphioplus sp. aff. *A. falcatus* Mortensen	..	..	..	..	s	s	s	s	..	
Amphipholis similis Mortensen	..	..	..	..	..	s	..	s d	..	
Amphipholis squamata (Delle Chiaje)	i	i s	i s d	i s	i	i s d	i s d	s d	i	Cosmopolitan

Amphipholis strata Mortensen	..	..	s	s	..	s	s	..	..	
Amphiura (Amphiura) acutisquama A. M. Clark	..	d	..	..	..	..	..	..	..	
Amphiura albella Mortensen	..	..	..	..	..	..	..	d vd	..	
Amphiura atlantica Ljungman	..	d	d	s	d	s d	d	..	..	W. Africa, St Helena
Amphiura candida Ljungman	..	..	..	..	..	..	..	..	i s	
Amphiura capensis Ljungman	i s	i s	i s d	i s	i	i s	s d	s	i	
Amphiura grandisquama natalensis Mortensen	..	..	..	..	..	..	..	d	..	
Amphiura incana Lyman	..	s	..	s	..	s	s	s	..	Mediterranean–W. Africa
Amphiura inhacensis Balinsky	..	..	..	..	..	..	..	..	i s	
Amphiura linearis Mortensen	..	..	..	..	..	..	..	s d	..	
Amphiura simonsi A. M. Clark	..	..	..	s	..	s	s d	s	..	
Amphiura uncinata Koehler	..	..	..	..	..	..	..	d	..	Zanzibar–Kei Is and Philippines
Amphiura sp. A.	..	..	..	..	..	..	d	..	..	
Amphiura (Fellaria) africana (Balinsky)	..	..	..	..	..	..	..	..	i	
Ophiocentrus dilatatus (Koehler)	..	..	..	..	..	..	..	..	i	Maldives, East Indies, N. Australia
Ophionephthys lowelli A. M. Clark	..	..	..	..	..	..	s	..	..	
Paracrocnida sacensis (Balinsky)	..	..	..	..	..	..	..	..	i	
AMPHIURIDAE : AMPHILEPIDINAE										
Amphilepis scutata Mortensen	..	..	..	..	..	..	..	d	..	

DISTRIBUTION TABLE FOR OPHIUROIDEA (*cont.*)

	Luderitz Bay area	Lambert's Bay area	Cape Town area	False Bay	Cape Agulhas area	Mossel Bay–Knysna area	Port Elizabeth area	Durban area	Lourenço Marques area	Other localities
OPHIACTIDAE										
Ophiactis abyssicola (M. Sars)	..	d	d vd	..	..	..	..	..	..	North and South Atlantic
Ophiactis carnea Ljungman	..	..	i s	i s	i s	i s	s d	i s d	i s	
Ophiactis delagoa Balinsky	..	..	..	..	..	..	..	..	i s	
Ophiactis hemiteles : Balinsky*	..	..	..	..	..	..	..	..	i	[N. Australia]
Ophiactis modesta Brock	..	..	..	..	..	..	..	..	i s	India–Hawaiian Is
Ophiactis nidarosiensis Mortensen		[Presumed to be from South Africa]								Norway
Ophiactis plana Lyman	..	..	d	..	d	..	d	i s d	i s	Deep tropical Atlantic, Indian and Pacific Oceans
Ophiactis savignyi (Müller & Troschel)	..	..	..	..	..	..	..	i s	i s	'Tropicopolitan'
OPHIACANTHIDAE										
Amphilimna cribriformis A. M. Clark	..	..	..	..	..	..	..	s d	d	
Anamphiura valida H. L. Clark	..	..	..	..	..	..	..	d	..	Zanzibar area
Ophiacantha baccata Mortensen	..	..	..	..	..	d	s	d	d	
Ophiacantha nerthepsila H. L. Clark	..	..	s	s	s	s	s	..	..	
Ophiacantha scutigera Mortensen	..	..	..	..	..	..	..	d	..	

Ophiacantha striolata Mortensen	..	..	d ?	..	..	..	s	d	..	
Ophiomitrella corynephora H. L. Clark	..	d	s d	..	..	s d	d	..	..	SW. Indian Ocean ridge
Ophiomitrella hamata Mortensen	..	..	..	..	..	..	..	d	..	
Ophioplinthaca sexradia Mortensen	..	..	..	..	..	..	s	..	..	
Ophiothamnus remotus Lyman	..	d	d	..	d	d	s d	d	..	
Ophiothamnus remotus forma *cordatus* Mortensen	..	..	..	..	d	..	..	d	..	
Ophiotreta durbanensis Mortensen	..	..	..	..	..	..	..	d	..	
OPHIOCOMIDAE										
Ophiocoma erinaceus Müller & Troschel	..	..	..	..	..	..	..	..	i	Tropical Indo-West Pacific
Ophiocoma pica Müller & Troschel	..	..	..	..	..	..	..	i ?	i	Tropical Indo-West Pacific
Ophiocoma pusilla (Brock)	..	..	..	..	..	..	..	..	i	Red Sea–Palao Is
Ophiocoma scolopendrina (Lamarck)	..	..	..	..	..	..	..	i	i	Tropical Indo-West Pacific
Ophiocoma valenciae Müller & Troschel	..	..	..	..	..	..	..	i	i	East Africa–Persian Gulf
Ophiocomella sexradia (Duncan)	..	..	..	..	..	..	..	..	i	Tropical Indo-West Pacific
Ophiomastix variabilis Koehler	..	..	..	..	..	..	..	..	i	Tropical Indo-West Pacific
Ophiomastix venosa Peters	..	..	..	..	..	..	..	..	i	Mombasa, Rodrigues, Aldabra
Ophiopsila bispinosa A. M. Clark	..	..	..	..	..	..	..	s d	..	
Ophiopsila seminuda A. M. Clark	..	..	..	s	..	d	s d	..	..	

* I am uncertain whether or not the two specimens from Inhaca which J. B. Balinsky (1957) referred to *Ophiactis hemiteles* H. L. Clark, 1915 are really conspecific with the holotype and only recorded specimen, from Torres Strait, northern Australia.

Distribution Table for Ophiuroidea (*cont.*)

	Luderitz Bay area	Lambert's Bay area	Cape Town area	False Bay	Cape Agulhas area	Mossel Bay–Knysna area	Port Elizabeth area	Durban area	Lourenço Marques area	Other localities
OPHIONEREIDAE										
Ophionereis australis (H. L. Clark)	..	..	..	..	..	..	..	s d	i d	
Ophionereis dubia (Müller & Troschel)	..	..	..	s	i	i	s	s d	s	Tropical Indo-West Pacific
Ophionereis porrecta Lyman	..	..	..	..	i	..	i s	s d	i	Vema Seamount, tropical Indo-West Pacific
Ophionereis vivipara Mortensen	..	..	..	..	..	..	..	..	i	Mauritius
OPHIODERMATIDAE										
Cryptopelta aster (Lyman)	..	..	d	..	d	d	s	d	..	
Ophiarachnella capensis (Bell)	..	..	i	i	i	i	i s	s	..	Vema Seamount
Ophiochasma nitida Hertz	..	..	..	..	..	d	..	..	..	
Ophioderma wahlbergi Müller & Troschel	s	..	i s	i s	i	..	..	s ?	..	
Ophiopeza fallax Peters	..	..	..	..	..	..	..	i	i	Mauritius–Philippines
OPHIOLEUCIDAE										
Ophiernus vallincola Lyman	..	..	vd	..	..	..	..	..	..	Azores, Crozet Is, Arabian Sea, E. Central America
Ophiocirce inutilis Koehler	..	..	..	..	..	..	..	..	d	East Africa, Red Sea, East Indies, Philippines
Ophiopallas paradoxa Koehler	..	..	..	..	..	..	..	..	d	East Indies, Philippines

8 OPHIURIDAE										
Amphiophiura trifolium Hertz	..	..	..	..	..	..	..	..	vd	East Africa, Maldives
Astrophiura permira Sladen	..	..	d	..	..	d	..	d	..	Madagascar, Japan
Dictenophiura anoidea H. L. Clark	..	d	s d	s	s	s d	s d	..	..	
Ophiocten latens Koehler	..	..	vd	..	..	..	..	..	..	North Atlantic
Ophiolepis cincta Müller & Troschel	..	..	..	..	..	..	..	i	i	Tropical Indo-West Pacific
Ophiomisidium pulchellum (W. Thomson)	..	..	s	..	..	d	..	..	..	Biscay, SE. U.S.A., Brazil, NW. Africa
Ophiomusium lymani W. Thomson	..	..	vd	..	..	..	..	..	..	Cosmopolitan deep-sea
Ophiura (*Ophiura*) *affinis simulans* (Mortensen)	..	d	s d	s	..	s d	s d	..	..	
Ophiura affinis simulans forma *microplax* (Mortensen)	..	..	..	..	..	..	i	..	..	
Ophiura flagellata (Lyman)	..	..	vd	..	..	..	..	..	..	Almost cosmopolitan deep-sea
Ophiura kinbergi Ljungman	..	..	..	..	..	..	..	s	..	Red Sea–Hawaiian Is and SE. Australia
Ophiura trimeni Bell	..	d	d	..	d vd	d	..	..	..	
Ophiura (*Ophiuroglypha*) *costata* (Lyman)	d	d	s d	..	d	d	d	..	..	
Ophiura costata tumida Mortensen	..	..	..	..	..	..	..	d	..	
Ophiura irrorata (Lyman)	..	d	vd	..	vd	..	..	..	..	Cosmopolitan deep-sea

Key to the Ophiuroidea

1 Arms covered with opaque skin and flexible dorsoventrally, branching or simple . 2

– Arms covered with segmental, normally undivided, dorsal and ventral plates (rarely obscured by skin and/or granules in the Ophiomyxidae and some Ophiotrichidae) and flexible mainly horizontally (again excepting in some Ophiotrichidae) ; arms simple 12

2 Belts of minute hyaline hooks present across each arm segment dorsally and laterally, except sometimes on the more proximal ones (Fig. 86) ; arms branching (except in *Astrothorax*). GORGONOCEPHALIDAE 3

– No belts of hooks, though the ventrally placed arm spines often become hook-like distally (Fig. 87) ; arms simple (in the genera represented) 9

3 Arms simple . . . . . ***Astrothorax papillata*** (H. L. Clark, 1923) (p. 132)

– Arms branched 4

4 A continuous fringe of oral papillae all around the edge of the mouth frame, even in the distal notches (Fig. 89) 5

– Oral papillae absent from the radial notches (Fig. 88) 7

5 Number of arm segments between consecutive forks varying from 7 to 13 in most cases, up to 20 distally ; arm spines present from the second segment ; no transverse belts of plates partly underlying the granules on the distal arm segments ***Astroconus capensis*** Mortensen, 1933a (p. 132)

– Number of segments between forks fewer and mostly varying by only ±1 (i.e. 6–8), or up to *c.* 11 distally ; larger specimens at least often with few, if any, spines before the first fork ; transverse series of flat plates (bearing the hooks) sometimes present under the granulation across each distal segment (Fig. 90) . . 6

6 Coarse armament of radial ribs consisting of large rounded tubercles (Fig. 91) ***Astrocladus euryale*** (Retzius, 1783) (p. 131)

– Ribs armed with 'moderately sized conical tubercles' (Fig. 92) ***Astrocladus africanus*** Mortensen, 1933a (p. 131)

7 No complete belts of hooks present before about the sixth fork ; only three to six segments present between successive main forks ***Astrocladus hirtus*** Mortensen, 1933a (p. 132)

– Complete hook belts present from the first or second fork ; often more than six segments between successive forks 8

8 Edge of disc reinforced interradially by a row of subcutaneous marginal platelets ; number of segments between forks very variable, 10–33 ; mouth papillae restricted to the apices of the jaws ***Gorgonocephalus pectinatus*** Mortensen, 1933a (p. 133)

– No marginal disc platelets ; four to eight arm segments between main forks ; oral papillae present each side of the apices of the jaws ***Astroboa nuda*** (Lyman, 1874) (p. 130)

9 Radial ribs and remaining disc completely smooth and naked, well marked off from the slender rounded arms (Fig. 93) ; arm spines more or less numerous, four to ten, the lowest becoming elongated. ASTERONYCHIDAE ***Asteronyx loveni*** Müller & Troschel, 1842 (p. 129)

– Radial ribs either armed with spines or tubercles and disc otherwise smooth (Fig. 94), or the whole surface granular (Fig. 95) ; arms stout and often armed with tubercles above, usually merging with the disc ; two to four arm spines, usually three, all short 10

10 Disc appearing granular. ASTEROSCHEMATIDAE ***Asteroschema capensis*** Mortensen, 1925 (p. 130)

– Disc smooth, radial ribs with spaced tubercles or spines. EURYALIDAE . . 11

11 Four or five rounded plates in a semicircle in each ventral interradius distal to the adoral shields (Fig. 96) ; a pair of conspicuous tubercles each side of the dorsal mid-line on each arm segment ***Asterostegus tuberculatus*** Mortensen, 1933a (p. 128)

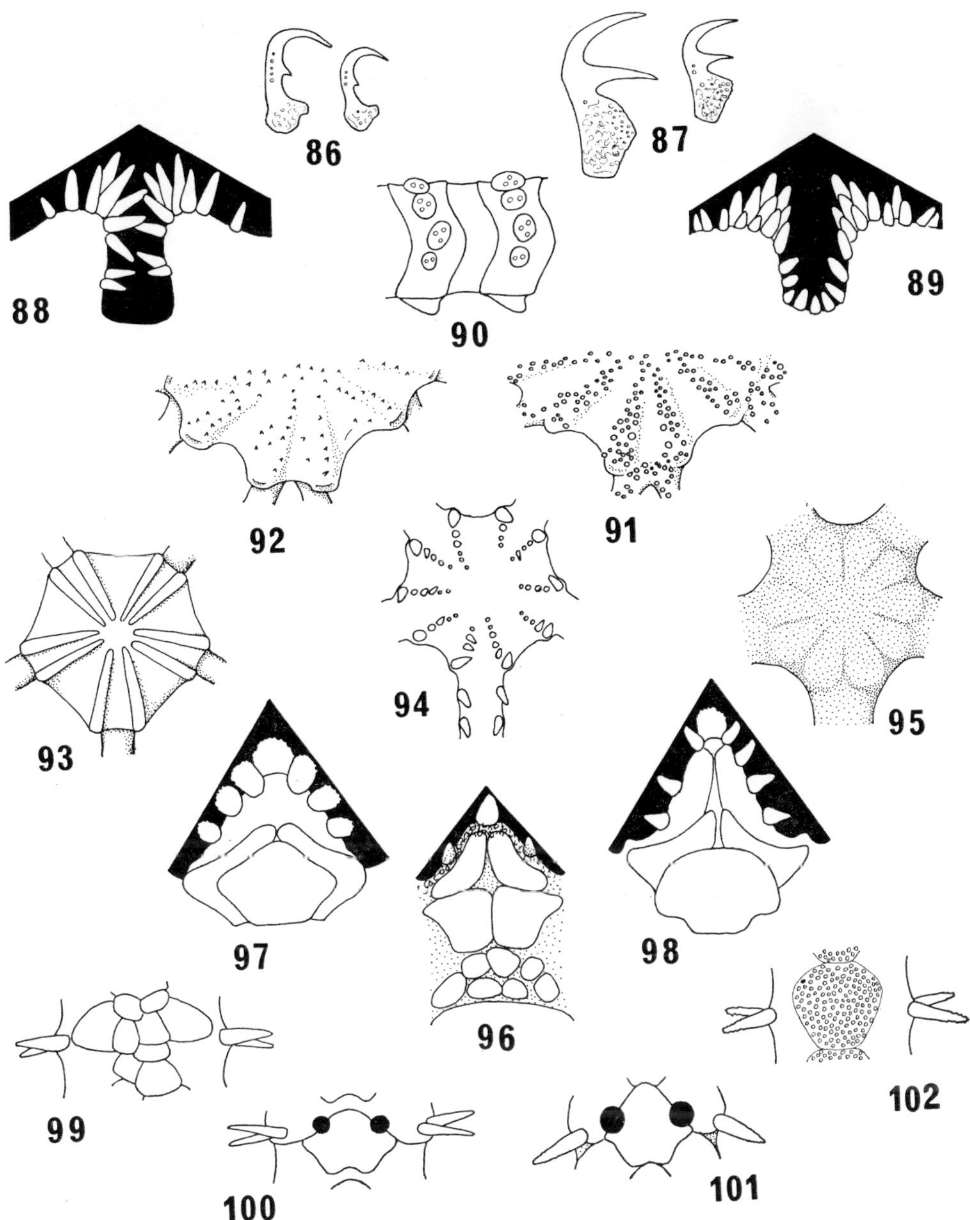

FIGS 86–102 Ophiuroidea. 86: hook from arm belt of *Gorgonocephalus pectinatus*. 87: hook-like modified arm spine of *Asterostegus tuberculatus*. 88, 89: two jaws of *Gorgonocephalus pectinatus* and *Astrocladus euryale*. 90: distal arm segments of *Astrocladus euryale* (side view). 91, 92: discs showing radial ribs of *Astrocladus euryale* and *A. africanus*. 93, 94, 95: discs of *Asteronyx loveni*, *Astroceras spinigerum* and *Asteroschema capensis*. 96: ventral interradius of *Asterostegus tuberculatus*. 97, 98: single jaws of *Ophiomyxa tenuispina* and *O. bengalensis*. 99, 100: dorsal and ventral arm plates of *Ophiomyxa tenuispina*. 101, 102: ventral and dorsal arm plates of *Ophiomyxa vivipara capensis*.

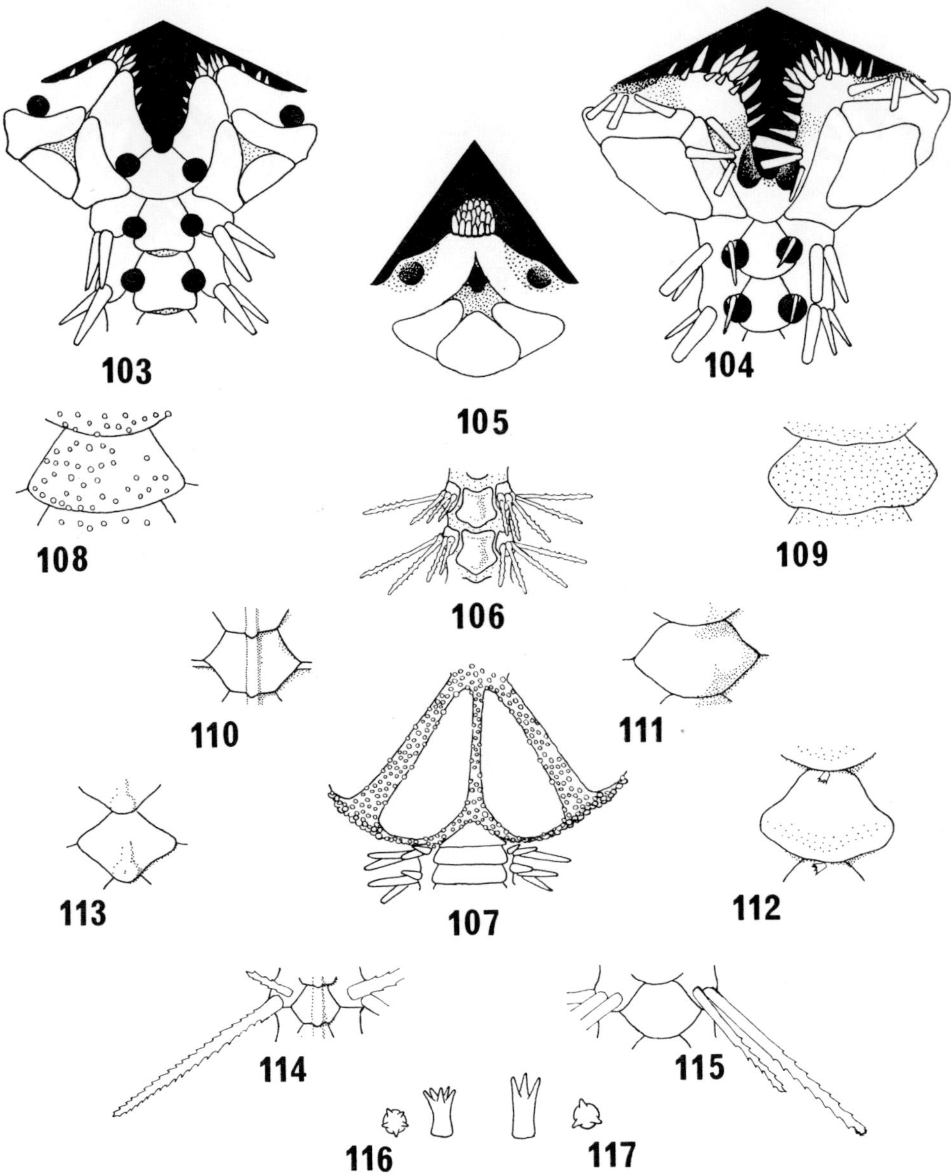

FIGS 103–117 Ophiuroidea. 103, 104: arm base and adjoining jaws of *Ophioscolex inermis* and *O. dentatus*. 105: single jaw of *Ophiothrix fragilis*. 106: dorsal arm plates of *Ophiogymna fulgens*. 107: part of disc of *Ophiocnemis marmorata*. 108, 109, 110, 111, 112, 113: dorsal arm plates of *Macrophiothrix demessa*, *M. hirsuta cheneyi*, *Ophiothrix* (*Acanthophiothrix*) *proteus*, *O. aristulata*, *O. echinotecta* and *O. fragilis*. 114, 115: arm segments of *Ophiothrix* (*Acanthophiothrix*) *proteus* and *O. aristulata* (dorsal view). 116, 117: disc stumps of *O. echinotecta* and *O. fragilis* f. *triglochis*.

11 Ventral interradii of disc naked ; only two spines or tubercles on each arm segment dorsally (Fig. 94) . . . . ***Astroceras spinigerum*** Mortensen, 1933a (p. 128)

12 Disc and arms covered by thick skin, the underlying plates only evident when dried or if the skin is dissolved with bleach. OPHIOMYXIDAE 13

– Disc with distinct scales or armament of granules, spines, spinelets or stumps, arm plates usually distinct wet or dry. 17

13 Teeth and usually also the oral papillae markedly broadened with serrated edges (Fig. 97) ; no tentacle scales 14

– Oral papillae and teeth all spiniform ; tentacle scales usually present . . . 16

14 Oral papillae tapering to a point (Fig. 98) ***Ophiomyxa bengalensis*** Koehler, 1899 (p. 134)

– Oral papillae broad and finely serrated like the teeth 15

15 The fragmented dorsal arm plates including an enlarged piece each side (Fig. 99) ; ventral arm plates markedly broader than long, with a wide distal concavity and spaced from each other (Fig. 100) ***Ophiomyxa tenuispina*** Mortensen, 1933a (p. 134)

– Dorsal arm plates mostly entire though delicate and fenestrated (Fig. 102) ; ventral arm plates about as broad as long or longer than broad, with only a slight distal concavity and the proximal ones at least contiguous (Fig. 101) ***Ophiomyxa vivipara capensis*** Mortensen, 1936 (p. 134)

16 No tentacle scales ; second oral tentacle pore superficial, outside the oral slit (Fig. 103) . . . . . ***Ophioscolex inermis*** Mortensen, 1933a (p. 135)

– A spiniform tentacle scale present on most pores ; second oral pore within the oral slit (Fig. 104) . . . . ***Ophioscolex dentatus*** Lyman, 1878 (p. 135)

17 Jaws armed superficially with only a compact cluster of apical tooth papillae, no oral papillae fringing the oral slits (Fig. 105). OPHIOTRICHIDAE . . 18

– Jaws armed both along the edges of the oral slits (or at least, e.g. in *Amphiura* and *Ophiactis*, at their distal ends) and at the apex with one to many papillae (e.g. Figs 123, 130, 157, 167, 186, 203, 212). 34

18 Disc scales and arm plates distinct, unless the disc armament is sufficiently dense to hide the scales, the skin transparent* 19

– Most of the disc and both sides of the arms obscured by thick skin, often with granules superimposed* 31

19 The bare radial shields huge, covering most of the upper side of the disc ; the scales of the narrow interradial and central areas bearing spherical granules (Fig. 107) ***Ophiocnemis marmorata*** (Lamarck, 1816) (p. 139)

– Radial shields with or without a covering of stumps or rugose granules and smaller in area than the scaled part of the upper side of the disc, which is usually armed with thorny stumps or serrated spinelets or spines, rarely naked . . . 20

20 Consecutive dorsal arm plates broad and broadly contiguous (Figs 108, 109) ; arms relatively long, eight to twenty times the disc diameter (d.d.) 21

– Dorsal arm plates much less than twice as long as broad, usually rhombic and only narrowly in contact (Figs 110–115) ; arm length usually four to eight times the d.d. 25

21 Only the ventral and peripheral disc scales armed with thorny stumps, the upper side naked. . . ***Ophiothrix (Keystonea) propinqua*** Lyman, 1861 (p. 145)

– Dorsal disc scales and usually also the radial shields armed with short thorny spinelets 22

22 Dorsal arm plates bearing scattered granules or (proximally) short thorny stumps (Fig. 108) ; ventral arm plates not broader than long ***Macrophiothrix demessa*** (Lyman, 1861) (p. 138)

* In dry specimens of *Ophiogymna fulgens*, which should run down to 31, the plates underlying the skin may be distinct; however, the isolation in skin of the successive dorsal and ventral arm plates serves to distinguish the genus (Fig. 106).

22 Dorsal arm plates unarmed, though their surface texture may be somewhat rugose (Fig. 109) ; ventral arm plates distinctly broader than long 23

23 Arms very long, usually fifteen to twenty times the d.d. ; dorsal arm plates trapezoidal with acute laterodistal angles ; colour patterned with small dark spots . . . . . ***Macrophiothrix longipeda*** (Lamarck, 1816) (p. 139)

– Arm length five to ten times the d.d. ; dorsal arm plates hexagonal or fan-shaped, lateral angles *c.* 90° or blunted ; colour pattern variegated, banded or with longitudinal stripes on the arms 24

24 Arms marked with a light, sometimes irregular, longitudinal midline, often defined by a dark blue line each side ; dorsal arm plates with blunted lateral angles . . . . ***Macrophiothrix hirsuta cheneyi*** (Lyman, 1861) (p. 138)

– No light longitudinal line along the arms, which are patterned purplish-brown in varying intensity, sometimes giving a banded effect ; dorsal arm plates with sharp lateral angles
Macrophiothrix aspidota (Müller & Troschel,1842) (p. 137)

25 Arms marked with a pair of dark longitudinal lines bordering a median light line, sometimes with two additional light lines outside these ; radial shields large and almost or completely bare 26

– Arms variously patterned but not with longitudinal lines ; radial shields sometimes bare but more often obscured by trifid stumps 28

26 Dorsal arm plates approximately hexagonal in shape with the distal end broadly truncated but often with a small median projection corresponding to the width of the sharp light-coloured keel running the whole length of the plate (Fig. 110) ; disc armed with trifid stumps and rugose spinelets ; arm spines needle-like with many very fine serrations, more than 20 each side on the longest, which equal in length five or even six arm segments (Fig. 114)
Ophiothrix (Acanthophiothrix) proteus Koehler, 1905 (p. 142)

– Dorsal arm plates rhombic or fan-shaped with the distal angle rounded, no sharp keel though the median distal part may be swollen (superficially giving a keel-like effect) (Fig. 111) ; disc armed predominantly with spines or spinelets, though small specimens (d.d. < 7 mm) usually have some trifid stumps, at least peripherally ; arm spines mostly truncated at the tip and with coarser thorns, rarely as many as 20 each side and in length not usually more than four times as long as the segment (Fig. 115) 27

27 Three longitudinal light lines on the arms, the outer ones sometimes not well defined ; littoral Mozambique . . ***Ophiothrix trilineata*** Lütken, 1869 (p. 145)

– One light longitudinal line, the pink or red stripes bordering it often poorly marked (at least in preserved specimens) ; from deep-water (100+ metres) off Cape Province and the west coast of South Africa
Ophiothrix aristulata Lyman, 1879 (p. 142)

28 A small thorny granule at the junction between two successive dorsal arm plates on most segments (Fig. 112) ; disc armed with stellate multi-pointed short stumps (Fig. 116) ; dorsal arm plates each marked with a light chevron on the distal edge ; ventral arm plates with the distal edge straight
Ophiothrix echinotecta Balinsky, 1957 (p. 143)

– Dorsal arm plates unarmed ; disc armament consisting of rugose spines, spinelets or trifid (Fig. 117) to multifid stumps ; no light-coloured chevrons on the arms ; most ventral arm plates with the distal edge distinctly concave 29

29 Disc beautifully patterned with dark lines on the radial shields, especially emphasizing their abradial edges, similar dark lines running transversely across each arm segment, often continued along the uppermost spine (Fig. 118)
Ophiothrix foveolata Marktanner-Turneretscher, 1887 (p. 143)

– No such linear patterns
Ophiothrix fragilis (Abildgaard *In* : Müller, 1789)*30 (p. 144)

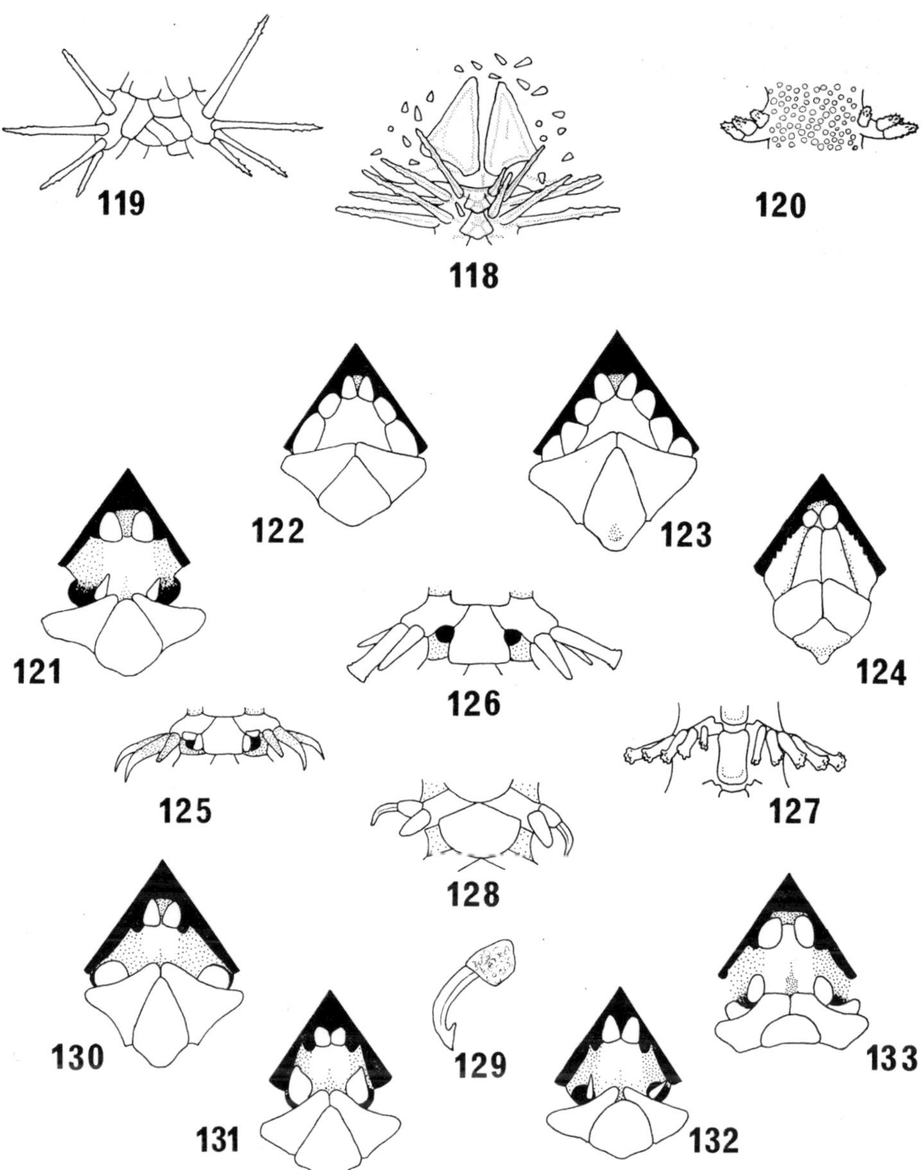

FIGS 118–133 Ophiuroidea. 118: arm base and adjoining disc of *Ophiothrix foveolata* (dorsal view). 119, 120: arm segments of *Ophiogymna capensis* and *Ophiothela nuda* (dorsal view). 121, 122, 123, 124: jaws of *Amphiura capensis*, *Amphipholis strata*, *Amphioplus* (*Lymanella*) *integer* and *Amphilycus scripta*. 125: proximal arm segments of *Amphiura uncinata* (ventral view). 126: middle arm segments of *A. atlantica* (ventral view). 127, 128: middle arm segments of *Amphiura* (*Fellaria*) *africana* and *A. simonsi* (dorsal view). 129: enlarged middle arm spine of *Amphiura simonsi*. 130, 131, 132, 133: jaws of *Amphiura uncinata*, *A. albella*, *A. atlantica* and *A. simonsi*.

30 Dorsal disc armament including many rugose spines or spinelets, with stumps often restricted to the periphery ; radial shields often completely naked
Ophiothrix fragilis forma ***pentaphyllum*** (Pennant, 1777)* (p. 144)

– Disc armament limited to trifid or multifid stumps, extending also on to the radial shields, though less densely there
Ophiothrix fragilis forma ***triglochis*** Müller & Troschel, 1842 (p. 144)

31 Arms very long, more than ten times the d.d., flexing more horizontally than vertically and covered with thick smooth skin ; dorsal and ventral arm plates present under the skin, though the dorsal ones at least are often more or less fragmented (Fig. 119) ; arm spines relatively long, the longer ones easily exceeding the segment in length 32

– Arms moderate in length or short, capable of rolling downwards, correlated with an epizoic habit, and covered with thick skin often itself obscured by a more or less dense coating of granules (Fig. 120) ; dorsal and ventral arm plates absent (though paired dorsal rudiments occur on a few basal segments, revealed by treatment with bleach when the underlying vertebrae are also exposed) ; arm spines short and stubby, barely exceeding the segment in length on most of the arm 33

32 Dorsal arm plates mostly entire . ***Ophiogymna fulgens*** (Koehler, 1905) (p. 140)

– Dorsal arm plates fragmented† . ***Ophiogymna capensis*** (Lütken, 1869) (p. 140)

33 Normally six-armed, fissiparous and small, d.d. usually 2–3 mm ; peripheral disc armament more often granuliform than spiniform
Ophiothela danae Verrill, 1869 (p. 141)

– Five-armed, d.d. often > 5 mm ; disc armament usually including peripheral spinelets . . . . . ***Ophiothela nuda*** (H. L. Clark, 1923) (p. 141)

34 Each jaw armed apically with a symmetrical pair of rounded or rectangular papillae and one to three other papillae each side, when only one then it is widely separated from the infradental papillae (Figs 121–124).‡ AMPHIURINAE 35

– Each jaw armed apically with one to many papillae (rarely none), only occasionally a more or less symmetrical pair on some jaws, often more than three distal papillae each side 59

35 Sexually dimorphic, the male dwarfed and living upturned on the underside of the normal-sized female, mouth to mouth ; the female with the jaws and adjacent parts slightly sunken, the oral papillae fused and irregular, often forming a serrated flange along the edge of the oral plate while the infradental papillae are not always symmetrical on all the jaws (Fig. 124) ; oral shields much smaller than the adorals and hollowed ; in life commensal on the underside of clypeasteroid echinoids . . . ***Amphilycus scripta*** (Koehler, 1904) (p. 147)

– Not sexually dimorphic, jaws not sunken and oral papillae normally perfectly symmetrical ; oral shields similar in size to the adorals or larger, usually flat ; mostly free-living, not commensal with clypeasteroids 36

* A case concerning the nomenclature of this species is still pending with the International Commission.

† According to Mortensen (1933a) another difference between *Ophiogymna fulgens* and *O. capensis* is the presence of a distal lobe to the adoral shields only in *O. capensis*. However, the holotype of *Placophiothrix phrixa* H. L. Clark, 1939 from the Gulf of Aden, which nominal species I referred to the synonymy of *O. fulgens* in 1967 (pp. 640–641) because of its entire dorsal arm plates, does have distal lobes to the adorals. The absence of *O. capensis* from any bona fide collections from South Africa leads one to suspect that the locality 'Cap' on the two syntypes of Lütken did not refer to the Cape of Good Hope.

‡ South African specimens of *Amphilimna* (Ophiacanthidae) and *Ophiopsila* (Ophiocomidae) may simulate amphiurids in having a degree of symmetry in the apical papillae when these number only two. They can easily be distinguished by the prolonged lateral angles on at least the second ventral arm plate (the one on the first arm segment) distal to the tentacle pores and by peculiarities of some of the tentacle scales. Conversely, *Amphilycus* often has rather irregular infradental papillae for an amphiurid.

36 Only two oral papillae each side of each jaw, the infradental one and a distal papilla (rarely two) separated by a space or diastema revealing the oral tentacle scale on the side of the oral plate within the oral slit (Figs 121, 130–133) . 37
– Three or four oral papillae in a continuous series (Figs 122, 123) 50
37 Disc scales obscured by skin and bearing scattered spinelets ; no tentacle scales ***Ophiocentrus dilatatus*** (Koehler, 1905) (p. 159)
– No disc spinelets, disc scales distinct, at least on the upper side, rarely lacking ; one or two tentacle scales, rarely none 38
38 Two tentacle scales 39
– One tentacle scale or none 43
39 Disc partially skin-covered below, the scaling incomplete 40
– Disc fully scaled 41
40 Proximally six or more arm spines, the middle ones on the basal segments ending in a glassy hook (Fig. 125) ; isolated scales present in the ventral skin of the disc ; distal oral papilla broad and semicircular (Fig. 130) ***Amphiura uncinata*** Koehler, 1904 (p. 158)
– Proximally only four or five arm spines, none hooked ; the disc scaling stopping abruptly on the ventral side ; distal oral papilla leaf-like (Fig. 131) ***Amphiura albella*** Mortensen, 1933a (p. 153)
41 Proximally about five arm spines ; the extremely fine disc scales pointed, except around the radial shields ; dorsal arm plates with a median distal angle (Fig. 134) ***Amphiura acutisquama*** A. M. Clark, 1952 (p. 153)
– Proximally seven or eight arm spines ; disc scales smooth ; dorsal arm plates with distal edge straight or evenly convex 42
42 Disc scales coarse and thick ; arm spines shorter than the segment ; ventral arm plates broader than long, the distal edge medially concave ***Amphiura incana*** Lyman, 1879 (p. 156)
– Disc scales fine and smooth ; arm spines slightly exceeding the segment in length ; ventral arm plates as long as or longer than broad, the distal edge straight or slightly convex . . . . ***Amphiura candida*** Ljungman, 1867 (p. 154)
43 No tentacle scales* 44
– One tentacle scale* 45
44 Disc scaled above, though bare below ; the upper and lower arm spines tapering and smooth, except that the second from lowest has an axe-headed tip (Fig. 126) ; dorsal arm plates broader than long ; distal oral papilla slender and spiniform (Fig. 132) . . . . ***Amphiura atlantica*** Ljungman, 1867 (p. 153)
– Disc naked above and below (often lost) ; all the arm spines with expanded rugose tips (Fig. 127) ; dorsal arm plates all much longer than broad ; distal oral papilla thick and rounded . ***Amphiura (Fellaria) africana*** (Balinsky, 1957) (p. 158)
45 Two distal oral papillae (Fig. 133) 46
– Only one distal oral papilla 47
46 Distal oral papillae not longer than broad (Fig. 133) ; second from lowest arm spine modified beyond the arm base into a conspicuous glassy hook (Figs 128, 129) ***Amphiura simonsi*** A. M. Clark, 1952 (p. 157)
– Distal oral papillae spiniform ; second arm spine not hooked ***Amphiura*** sp. A† (p. 158)
47 Lowest arm spine markedly longer than the rest 48
– Lowest arm spine similar in length to the rest or only slightly longer . . . 49
48 Radial shields long, narrow and bar-shaped, l : br > 5 : 1, well separated distally and almost parallel (Fig. 141) . ***Amphiura linearis*** Mortensen, 1933a (p. 157)
– Radial shields tapering proximally, l : br *c.* 3 : 1, approximating and only just separated distally (Fig. 142) ***Amphiura grandisquama natalensis*** Mortensen, 1933a (p. 155)

* *Amphiura simonsi* often has no tentacle scale on up to *c.* 12 of the proximal arm segments.

† This derives from a single poorly preserved *Amphiura* collected at 34°15′S 25°50·5′E (near Port Elizabeth) in 108 metres.

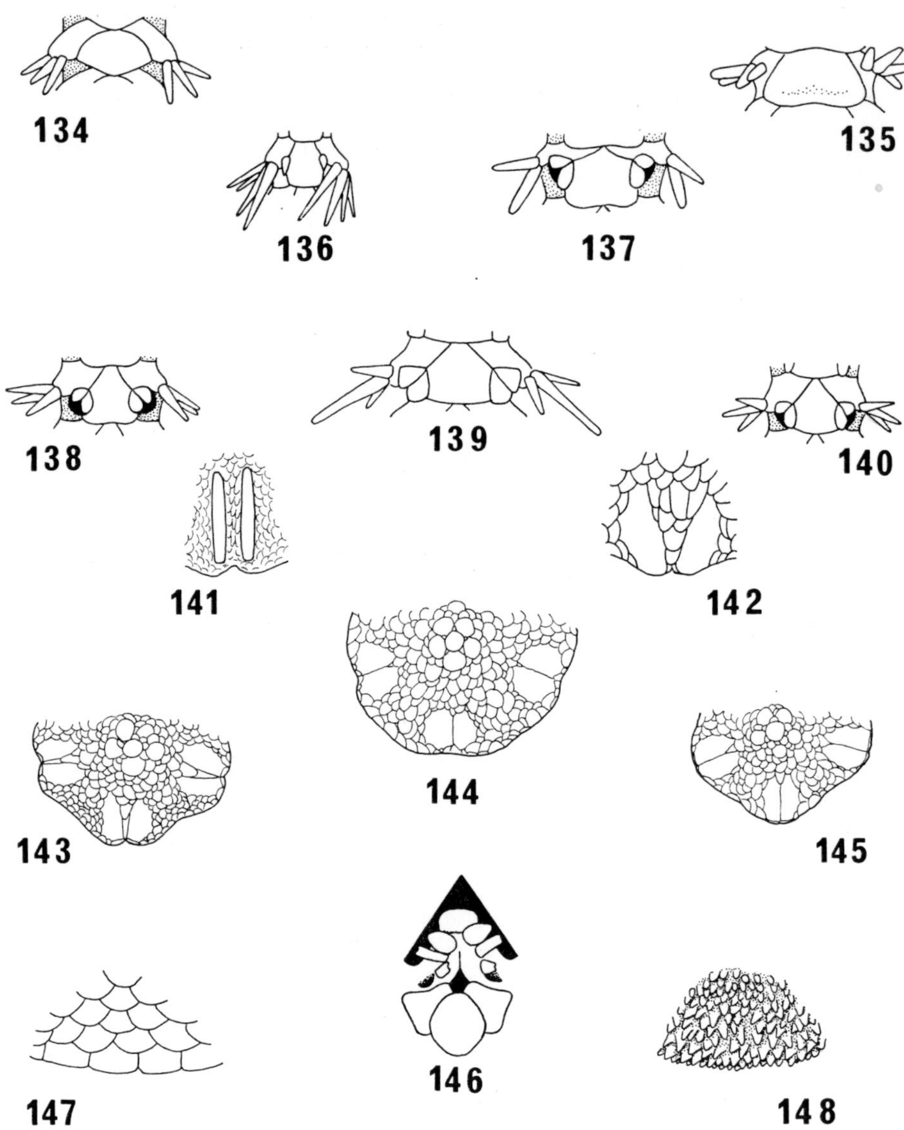

FIGS 134–148 Ophiuroidea. 134, 135: dorsal arm plates of *Amphiura acutisquama* and *Paracrocnida sacensis*. 136, 137, 138, 139, 140: arm segments of *Amphiura linearis*, *Amphioplus* (*Lymanella*) *integer*, *Amphipholis squamata*, *A. strata* and *A. similis* (ventral view). 141, 142: radial shields of *Amphiura linearis* and *A. grandisquama natalensis*. 143, 144, 145: outline of disc with rosette of *Amphiura capensis*, *Amphipholis strata* and *A. similis*. 146: jaw of *Ophionephthys lowelli*. 147, 148: interradial edge of disc of *Amphipholis strata* and *Paracrocnida sacensis* (ventral view).

49 Disc delicate centrally and interradially, the dorsal scaling extremely fine, no distinct rosette in the middle ; oral shields slightly broader than long
Amphiura inhacensis Balinsky, 1957 (p. 156)

– Disc not particularly delicate, dorsal scaling well developed and appearing moderately coarse, especially in smaller specimens (d.d. < 5 mm) ; rosette often distinct (Fig. 143) ; oral shields as long as broad or longer
Amphiura capensis Ljungman, 1867 (p. 155)

50 Three oral papillae each side of each jaw* 51

– Four oral papillae each side* 54

51 Both distal oral papillae based on the oral plate and spiniform in shape (Fig. 146) ; disc with exceedingly fine scales and partly naked below, easily lost
Ophionephthys lowelli A. M. Clark, 1974 (p. 159)

– The distalmost oral papilla extremely broad and opercular, based partly on the adoral shield (Fig. 122) ; disc fully scaled 52

52 Primary disc plates indistinguishable in shape from the other disc scales (at least at d.d. > 2 mm) ; oral shields broader than long or as long as broad ; tentacle scales relatively small, not fully covering the pores (Fig. 137)
Amphipholis squamata (Delle Chiaje, 1829) (p. 151)

– Rosette conspicuous among the disc scales ; oral shields as long as broad or longer ; tentacles scales moderately or very large (Figs 138, 139) 53

53 Outermost row of ventral disc scales just below the edge of the disc abruptly larger and squarer than the adjacent scales proximal to them (Fig. 147) ; rosette plates only separated at their corners and radial shields very broad, rounded abradially (Fig. 144) ; tentacle scales very large (Fig. 138)
Amphipholis strata Mortensen, 1933a (p. 152)

– Outermost ventral disc scales not conspicuously different from the adjacent ones ; plates of rosette separated from each other all round already by d.d. 3 mm and radial shields not very broad, their sides often almost parallel (Fig. 145) ; tentacle scales moderately large (Fig. 139) ***Amphipholis similis*** Mortensen, 1933a (p. 151)

54 Ventral disc scales small but markedly thickened and projecting to give a rough surface (Fig. 148) ; arm spines seven or more ; dorsal arm plates with median distal concavity (Fig. 135) . . ***Paracrocnida sucensis*** (Balinsky, 1957) (p. 160)

– Ventral disc scales smooth (only those of the uppermost row, along or just above the ambitus, sometimes with fine spinose projections) ; only three or four arm spines for most of the arm, sometimes five or six basally ; dorsal arm plates with distal edge convex 55

55 Radial shields contiguous for almost their whole length (Figs 149, 150) ; the two tentacle scales very large (Fig. 137) 56

– Radial shields only contiguous distally (Fig. 151) ; if two tentacle scales are present then they are small or moderate in size 57

56 Arm spines all tapering, superficially appearing pointed ; disc scaling moderately fine, about 13 scales between adjacent radial shields across the interradius in larger specimens (d.d. *c.* 5 mm) ; spinose projections often present on the marginal row of scales (Fig. 149)
Amphioplus (Lymanella) furcatus Mortensen, 1933a (p. 149)

– Middle arm spines stout, not markedly tapering, blunt at the tip ; disc scaling coarse, only 9–11 scales between the radial shields at d.d. *c.* 5 mm ; no spinose projections on the disc margin (Fig. 150)
Amphioplus (Lymanella) integer (Ljungman, 1867) (p. 149)

* Young specimens of *Amphioplus integer* (d.d. < 3 mm) may have the fourth (outermost) oral papilla hidden behind the somewhat enlarged third papilla, or even undeveloped, and so may run down to 51 and be confused with *Amphipholis similis*; the latter species has the third papilla even broader, well over twice as broad as long normally and the tentacle scales are only moderate in size, while the arm spines are all shorter than the segments.

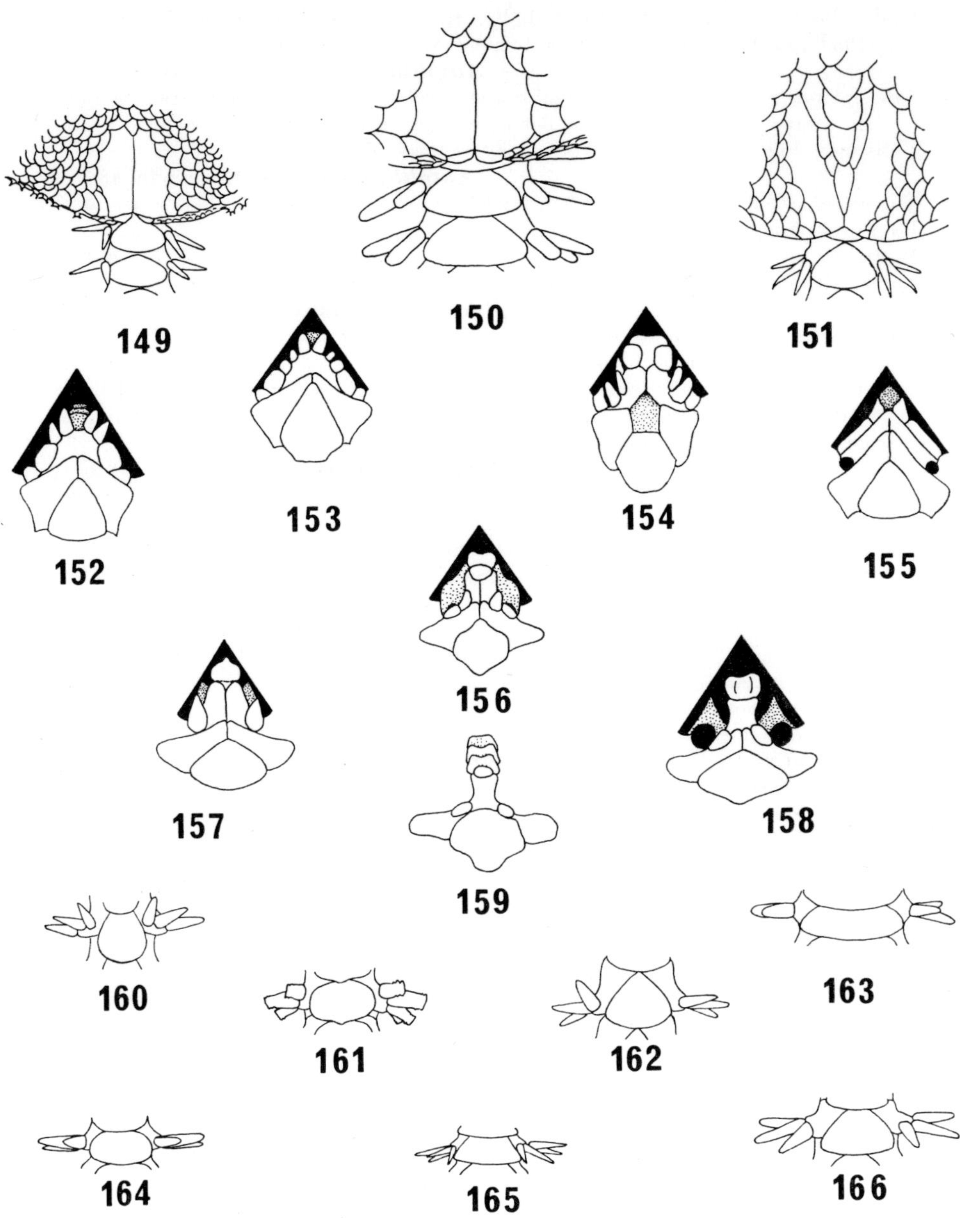

FIGS 149–166 Ophiuroidea. 149, 150, 151: arm base and adjoining disc of *Amphioplus* (*Lymanella*) *furcatus*, *A.* (*L.*) *integer* and *A.* (*Unioplus*) *falcatus* (dorsal view). 152, 153, 154, 155, 156, 157, 158, 159: jaws of *Amphioplus* (*Unioplus*) *falcatus*, *A. pectinatus*, *A.* sp. aff. *A. falcatus*, *Amphilepis scutata*, *Ophiactis savignyi*, *O. plana*, *O. carnea* and *O. delagoa*. 160, 161, 162, 163, 164, 165, 166: arm segments of *Ophiactis nidarosiensis*, *O. savignyi*, *O. plana*, *O. modesta*, *O. carnea* (oval and quadrangular forms) and *O. delagoa* (dorsal view).

57 One tentacle scale ; only three true oral papillae plus the almost superficial oral tentacle scale (Fig. 152) ***Amphioplus (Unioplus) falcatus*** Mortensen, 1933a (p. 150)

– Two tentacle scales ; four, sometimes five, superficial oral papillae . . . 58

58 All oral papillae rounded ; adoral shields interradially contiguous (Fig. 153) ***Amphioplus (Amphioplus) pectinatus*** Mortensen, 1933a (p. 148)

– Two middle oral papillae finger- or spine-like (Fig. 154) ; adoral shields broadly separated interradially . ***Amphioplus*** sp. aff. ***A. falcatus*** Mortensen (p. 150)

59 No infradental papilla or papillae, though the inner one of the two (rarely three) lateral oral papillae is only slightly offset from the relatively narrow lowermost tooth (Fig. 155) ; second oral tentacle pore superficial and scaleless. AMPHILEPIDINAE . . . . ***Amphilepis scutata*** Mortensen, 1933a (p. 146)

– Teeth usually more or less broadened with one or a cluster of infradental papillae superficial to them ; second oral tentacle usually inset into the oral slit but, if superficial, then armed with one or more scales (Figs 213–217) . . . 60

60 A single broad infradental papilla (rarely none at all) separated by a space or diastema from the one, sometimes two, distal oral papillae each side (Figs 156–159). OPHIACTIDAE 61

– If only a single infradental papilla occurs, it is not broader than long but similar to and in series with the other papillae ; a diastema only present in *Anamphiura* (here referred to the Ophiacanthidae) in which there are two or three somewhat irregular infradental papillae 68

61 Six arms, sometimes five or seven, most individuals fissiparous (liable to split and regenerate) 62

– Normally five arms 65

62 Two distal oral papillae each side (Fig. 156) ; radial shields large . . . 63

– One distal oral papilla (Fig. 157) ; radial shields small 64

63 Dorsal arm plates elongate oval (Fig. 160) ; arm spines tapering to points and not noticeably rugose ; no green in colour and radial shields not conspicuously patterned . . . . . ***Ophiactis nidarosiensis*** Mortensen, 1920 (p. 163)

– Dorsal arm plates elliptical but with a slight median distal lobe defined by a pair of small indentations (Fig. 161) ; arm spines short, broadly truncated at the tips and rugose ; colour green and white, radial shields conspicuously marked with dark patches . . . ***Ophiactis savignyi*** (Müller & Troschel, 1842) (p. 164)

64 Dorsal arm plates fan-shaped, less than twice as broad as long, barely contiguous (Fig. 162) . . . . ***Ophiactis plana*** Lyman, 1869* (p. 163)

– Dorsal arm plates broad elliptical, at least twice as broad as long, broadly contiguous (Fig. 163) . . . . . ***Ophiactis modesta*** Brock, 1888 (p. 163)

65 Two distal oral papillae each side 66

– Only one distal oral papilla 67

66 Dorsal arm plates fan-shaped, barely contiguous ; disc fully scaled below ; oral shields rhombic, oral papillae rather irregular ***Ophiactis abyssicola*** (M. Sars, 1861) (p. 161)

– Dorsal arm plates elliptical, broadly in contact ; disc with only isolated scales in the skin ventrally ; oral shields elliptical ; oral papillae regular ***Ophiactis hemiteles*** H. L. Clark, 1915 (p. 162)

67 Infradental papilla usually lacking, the broad lowest tooth superficial, often trilobed (Fig. 158) ; dorsal arm plates very variable in shape, oval or like a broad trapezium (Figs 164, 165), usually contiguous for more than half their breadth ***Ophiactis carnea*** Ljungman, 1867 (p. 161)

– Infradental papilla developed (at least in the type material), distinctly smaller than the lowest tooth (Fig. 159) ; dorsal arm plates fan-shaped, tapering proximally and only contiguous for about half their width (Fig. 166) ***Ophiactis delagoa*** Balinsky, 1957 (p. 162)

* The specimens from Inhaca which J. B. Balinsky referred to *Ophiactis lymani* (otherwise known from the West Indies and St Helena) and *O. parva* (from the Red Sea and Persian Gulf) may be better placed in *O. plana*.

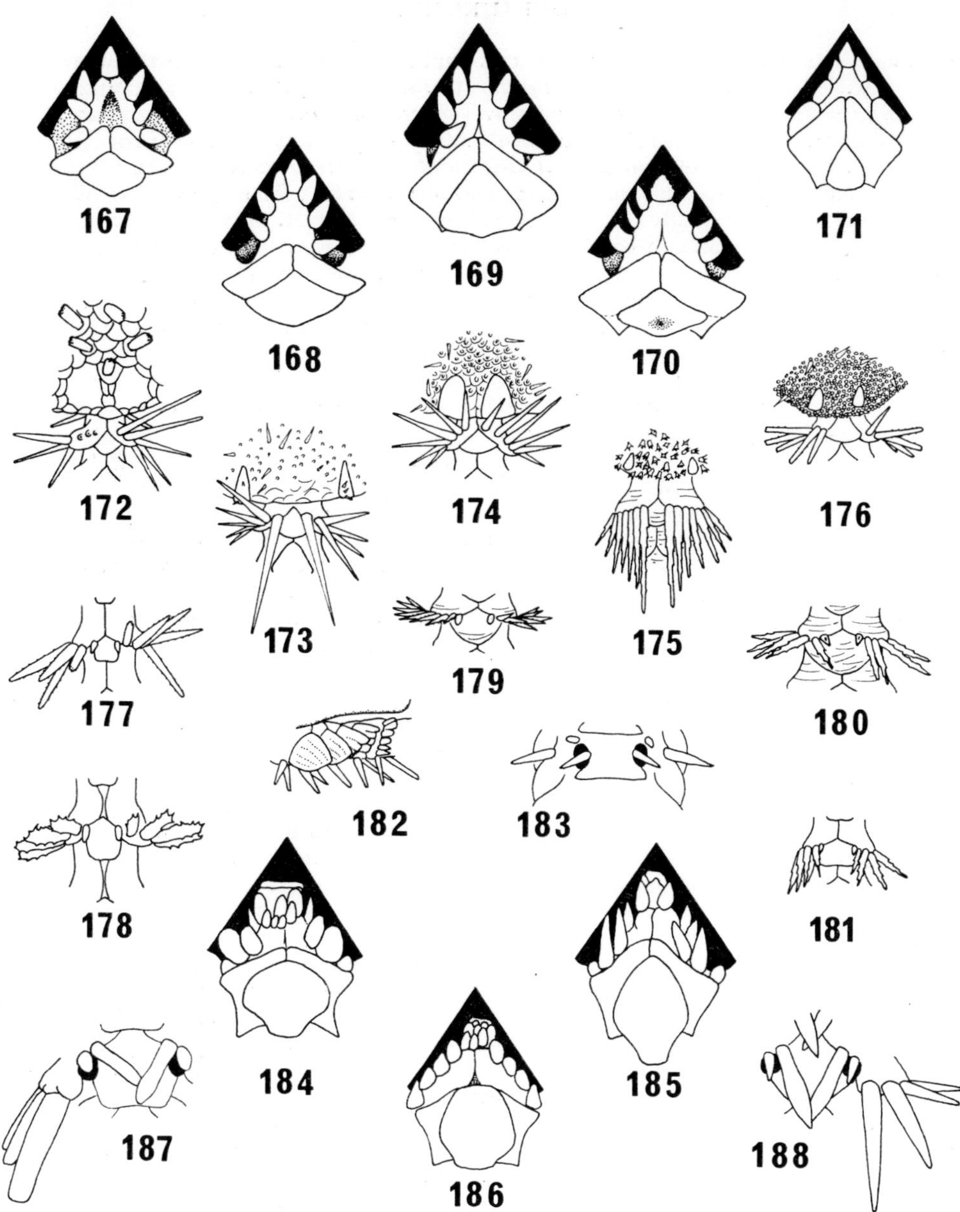

Figs 167–188 Ophiuroidea. 167, 168, 169, 170, 171: jaws of *Ophiomitrella corynephora*, *O. hamata*, *Ophiacantha striolata*, *O. baccata* and *Ophiothamnus remotus*. 172, 173, 174, 175, 176: arm base and adjoining disc of *Ophiomitrella corynephora*, *Ophiacantha nerthepsila*, *O. scutigera*, *O. striolata* and *Ophiotreta durbanensis*. 177, 178, 179, 180, 181: arm segments of *Ophiomitrella corynephora*, *O. hamata*, *Ophiacantha scutigera*, *O. striolata* and *O. baccata* (ventral view). 182: basal arm segments of *Amphilimna cribriformis* (side view) and 183: (ventral view). 184, 185, 186: jaws of *Ophiopsila seminuda*, *O. bispinosa* and *Ophiocoma valenciae*. 187, 188: middle arm segments of *Ophiopsila seminuda* and *O. bispinosa* (ventral view).

68 Teeth and single apical oral papilla narrow and pointed, usually with three similar oral papillae each side (rectangular only in *Ophiothamnus*) (Figs 167–171); lateral arm plates large and even the proximal ones partially contiguous dorsally, separating the consecutive dorsal arm plates (Figs 172–175); arm spines conspicuously long (except in *Ophioplinthaca*). OPHIACANTHIDAE (part) . 69

– Teeth either narrow or rounded and more or less broadened, often with more than three papillae of various shapes each side (Figs 184, 185, 214–217); if the lateral arm plates are broadly contiguous then most of the arm spines are reduced and more or less appressed, the radial shields are well developed and the disc lacks armament (except in *Ophiotreta*) 76

69 Radial shields relatively small and the two shields of each pair separated (Figs 172–175) except in very small specimens, d.d. $<$ 4 mm 70

– Radial shields large and at least partially contiguous radially 75

70 Disc scaling distinguishable, the armament consisting of coarse, spaced, short, rugose-tipped cylindrical stumps (Fig. 172) 71

– Disc scaling obscured by the armament, which consists of either numerous fine trifid or multifid stumps, or granules and sporadic spinelets (Figs 173–175) . 72

71 Arm spines very finely serrated, superficially appearing smooth, the longer ones markedly longer than the segment (Fig. 177); jaws sunken ventrally (Fig. 167) ***Ophiomitrella corynephora*** H. L. Clark, 1923 (p. 169)

– Arm spines distinctly serrated, none longer than the segment (Fig. 178); ventral surface of jaws flat (Fig. 168) . ***Ophiomitrella hamata*** Mortensen, 1933a (p. 170)

72 Disc covered with rugose granules usually interspersed with some small spinelets (Figs 173, 174) 73

– Disc covered with uniform trifid, sometimes multifid, stumps (Fig. 175) . . 74

73 Radial shields narrow, usually rib-like, the two of each pair separated by about the breadth of the arm (Fig. 173); ventral arm plates uniform in texture ***Ophiacantha nerthepsila*** H, L. Clark, 1923 (p. 168)

– Radial shields broad, flat and separated by less than the arm breadth (Fig. 174); ventral arm plates with concentric striations near their distal edges (Fig. 179) ***Ophiacantha scutigera*** Mortensen, 1933a (p. 168)

74 Third (distalmost) oral papilla similar to the other two and rather thorny (Fig. 169) (or a fourth papilla present); ventral arm plates with distal edge convex, all arm plates striated transversely (Fig. 180) ***Ophiacantha striolata*** Mortensen, 1933a (p. 168)

– Third oral papilla distinctly stouter and blunter than the other two and almost smooth, only very finely rugose (Fig. 170) (or a fourth papilla present); ventral arm plates with distal edge slightly concave, arms not striated (Fig. 181) ***Ophiacantha baccata*** Mortensen, 1933a (p. 167)

75 Five arms; disc scales usually bearing scattered spinelets; distalmost oral papilla very broad operculiform (Fig. 171) ***Ophiothamnus remotus*** Lyman, 1878 (p. 170)

– Six arms; disc scales with scattered granules; distalmost oral papilla papilliform ***Ophioplinthaca sexradia*** Mortensen, 1933a (p. 170)

76 Disc armed above both with fine granules concealing the scaling (except for the very small radial shields) and with scattered spinelets (Fig. 176) ***Ophiotreta durbanensis*** (Mortensen, 1933a)* (p. 171)

* I support Koehler and H. L. Clark in regarding *Ophiotreta* as generically distinct from *Ophiacantha*. Certainly *O. durbanensis* shows more markedly superficial resemblance to the Chilophiurida (notably the families Ophiocomidae and Ophiodermatidae) than to most other Ophiacanthids, having continuous disc granulation, spearhead-shaped oral shields, numerous oral papillae leading distally to a partially inset papilla or oral tentacle scale on the inclined first ventral arm plate, and contiguous dorsal arm plates. ***Ophiotreta matura*** (Koehler, 1904), collected off northern Natal (27°45′S 32°44·8′E, in 492–700 metres) only in May, 1975 and previously known from the Gulf of Aden and the East Indian area, runs down here. It is distinguished from *O. durbanensis* by the dense covering of elongated spinelets on the disc and the single very large tentacle scales.

76 Disc scaling naked or covered with close granules or scattered spinelets but not both together 77

77 A curved flange on each side of the first one or several arm segments opposite the genital slits above one or two arm spines (Fig. 182); disc armed with spaced scattered spinelets 78

– Basal arm segments without flanges, the spines all separate; disc naked or granule-covered, rarely with fine spinelets 79

78 Primary rosette conspicuous, covering most of the disc, radial shields short and broad; only one arm segment with a lateral flange each side above the two lowest arm spines . . . ***Anamphiura valida*** H. L. Clark, 1939 (p. 166)

– Rosette indistinct, radial shields long and narrow; several arm segments with flanges below which only one spine is free ***Amphilimna cribriformis*** A. M. Clark, 1974 (p. 165)

79 Apex of each jaw armed with three to many tooth papillae, usually a cluster superficial to the tooth row (Figs 184–186)*. OPHIOCOMIDAE . . . 80

– No tooth papillae, though the two proximalmost oral papillae may be infradental in position 89

80 Two tentacle scales of which at least the inner one on the arms becomes long and sword-like, aligned obliquely across the ventral plate forming a cross with the corresponding scale (Figs 187, 188) 81

– One or two oval and unspecialized scales 82

81 Only the inner scale spiniform (Fig. 187); distal oral papillae flat papilliform with rounded tips (Fig. 184) . . ***Ophiopsila seminuda*** A. M. Clark, 1952 (p. 178)

– Both scales becoming spiniform on the arms (Fig. 188); distal oral papillae also spiniform (Fig. 185) . . . ***Ophiopsila bispinosa*** A. M. Clark, 1974 (p. 177)

82 Six-armed, fissiparous . . . ***Ophiocomella sexradia*** (Duncan, 1887) (p. 175)

– Five armed 83

83 Disc with continuous granulation, at least dorsally 84

– Disc skin-covered, with or without spaced spines or spinelets 88

84 Disc granules coarse, leaving the proximal parts of the ventral interradii naked . 85

– Disc granules fine and completely covering both sides 86

85 Colour uniformly dark, above and below ***Ophiocoma erinaceus*** Müller & Troschel, 1842 (p. 173)

– Colour light on the under side, variegated dark and light or sometimes all dark above . . . . . ***Ophiocoma scolopendrina*** (Lamarck, 1816) (p. 174)

86 One tentacle scale; disc granules of ambital area distinctly elongated (Fig. 189); colour mid-brown, arms greenish in life ***Ophiocoma valenciae*** Müller & Troschel, 1842 (p. 175)

– Two tentacle scales; disc granules all spherical 87

87 No clavate arm spines; disc beautifully patterned with many light radiating lines (gold in life) on a dark background ***Ophiocoma pica*** Müller & Troschel, 1842 (p. 173)

– One to three consecutive segments of each arm, about a third of the length from the disc, with the second from uppermost spine markedly clavate at the tip, especially in wet specimens (Fig. 190); colour dark brown, disc markings, if any, consisting of darker spots or lighter rings, sometimes both ***Ophiocoma pusilla*** (Brock, 1888) (p. 174)

88 Disc spines few or absent, colour predominantly light, disc marked with a conspicuous dark linear pattern; some of the uppermost arm spines markedly clavate at the tips (Fig. 191) . . ***Ophiomastix venosa*** Peters, 1851 (p. 176)

– Disc spines fairly numerous; colour predominantly dark, sometimes with irregular light markings; arm spines regularly alternating two and three, the uppermost one of each three markedly enlarged but hardly, if at all, clavate at the tip (Fig. 192) . . . . ***Ophiomastix variabilis*** Koehler, 1905 (p. 176)

* Small specimens of *Ophiopsila*, d.d. <4 mm, usually have only two infradental papillae; the sword-like tentacle scales on the arms should distinguish them.

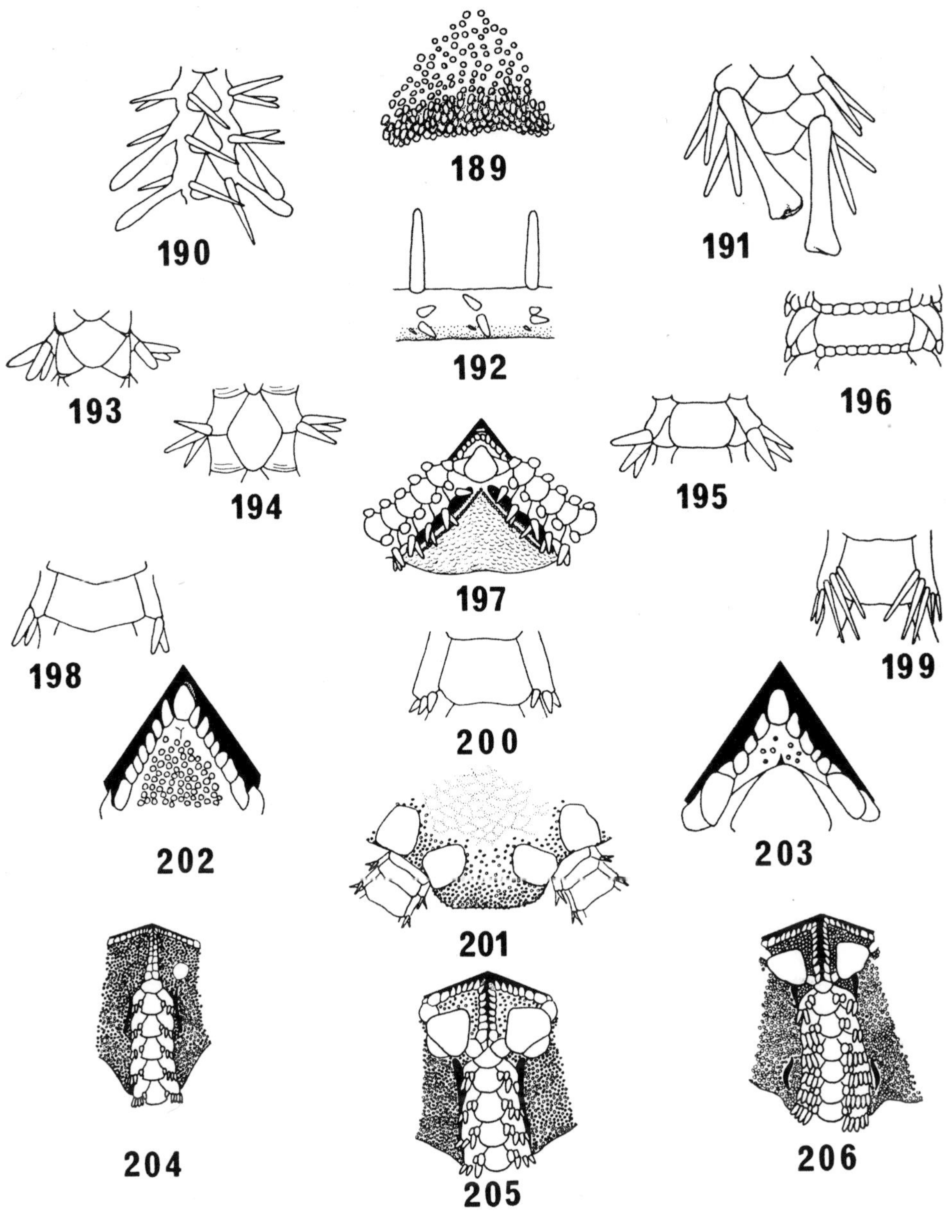

FIGS 189–206 Ophiuroidea. 189: periphery of disc of *Ophiocoma valenciae* (ventral view). 190, 191: arm segments of *Ophiocoma pusilla* and *Ophiomastix venosa* (dorsal view). 192: arm segments of *Ophiomastix variabilis* (side view). 193, 194, 195, 196: arm segments of *Ophionereis dubia, O. vivipara, O. australis* and *Ophiolepis cincta* (dorsal view). 197: edge of arm and genital slit of *Ophionereis porrecta* (ventral view). 198, 199, 200: arm segments of *Ophiocirce inutilis, Ophiopallas paradoxa* and *Ophiarachnella capensis* (dorsal view). 201: part of disc of *Ophiernus vallincola* (dorsal view). 202, 203: jaws of *Ophiocirce inutilis* and *Ophiopallas paradoxa*. 204, 205, 206: arm base and adjoining jaws of *Cryptopelta aster, Ophiarachnella capensis* and *Ophioderma wahlbergi* (ventral view).

89 A pair of supplementary dorsal arm plates on each segment, at least proximally (Fig. 193); disc overlying the bases of the arms; arm spines well developed and erect. OPHIONEREIDAE 90

– If any supplementary dorsal arm plates are present, then they are numerous and encircle the main dorsal plates (Fig. 196); disc usually more or less fused around the arm bases; arm spines often short and appressed 93

90 No genital papillae present; disc scaling very fine 91

– Genital papillae present (Fig. 197); disc scaling moderately coarse . . . 92

91 Colour pattern on disc reticular, usually with a well-marked dark Y opposite the base of each arm on a light background; disc scaling very fine but continuous; dorsal arm plates widest at the proximal end of the segment and supplementary plates running the whole length of the segment (Fig. 193)
Ophionereis dubia (Müller & Troschel, 1842) (p. 179)

– Disc with a large brown patch in the centre; disc scaling very reduced, discontinuous; dorsal arm plates widest in the middle of the segment and supplementary plates only bordering their distal halves (Fig. 194)
Ophionereis vivipara Mortensen, 1933b (p. 180)

92 Supplementary dorsal arm plates small and only developed on the proximal segments (Fig. 195) . . **_Ophionereis australis_** (H. L. Clark, 1923) (p. 179)

– Supplementary arm plates developed for most of the arm length
Ophionereis porrecta Lyman, 1865 (p. 180)

93 Disc covered partially or completely by a close coating of granules, though the central and peripheral scaling or skin may be distinguishable 94

– No granules on the disc, scaling distinct (rarely with spaced tubercles or spinelets present). OPHIURIDAE 100

94 Fragile deep-water species; many arm spines equalling the segment in length (Fig. 198). OPHIOLEUCIDAE 95

– Stout species, from intertidal to deep-water, often large, d.d. 15 mm or more; arm spines all very short and more or less appressed, equal in length to about half the segment, or less (Fig. 200). OPHIODERMATIDAE . . . 97

95 Granule covering incomplete, sparse peripherally and absent centrally (Fig. 201); radial shields exposed . . . **_Ophiernus vallincola_** Lyman, 1878 (p. 185)

– Granule covering complete in life (though easily rubbed off in preserved specimens); radial shields concealed 96

96 Only two arm spines (Fig. 198); seven or eight oral papillae each side of the jaw (Fig. 202) . . . . . . **_Ophiocirce inutilis_** Koehler, 1904 (p. 185)

– Numerous arm spines (Fig. 199); about six oral papillae each side (Fig. 203)
Ophiopallas paradoxa Koehler, 1904 (p. 186)

97 Granulation extending on to the oral shields, though the central parts of one or two may be exposed (Fig. 204); one tentacle scale
Cryptopelta aster (Lyman, 1879) (p. 182)

– Oral shields all bare; two tentacle scales. 98

98 Two short genital slits each side of each ventral interradial area (Fig. 206)
Ophioderma wahlbergi Müller & Troschel, 1842 (p. 183)

– Only a single genital slit each side (Fig. 205) 99

99 Radial shields concealed by granulation . . **_Ophiopeza fallax_** Peters, 1851 (p. 184)

– Radial shields naked (though the granulation may encroach around the edge) . 100

100 Up to seven arm spines; a broad supplementary oral shield present, contiguous with the distal side of the main shield (Fig. 205)
Ophiarachnella capensis (Bell, 1888) (p. 182)

– Up to ten arm spines; supplementary oral shields largely covered (or concealed) by the granulation . . . . . **_Ophiochasma nitida_** Hertz, 1927 (p. 183)

101 Disc extending into a flat pentagon, largely by extreme modification and extension of several proximal lateral arm plates, its straight interradial edges fringed by a

continuous series of realigned arm spines (Fig. 207); the free parts of the arms very reduced . . . . . . ***Astrophiura permira*** Sladen, 1879 (p. 188)

101 Disc not extended beyond its normal limits (though the first one or two lateral arm plates may be somewhat enlarged without contrasting in form with the succeeding plates) 102

102 Only one tentacle scale, the pores being abruptly lost after the first two to five arm segments (Figs 210, 211) 103

– Two or usually more scales on the proximal pores, though often only one on the arm pores, the pores either gradually reducing in size on the proximal segments or extending for the whole arm (Figs 212, 213) 104

103 First ventral arm plate (flanked by the adoral shields) large and with a well-developed superficial 'oral' tentacle pore and scale each side of it, resembling the plates and pores on the first three or four proper arm segments (Fig. 210); primary rosette occupying the whole centre of the disc and each of its plates liable to bear a median tubercle (Fig. 208); size small, d.d. up to only 5 mm ***Ophiomisidium pulchellum*** (W. Thomson, 1877) (p. 190)

– Oral tentacles not visible externally, the first ventral arm plate reduced in comparison with the second and third ones, which are the only other ventral arm plates developed, accordingly only two pairs of tentacle pores present (Fig. 211); disc with multiple plates centrally, though the rosette may be distinct; upper side bearing many small tubercles; size often large, d.d. exceeding 20 mm ***Ophiomusium lymani*** W. Thomson, 1873 (p. 191)

104 Dorsal arm plates ringed by regularly arranged small secondary plates (Fig. 196); second oral tentacle pore concealed within the oral slit ***Ophiolepis cincta*** Müller & Troschel, 1842 (p. 189)

– No secondary dorsal arm plates; second oral tentacle pore more or less completely exposed on the ventral face outside the oral slit, though its abradial scales may be in series with the distal oral papillae (Figs 213–217) 105

105 Oral shields with a huge broad rounded distal lobe covering much of the ventral interradius, the proximal part of the shield with a small constricted angle ***Amphiophiura trifolium*** Hertz, 1927 (p. 187)

– Oral shields not markedly broader distally than proximally 106

106 Disc elevated above the arm bases with a swollen midradial plate, indented or creased outwardly, distal to each pair of radial shields and bearing a secondary arm comb each side opposing the main combs (Fig. 220) ***Dictenophiura anoidea*** H. L. Clark, 1923 (p. 188)

– No such median plate and no secondary arm combs 107

107 Disc not distinctly indented opposite the base of each arm* (Fig. 219); oral shields extremely broad and short and at the same time only two scales on the superficial oral tentacle pores . . ***Ophiocten latens*** Koehler, 1906 (p. 189)

– Disc more or less indented opposite the arm bases* (Figs 218, 220–223); if the oral shields are broader than long then there are multiple scales on the superficial oral pores 108

108 Only two (rarely three) tentacle scales on the second oral pore (Fig. 214) and those of the proximal arm segments ***Ophiura (Ophiura) affinis simulans*** (Mortensen, 1936) (p. 192)

– Five or more tentacle scales on the second oral pores and often also on the proximal arm pores 109

109 Tentacle pores gradually reducing in size along the arms, distinct (or their scale or scales are) for at least the proximal half of the arm; arm spines as long as or longer than the segment (Fig. 224) 110

* In *Ophiura affinis simulans*, which was formerly included in *Ophiocten*, the indentation opposite the arm bases may not be well marked; the elongate oral shields should serve to differentiate it from *Ophiocten latens*, which also comes from much deeper water.

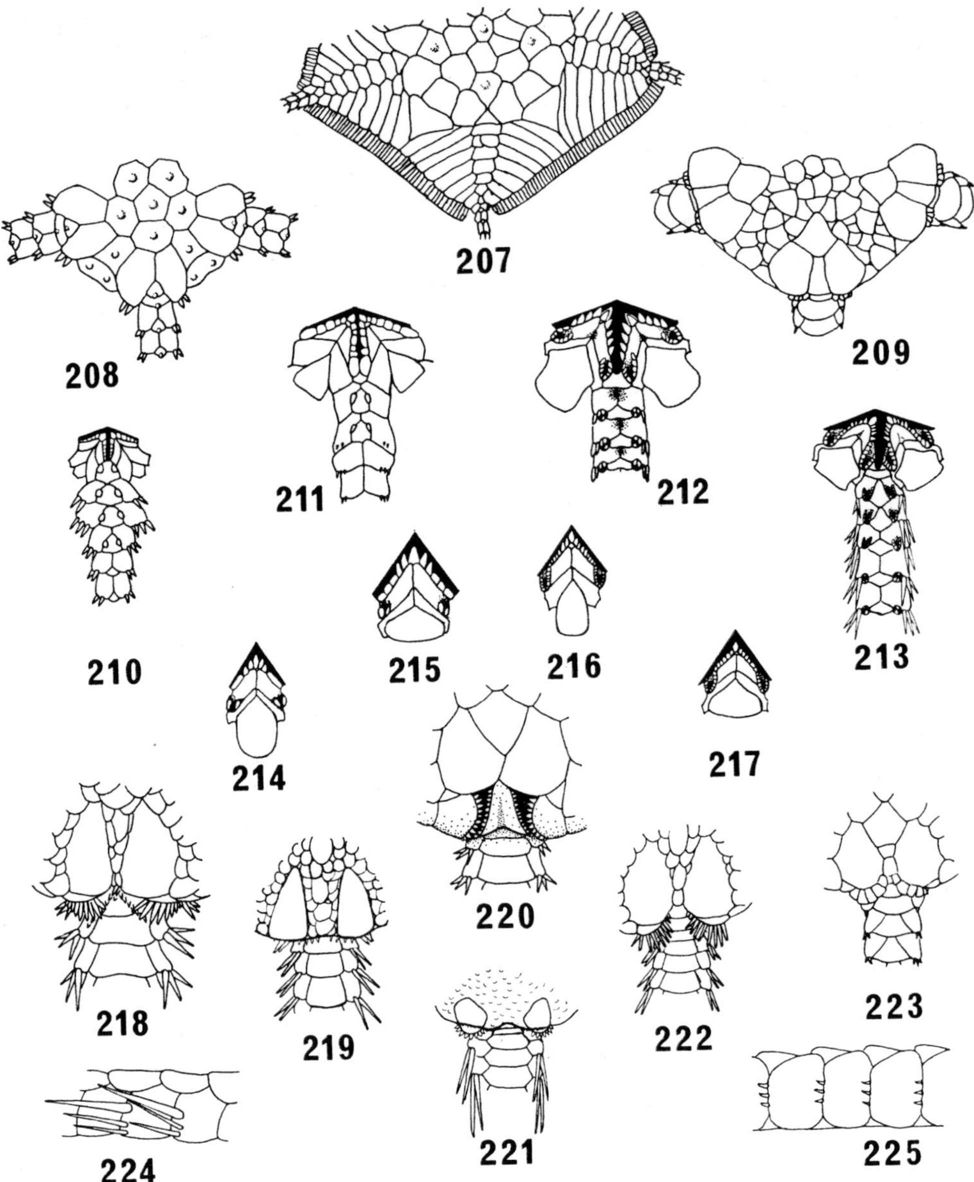

FIGS 207–225 Ophiuroidea. 207, 208, 209: half of disc and arm bases of *Astrophiura permira*, *Ophiomisidium pulchellum* and *Ophiura costata* (dorsal view). 210, 211, 212, 213: arm base and adjoining jaws of *Ophiomisidium pulchellum*, *Ophiomusium lymani*, *Ophiura kinbergi* and *O. trimeni*. 214, 215, 216, 217: jaws of *Ophiura affinis simulans*, *Ophiocten latens*, *Ophiura costata costata* and *O. irrorata*. 218, 219, 220, 221, 222, 223: arm base and adjoining disc of *Ophiura affinis simulans*, *Ophiocten latens*, *Dictenophiura anoidea*, *Ophiura flagellata*, *O. kinbergi* and *O. costata tumida* (dorsal view). 224, 225: arm segments of *Ophiura trimeni* and *O. costata tumida* (side view).

109 Tentacle pores indistinct from about either the third or the tenth segment; arm spines all much shorter than the segment (Fig. 225) 112

110 Uppermost arm spine much the longest and stoutest (Fig. 224); oral shields often broader than long . . . ***Ophiura* (*Ophiura*) *trimeni*** Bell, 1905b (p. 194)

– Arm spines similar; oral shields usually longer than broad 111

111 Radial shields very short and widely separated (Fig. 221); no hollows between the basal ventral arm plates . ***Ophiura* (*Ophiura*) *flagellata*** (Lyman, 1878) (p. 193)

– Radial shields moderately large, longer than broad and just contiguous or nearly so (Fig. 222); conspicuous deep hollows present between several of the basal ventral arm plates (Fig. 212) ***Ophiura (Ophiura) kinbergi*** Ljungman, 1867 (p. 194)

112 Oral shields broader than long* (Fig. 217); disc scales numerous, mostly small, overlapping and irregular ***Ophiura (Ophiuroglypha) irrorata*** (Lyman, 1878) (p. 197)

– Oral shields about as long as broad or longer* (Fig. 216); disc covered with coarse plates and relatively few small interstitial ones (Fig. 209) . . 113

113 Arm combs separated radially by more than half the breadth of the first free arm segment (Fig. 209); the disc plates not markedly convex and the dorsal arm plates also almost flat in profile
Ophiura (Ophiuroglypha) costata costata (Lyman, 1878) (p. 195)

– Arm combs approximating midradially (Fig. 223); the larger disc and dorsal arm plates markedly swollen, distinctly convex in side view (Fig. 225)
Ophiura (Ophiuroglypha) costata tumida Mortensen, 1933 (p. 196)

Suborder EURYALINA Lamarck†

Ophiuroids lacking continuous scaling over the disc (which is liable to considerable shrinkage interradially in preservation), the skin-covering very thick and naked or with embedded granules or fine spinelets, sometimes coarser scattered armament; radial shields usually rib-like, the hinge at their distal ends with the outer end of the genital plate often conspicuous externally; oral shields reduced or absent, except for the madreporite; arms distally attenuated and capable of coiling downwards, sometimes branching repeatedly, completely covered with thick skin so as to hide the superficial underlying plates, of which no entire median dorsal plates are found, ventral plates more or less reduced, contiguous only on the basal segments, if at all, and the spine-bearing lateral arm plates ventrolateral in position with the spines (sometimes misnamed tentacle scales) more or less appressed, aligned mainly or entirely on the ventral side and usually modified into hooks distally; no true tentacle scales restricted to the small pores present.

* In young specimens of *Ophiura costata* with d.d. < 5 mm the oral shields are broader than long; the few relatively huge plates covering the disc should serve to distinguish them.

† After vainly attempting to compose a satisfactory diagnosis for Matsumoto's order Phrynophiurida to include the family Ophiomyxidae together with the 'euryalid' families, as adopted by H. L. Clark, Koehler, Fell (latterly) and Spencer and Wright, I propose to use here instead the five *sub*ordinal divisions of recent Ophiuroidea used by Spencer & Wright. This also has the merit of producing better fragmentation into smaller groups of families. Although, at present, my knowledge of the range of form within the family Ophiomyxidae is limited, I am inclined to follow Mortensen (and Djakonov) in regarding the Ophiomyxidae as more closely related to the families of the order Ophiurida than to the euryalid families. A comparison of some of the ophiomyxids included here, notably *Ophioscolex dentatus*, with the non-euryalids shows many similarities in the relative proportions of disc and arms, the well-developed oral shields and their relations with the adorals, the continuous scaling over the disc (even though obscured by skin in wet specimens), the horizontal flexure of the arms, the prominence and lateral position of the lateral arm plates with their conspicuous laterally projecting spines, the development of superficial dorsal arm plates (though fragmented) and the large tentacle pores provided with tentacle scales. Although it must be admitted that these are superficial characters, none of them are shared by the euryalids.

Family **EURYALIDAE** Gray

See : Spencer & Wright, 1966 : U92.
[With synonym Trichasteridae Döderlein.]

A family of Euryalina with a moderately large disc and simple, sometimes branching, arms (simple in the species represented), often armed dorsolaterally with one or more short blunt spines on each side of the proximal segments ; jaws armed with a single row of teeth apically and sometimes a few granules each side, usually within the oral slit ; vertebrae of arms very short and with the ventral longitudinal groove more or less extensively floored over, at least, on the more distal segments ; no belts of microscopic hooks across the arms dorsally, though the distal arm spines are modified into ventrolateral hooks ; distal segments somewhat attenuated, as long as or longer than broad ; tubular gonads present in at least the basal part of the arms dorsally.

Asterostegus tuberculatus Mortensen

Asterostegus tuberculatus Mortensen, 1933a : 298–300, figs 25, 26, pl. 19, figs 4, 5.

D.d. 23 mm (only one specimen recorded) ; a.l./d.d. *c*. 15/1.

A species of Euryalidae with simple arms of considerable length ; radial ribs bearing rounded tubercles of which three to five form an irregular group at the distal end ; a few small tubercles also present along the disc margin interradially ; no distinct oral shields present but a group of rounded plates outside each pair of adoral shields ; jaws armed with some fine granules laterally each side of the row of conical teeth but no true oral papillae ; upper side of each proximal arm segment with a dorsolateral oblong plate each side bearing usually two short rugose capitate spines or tubercles aligned transverse to the arm axis ; ventral arm plates fairly well developed but only the proximal ones contiguous ; lateral arm plates armed with two capitate spines like those on the dorsolateral plates.

TYPE LOCALITY. Off Durban (29/31), 382 metres.

Astroceras spinigerum Mortensen

Astroceras spinigerum Mortensen, 1933a : 296–297, fig. 23, pl. 18, figs 8, 9.

D.d. up to 13 mm ; a.l./d.d. *c*. 6/1.

A species of Euryalidae with simple arms of moderate length ; radial ribs armed with three to five short cylindrical, rugose-tipped spines, the outermost one spaced and enlarged, disc otherwise naked, including the area distal to the adoral shields on the ventral side ; jaws armed with broad teeth, though the superficial one may be more conical, laterally bearing several low tubercular papillae each side within the oral slit ; proximal arm segments with a single short cylindrical, rugose-tipped dorsolateral spine each side mounted on a small plate ; ventral arm plates small, all separated, with the lateral arm plates meeting between them in the proximal part of the arm, these bearing two spines each, similar to the dorsolateral spines.

Type locality. Off Durban, 410 metres.

Southern African records. 29/31/d ; 25/34/d ; 112–410 metres.

Nature of bottom. Sand with sponges ; dark sandy mud.

Family **ASTERONYCHIDAE** Müller & Troschel

See : Matsumoto, 1917 : 28 (as subfamily) ; Spencer & Wright, 1966 : U91.

A family of Euryalina with the disc relatively large and arms long, simple, distally coiling and without conspicuous dorsolateral armament ; jaws with clustered armament apically in adults ; vertebrae very short and with a continuous longitudinal ventral groove ; no belts of microscopic hooks across the arms dorsally ; distal arm segments short ; gonads restricted to the disc ; at least three arm spines on each lateral plate.

Asteronyx loveni Müller & Troschel

Asteronyx loveni Müller & Troschel, 1842 : 119, pl. 10, figs 3–5 ; H. L. Clark, 1923 : 314–315 ; Mortensen, 1933a : 300–301 ; A. M. Clark, 1952 : 199, 212.

D.d. up to *c.* 50 mm ; a.l./d.d. 6–17/1, increasing with size, usually *c.* 10/1.

A species of Asteronychidae with arms usually of unequal length in adults ; disc and arms covered with thick smooth skin, the very long radial ribs naked ; jaws armed apically with a cluster of irregular teeth and papillae, though small specimens may have only a single vertical row ; three to nine arm spines placed ventrolaterally and short and hook-like except that on large specimens the lowest spine on the more distal segments of the larger arms is markedly elongated.

Type locality. SW. Sweden.

Southern African records. 30/14/d ; 31/16/d ; 33/16/vd ; 34/17/vd ; 457–1830 metres.

Nature of bottom. Green mud and sand (epizoic on pennatulids, gorgonians, etc.).

Family **ASTEROSCHEMATIDAE** Verrill

See : Matsumoto, 1917 : 29 (as subfamily) ; Spencer & Wright, 1966 : U91.

A family of Euryalina with a relatively small disc and stout, simple, distally coiling arms, without conspicuous dorsolateral armament ; jaws armed apically with a vertical series of teeth, usually conical, and sometimes some low granuliform papillae each side ; vertebrae very short and with a longitudinal ventral furrow ; no belts of microscopic hooks across the arms dorsally, distal arm segments short ; gonads extending at least midway along the arms dorsally ; only two spines on each lateral arm plate.

Asteroschema capensis Mortensen

Astroschema [sic] *capensis* Mortensen, 1925 : 152–154, fig. 5.

D.d. 8 mm (only one specimen recorded) ; a.l./d.d. *c.* 6/1.

A species of Asteroschematidae with the disc and fairly short arms covered all over by a uniform granular coating, slightly coarser on the radial ribs which are broad distally ; jaws armed only with teeth ; arms annulated proximally ; lateral arm plates bearing two very short similar rugose-tipped spines.

TYPE LOCALITY. Off Umvoti River, Natal (no depth).

SOUTHERN AFRICAN RECORDS. 29/31/[?]s ; 24/35/d ; 110–132 metres.

NATURE OF BOTTOM. Rock and shelly sand ; coarse sand and rock (epizoic on gorgonians).

Family **GORGONOCEPHALIDAE** Ljungman

See : Spencer & Wright, 1966 : U91.

A family of Euryalina with a large disc and branching or sometimes long simple arms, capable of coiling downwards, smooth dorsolaterally or with low tubercles but these are more common medially ; jaws armed with more or less numerous spiniform papillae clustered especially near the apex where there is no distinction between teeth and other papillae, sometimes also papillae fringing the distal edges of the oral slits ; vertebrae with a longitudinal ventral groove ; belts of microscopic F-shaped hooks developed across each arm segment dorsally, for much of the arms ; distal arm segments short ; gonads restricted to the disc ; usually two or three arm spines on each lateral arm plate, converting to hooks on the more distal segments.

Astroboa nuda (Lyman)

Astrophyton nudum Lyman, 1874, *Bull. Mus. comp. zool. Harv.* **3** : 251–252, pl. 4, figs 4, 5.
Astroboa nuda : Döderlein, 1911, *Abh. bayer. Akad. Wiss.* **2**. Suppl. 5 : 86–88.
Astroboa nigra Döderlein, 1911 : 83–86, pl. 9, fig. 9.
Astroboa nuda var. *nigra* : Döderlein, 1927, *Abh. bayer. Akad. Wiss.* (Math. nat.) **31** (6) : 44 ; J. B. Balinsky, 1957 : 2–3 ; B. I. Balinsky *In* : Macnae & Kalk, 1969 : 99, 106, 130, fig. 28.

D.d. up to *c.* 100 mm ; a.l. up to *c.* 400 mm.

A species of Gorgonocephalidae with branching arms ; disc coated with fine granules, slightly coarser on the radial ribs, superficially appearing naked, no belt of platelets reinforcing the margin interradially ; jaws bearing oral papillae lateral to their apices but not extending around the distal notches of the oral slits to produce a continuous fringe ; arms coated with small flat superficial platelets ; usually four to eight segments between successive forks of the main arm branches ; complete belts of hooks across the arms even at their bases, arm spines only developed after about the fourth fork, except in small specimens (d.d. $< c.$ 20 mm). Colour (forma *nigra*) black above, lighter below.

TYPE LOCALITY. Philippines (*A. nuda*) ; Zanzibar (forma *nigra*).

Southern African record. 26/33/s ; 1–2 metres.

Nature of bottom. Coral.

Genus *ASTROCLADUS* Verrill

See : Döderlein, 1911, *Abh. bayer. Akad. Wiss.* **2**, Suppl. 5 : 40.

A genus of Gorgonocephalidae with branching arms ; disc armed with flat or conical more or less conspicuous tubercles, no belts of marginal platelets ; oral slits armed with papillae all around the edge, even in the distal notches, except in *A. hirtus* ; usually more arm segments before the first fork than between the first and second forks, distally not more than 11 segments between successive forks ; arm spines very small, usually starting after the first fork but sometimes small ones present from the segments bearing the second or third pores.

Astrocladus africanus Mortensen

Astrocladus africanus Mortensen, 1933a : 291–293, fig. 20, pl. 17, figs 1, 2.

D.d. 58 mm (only one specimen recorded).

A species of *Astrocladus* with the centre of the disc and the radial ribs armed with irregular small conical tubercles, otherwise with a smooth coat of minute flat platelets, ventral interradii with numerous fine short spinelets ; arms also coated with flat platelets ; arm forks proximally close together with only seven or eight segments between them ; belts of hooks only complete on the distal branches, though there are scattered hooks proximally ; arm spines starting at or just before the first fork.

Type locality. Unknown.

Astrocladus euryale (Retzius)

Asterias euryale Retzius, 1783, *K. svenska VetenskAkad. Handl.* **4** : 243.
Gorgonocephalus verrucosus : Lyman, 1882 : 262–263 ; Bell, 1905b : 260.
Astrocladus euryale : H. L. Clark, 1923 : 319 ; Mortensen, 1933a : 293–296, figs 21, 22, pl. 18, fig. 7 ; A. M. Clark, 1952 : 199 ; Day, Millard & Harrison, 1952 : 396 ; Morgans, 1962 : 315, 324 ; Day, Field & Penrith, 1970 : 80 ; A. M. Clark, 1974 : 440–441, pl. 3, figs 1, 2.

D.d. up to *c.* 75 mm.

A species of *Astrocladus* with radial ribs armed with isolated moderate to large rounded tubercles, the whole disc otherwise with a smooth coat of minute flat platelets ; arms similarly coated and often also with some median tubercles, arm forks proximally with six to nine segments between them ; complete belts of hooks occurring proximally to a varying extent, in larger specimens at least mounted on transverse series of small plates concealed under the skin ; arm spines starting either at the first fork or earlier.

Type locality. Cape of Good Hope.

Southern African records. 34/18/FB/s ; 34/19/s ; 34/20/s ; 34/21/s ; 34/22/s ; 34/23/s ; 34/25/s ; 33/26/s ; 33/27/s ; 32/28/s ; 13–90 metres.

Nature of bottom. Usually rock ; in one case grey sand.

Astrocladus hirtus Mortensen

Astrocladus hirtus Mortensen, 1933a : 288–290, fig. 19, pl. 19, figs 1–3.
Astrocladus hirtus var. *reticulatus* Mortensen, 1933a : 290–291, pl. 18, figs 5, 6.

D.d. up to *c.* 25 mm.

A species of *Astrocladus* with a finely granular disc coating, except on the radial ribs where it is more papilliform, some coarser papillae scattered sparsely between the radial ribs radially and interradially, ventral interradii with a finer granular coating but some short slender spines towards the ambitus ; oral slits lacking papillae around the distal notches ; arms with fine papilliform coating and some small flat platelets, only three to six segments between most successive forks ; complete belts of hooks from about the sixth fork ; arm spines starting at the second or third fork.

Type locality. Natal 'or perhaps' from Mozambique.

Southern African record. 29/31/s ; 24 metres.

Astroconus capensis Mortensen

Astroconus capensis Mortensen, 1933a : 285–288, fig. 18a–d, pl. 18, figs 3, 4.

D.d. *c.* 30 mm.

A species of Gorgonocephalidae with branching arms ; disc appearing finely granular with some small conical tubercles towards the ambitus but no belt of underlying platelets, radial ribs bearing several somewhat larger conical tubercles of moderate size; oral papillae fringing the distal notches of the oral slits as well as adapically ; about eight or nine arm segments both before the first fork and between the first and second forks, distally up to *c.* 20 segments between successive forks ; hook belts starting from about the third fork ; short arm spines developed from level with the second pair of pores.

Type locality. Off Natal (29/31) ; 420 metres.

Astrothorax papillata (H. L. Clark)

Astrothamnus papillatus H. L. Clark, 1923 : 316–318, pl. 20, figs 5, 6.
Astrothorax papillata : Mortensen, 1933a : 279–280, fig. 15 ; A. M. Clark, 1952 : 199.

D.d. up to 20 mm ; a.l. 100+ mm.

A species of Gorgonocephalidae with long simple arms ; disc covered above with coarse and fine granules intermixed, below with abruptly finer granules ; oral slits armed with numerous teeth and papillae, similarly spiniform ; arms covered with alternating bands of fine and coarse granules, the fine areas with numerous small hooks ; arm spines increasing from two at the level of the second pores to six or seven, their shape changing from thorny-tipped stumps proximally to F-shaped hooks distally.

Type locality. SW. of Cape Hangklip, 110 metres.

Southern African records. 34/14/d ; 34/18/d ; 33/27/s, vd ; 33/28/d ; 30/31/d ; 44–567 metres.

Nature of bottom. Green sand ; green mud.

Gorgonocephalus pectinatus Mortensen

Gorgonocephalus pourtalesi : Bell, 1905b : 259. [Non *Astrophyton pourtalesii* Lyman, 1875.]
Gorgonocephalus chilensis : H. L. Clark, 1923 : 318. [Non *Astrophyton chilense* Philippi, 1858.]
Gorgonocephalus pectinatus Mortensen, 1933a : 281–285, fig. 17, pl. 18, figs 1, 2.

D.d. up to *c.* 55 mm.

A species of Gorgonocephalidae with branching arms ; disc covering rather variable, in some specimens appearing almost naked interradially, in others with more or less numerous tubercles, either conical or almost spine-like, developed especially on the distinct belt of marginal platelets, slightly larger tubercles fairly numerous on the radial ribs ; jaws with papillae restricted to their apices ; only about eight arm segments before the first fork but 10 or 11 between the first and second ones, subsequent branches very variable in length with 10–33 segments between successive forks ; complete belts of hooks developed from about the second fork ; arm spines usually starting level with the second pair of pores. Colour deep pinkish-brown.

TYPE LOCALITY. Off East London, 357 metres.

SOUTHERN AFRICAN RECORDS. 33/17/d ; 34/18/s, d, vd ; 32/28/d ; 79–860 metres.

NATURE OF BOTTOM. Fine sand ; black specks ; rock.

Suborder OPHIOMYXINA Fell

Ophiuroids with a moderately large disc covered with thick skin, sometimes with fine scales embedded but only visible when dried or the skin dissolved, occasionally the scales bearing small spaced spinelets, more often the disc appearing quite naked (distorted or ruptured in many preserved specimens) ; radial shields usually flattened bar-like, separated by the width of the arm base, sometimes more or less reduced, their distal articulation with the genital plates concealed or inconspicuous ; oral shields usually distinct on all the jaw angles distal to the pairs of adoral shields to which they are occasionally fused ; arms simple, moderately long and slender, not markedly attenuated, usually more flexible horizontally than vertically, covered with thick skin obscuring the underlying plates, of which the dorsal ones are very delicate, or fragmented into two or more parts, or even lacking, though the ventral ones are better calcified and entire, though not usually contiguous, lateral plates bearing laterally aligned more or less conspicuous series of spines, visible from above as well as below, only occasionally some of them modified into hooks distally ; tentacle pores relatively large, sometimes armed with a tentacle scale.

Family **OPHIOMYXIDAE** Ljungman

See : Matsumoto, 1917 : 7 ; Spencer & Wright, 1966 : U89.

Diagnosis as for the suborder.

Genus *OPHIOMYXA* Müller & Troschel

See : Lyman, 1882 : 244 ; Mortensen, 1927 : 167.

A genus of Ophiomyxidae with only scattered glassy plates or C-shaped spicules concealed in the skin of the disc ; teeth very broad and flat with serrated edges, oral papillae usually similar to the teeth ; no tentacle scales.

Ophiomyxa bengalensis Koehler

Ophiomyxa bengalensis Koehler, 1897, *Annls Sci. nat.* (8) **4** : 363, pl. 9, figs 70, 71 ; 1922 : 17, pl. 5, figs 5, 6, pl. 92, fig. 1 ; Mortensen, 1933a : 306–309, fig. 31.

D.d. 16 mm ; a.l. 110 mm.

A species of *Ophiomyxa* with the oral shields semicircular except for a very short, broadly truncated, distal lobe ; teeth and the two to four oral papillae each side rather variable in shape, usually pointed and rarely very wide ; dorsal arm plates consisting of some irregular plates concealed in the skin, if present at all ; ventral arm plates better developed, broad with a deep median distal notch ; three slender, very finely serrated spines, the uppermost one stoutest, on most free arm segments.

Type locality. Andaman Is, 316–457 metres.

Southern African record. 29/31/d ; 250–550 metres.

Ophiomyxa tenuispina Mortensen

Ophiomyxa tenuispina Mortensen, 1933a : 304–306, fig. 30, pl. 19, fig. 27.

D.d. 12 mm (only one specimen recorded).

A species of *Ophiomyxa* with the oral shields broad rhombic except that the distal angle is broadly truncated ; teeth and the three oral papillae each side broad and flat, very finely serrated at the edge ; dorsal arm plates fragmented into irregular pieces medially but a larger piece each side (at least in the holotype) ; ventral arm plates distinctly broader than long, with a large median distal notch ; up to four slender pointed similar spines each side of the free arm segments.

Type locality. NE. of East London (32/28), 174 metres.

Ophiomyxa vivipara capensis Mortensen

Ophiomyxa vivipara (part) : Lyman, 1882 : 246 (St. 142 only).
Ophiomyxa vivipara : H. L. Clark, 1923 : 313 ; Mortensen, 1933a : 301–304, figs 27–29.
Ophiomyxa vivipara var. *capensis* Mortensen, 1936 : 242 ; A. M. Clark, 1952 : 199.

D.d. up to *c.* 20 mm ; a.l. up to 90 mm.

A subspecies of *Ophiomyxa* with the oral shields rather variable in form, rounded triangular, with a small distal lobe, or ovate ; teeth and the three to five oral papillae each side flat and serrated ; dorsal arm plates entire but extremely thin and transparent ; ventral arm plates as long as broad or longer, with the middle of the distal edge straight or only very slightly notched ; three or four arm spines each side of most free arm segments, the uppermost one stoutest but sometimes lacking on alternate segments.*

TYPE LOCALITY. Unspecified.

SOUTHERN AFRICAN RECORDS. 31/16/d; 32/16/d; 33/16/d; 34/17/d; 34/18/s, d; 34/17/d; 34/18/d; 33/28/d; 29/31/d; 27/32/d; 129–410 metres.

NATURE OF BOTTOM. Rock; coral and rock; black specks; clay and sand.

Genus *OPHIOSCOLEX* Müller & Troschel

See: Lyman, 1882: 232; Mortensen, 1927: 169.

A genus of Ophiomyxidae with a continuous coat of fine scales in the skin of the disc, usually visible when dried, sometimes armed with scattered spinelets; teeth and oral papillae all narrow, spiniform; tentacle pores sometimes armed with a spiniform scale.

Ophioscolex (Ophiolycus) dentatus Lyman

Ophioscolex dentatus Lyman, 1878, *Bull. Mus. comp. Zool. Harv.* **5**: 157, pl. 7, figs 184–186; 1882: 233, pl. 24, figs 4–6; Bell, 1905b: 259; H. L. Clark, 1923: 314; A. M. Clark, 1952: 199.
Ophioscolex (*Ophiolycus*) *dentatus*: Mortensen, 1933a: 309–312, figs 32–34.
Ophioscolex dentatus var. *spiniger* Mortensen, 1933a: 312–313, fig. 35.

D.d. up to 18 mm.

A species of *Ophioscolex* with the disc scales bare or with scattered small spinelets (forma *spiniger*); oral shields broad rhombic with a very short, broadly truncated, distal lobe; jaws with an apical cluster of similar pointed papillae and/or teeth and distally about four longer papillae, blunt or expanded at the tips, arising close to the second oral tentacle pore which is inclined into the oral slit; dorsal arm plates fragmented irregularly, sometimes split transversely into two main pieces in smaller specimens; ventral arm plates longer than broad with a rounded median distal angle or lobe; three arm spines on most free arm segments, the lowest one broader and flatter than the other two which are modified into double, F-shaped, hooks on the distal segments; one spiniform tentacle scale present. Colour dark red dorsally.

TYPE LOCALITY. Agulhas Bank, 275 metres.

SOUTHERN AFRICAN RECORDS. 31/16/d; 31/17/d; 32/16/d; 32/17/d; 33/16/d; 34/17/d; 34/18/d; 35/18/d; 33/28/d; 29/31/d; 27/32/d; 129–410 metres.

NATURE OF BOTTOM. Rock; polyzoa and rock; brown shell and rock; black specks.

* Mortensen distinguished his variety *capensis* on the grounds that Magellanic specimens of *O. vivipara* have two arm spines developed on the third and next few arm segments, with a single spine only on the first two segments, whereas he thought that South African specimens have a single spine for the first five or six segments. However, a specimen from the west coast of South Africa with d.d. 20 mm has two arm spines on at least one side of the third arm segments and I doubt if this character is significant. Even so, zoogeographically, I think it is likely that there is at least a subspecific difference between specimens from the Falkland–Magellan area and South Africa so I am continuing the use of the name *capensis*.

Ophioscolex (Ophioscolex) inermis Mortensen

Ophioscolex inermis Mortensen, 1933a : 313–315, fig. 36.

D.d. *c.* 8 mm (only one specimen recorded) ; a.l. *c.* 25 mm.

A species of *Ophioscolex* of which the disc covering is unknown ; oral shields relatively small compared with the adorals ; jaws with an apical cluster of small spiniform papillae and/or teeth, with a few similar oral papillae each side but proximal to the second oral tentacle pore which is superficial, not sunk into the slit ; dorsal arm plates extremely thin and transparent, entire or split into two transversely ; ventral arm plates about as long as broad, the distal edge flat or slightly concave ; three arm spines longer than the segment, the lowest one somewhat clavate ; no tentacle scales.

TYPE LOCALITY. Off Durban (29/31), 400 metres, sandy mud.

Suborder GNATHOPHIURINA Matsumoto

Ophiuroids with a moderately large disc covered with usually thin skin in which continuous scaling is either distinct or obscured by a superficial armament of spinelets, thorny stumps or granules, if this armament is dense, the skin rarely thick enough to obscure the scaling or the scaling very reduced ; radial shields flat and triangular, often conspicuous, usually more or less contiguous but sometimes well separated, occasionally obscured by thorny stumps, the distal articulation with the genital plates usually distinct externally without being conspicuous ; oral shields well developed on all the jaw angles ; arms simple, usually fairly slender and more flexible horizontally than vertically (except in a few small species specialized for epizoic life which have downwards flexing arms covered with thick skin as on the disc and dorsal arm plates reduced), dorsal and ventral arm plates usually well developed, successive ones contiguous for most of the arm ; arm spines erect, rarely more than one of them, if any, becoming hooked distally ; tentacle pores usually large and with one or two scales, sometimes none.

Family **OPHIOTRICHIDAE** Ljungman

See : Matsumoto, 1917 : 214 ; Spencer & Wright, 1966 : U102.

A family of Gnathophiurina with disc scales small, similar but for an enlarged central plate in juvenile specimens (d.d. < 2 or 3 mm), usually the dorsal ones obscured by an armament of thorny stumps, spines or granules or a covering of more or less thick skin, rarely naked ; radial shields more or less conspicuous unless covered by thorny stumps ; jaw structure very constant, oral shields broad rhombic with a conspicuous foramen proximal to them between the oral plates, teeth broad rectangular with a compact cluster of small rounded tooth papillae superficial to them, no oral papillae so that the second (oral) tentacle is unguarded ; arms stout or slender, sometimes very long, a.l. $> 10 \times$ d.d. but more often $5-8 \times$ d.d., rarely numbering six (in a fissiparous epizoic species), often more flexible vertically than most non-euryalid ophiuroids, especially in some epizoic species, successive dorsal and ventral arm plates usually contiguous (dorsal plates reduced in epizoic species); arm spines usually more or less serrated and terminally rugose, often glassy; usually only a single inconspicuous tentacle scale, if any.

Genus *MACROPHIOTHRIX* H. L. Clark

See: H. L. Clark, 1938, *Mem. Mus. comp. Zool. Harv.* **55** : 278; A. M. Clark, 1968, *Bull. Br. Mus. nat. Hist.* (Zool.) **16** : 284.

A genus of Ophiotrichidae with species often exceeding 20 mm d.d. and the arms of moderate or considerable length, often > 200 mm, disc soft and puffy, the fine scales covered with thorny stumps, sometimes short spinelets or rugose granules, usually granules also obscuring the radial shields; arms mainly flexible horizontally, the segments relatively broad; dorsal arm plates broad, breadth : length usually 2·0–2·3 : 1 and successive ones broadly contiguous; arm spines long, serrated (or sometimes smooth basally), glassy or sometimes opaque at the tip, especially if clavate; tentacle scale single.

Macrophiothrix aspidota (Müller & Troschel)

Ophiothrix aspidota Müller & Troschel, 1842 : 115.
Ophiothrix longipeda (part): H. L. Clark, 1923 : 340 (Delagoa Bay specimen). [Non *Ophiura longipeda* Lamarck, 1816.]
Macrophiothrix aspidota: H. L. Clark, 1938, *Mem. Mus. comp. Zool. Harv.* **55** : 284; A. M. Clark, 1968, *Bull. Br. Mus. nat. Hist.* (Zool.) **16** : 285–287, figs 3a, 4a, 5a, b, c, 7a.
? *M. aspidota*: J. B. Balinsky, 1957 : 18; Kalk, 1958 : 214; B. I. Balinsky *In*: Macnae & Kalk, 1969 : 99, 101, 106, 129.

D.d. up to *c.* 22 mm; a.l. up to *c.* 175 mm; d.d./a.l. 1/5–10, usually *c.* 1/8.

A species of *Macrophiothrix* with the disc scales covered with short stumps having usually five or six points but the radial shields almost completely naked; arms relatively short (for the genus), less than ten times as long as the d.d.; dorsal arm plates very short and broad, often with breadth : length nearly 2·5 : 1, fan-shaped or sometimes hexagonal, angular laterally but not acutely so; ventral arm plates broad hexagonal, the distal edge straight or slightly concave; up to only *c.* 8 arm spines, none markedly clavate. Colour (Balinsky) purplish-brown, indistinctly variegated but with no longitudinal stripes on the arms.

Type locality. 'Ostindien'.

Southern African record. 26/33/i. This record (the only one of *M. aspidota* from East Africa) depends on J. B. Balinsky's determination and I think needs confirmation. H. L. Clark (1938 : 239), under the heading of *M. longipeda*, mentions some specimens probably from Delagoa Bay with naked radial shields and arms shorter than usual (for *longipeda*, in which they are commonly 15–20 × d.d.).

Nature of bottom. Dead coral debris.

Macrophiothrix demessa (Lyman)

Ophiothrix demessa Lyman, 1861, *Proc. Boston Soc. nat. Hist.* **8** : 82.
Macrophiothrix mossambica J. B. Balinsky, 1957 : 18–20, fig. 7, pl. 3, figs 11, 12 ; Macnae & Kalk, 1969 : 129.
Macrophiothrix demessa : A. M. Clark, 1968, *Bull. Br. Mus. nat. Hist.* (Zool.) **16** : 289–291, figs 3e, f, 4h, 5h, 7e.

D.d. up to *c.* 20 mm ; a.l. up to *c.* 300 mm, d.d./a.l. 1/9–15.

A species of *Macrophiothrix* with the disc scales covered with short stumps ending in three to six points, similar ones extending also on to the radial shields ; arms relatively long ; dorsal arm plates with breadth : length usually *c.* 2 : 1, fan-shaped with blunt lateral angles, bearing small scattered short stumps or rugose granules ; ventral arm plates relatively narrow, not appreciably broader than long ; arm spines up to 14. Colour : 'arms banded with dark purple' (Balinsky, from a dry specimen).

Type locality. Hawaiian Is.

Southern African record. 26/33/i.

Macrophiothrix hirsuta cheneyi (Lyman)

Ophiothrix cheneyi Lyman, 1861, *Proc. Boston Soc. nat. Hist.* **8** : 84.
Ophiothrix aristulata (part) : Bell, 1905b : 258 ('Mossel Bay' sample).
? *Ophiothrix longipeda* (part) : H. L. Clark, 1923 : 340.
Macrophiothrix brevipeda H. L. Clark, 1938, *Mem. Mus. comp. Zool. Harv.* **55** : 290–292, fig. 20 ; 1939 : 91.
Macrophiothrix hirsuta : J. B. Balinsky, 1957 : 17–18 ; Kalk, 1958 : 207, 214 ; Macnae & Kalk, 1962 : 118 ; B. I. Balinsky *In* : Macnae & Kalk, 1969 : 99, 101.
Macrophiothrix hirsuta cheneyi : A. M. Clark, 1968, *Bull. Br. Mus. nat. Hist.* (Zool.) **16** : 296–298, figs 3k, 4n, 5n, 7j.

D.d. up to *c.* 20 mm ; d.d./a.l. 1/5–10, usually *c.* 1/7·5, though Balinsky's 1/9 may be a better figure based on fresh material.

A species of *Macrophiothrix* with the disc scales densely covered with short stumps ending in numerous points but only rugose grains on the radial shields ; dorsal arm plates relatively broad, breadth : length usually > 2 : 1, hexagonal or fan-shaped but the median part of the distal edge usually straight, slightly carinate, at least near the base of the arm, textured with very fine rugosities ; ventral arm plates broader than long, the distal side distinctly concave ; up to *c.* 10 arm spines, the middle ones somewhat clavate. Colour dark purple, arms with a median longitudinal yellow line bordered by two dark blue lines dorsally, often a distinct light median ventral line as well.

Type locality. Zanzibar.

Southern African records. 31/30/s ; 30/30/s ; 29/31/s ; 26/32/i ; 26/33/i ; 24/35/i ; 0–90 metres. [Although some of Bell's specimens are labelled as from Mossel Bay, I think they were far more likely to have been from Natal.]

Nature of bottom. Dead coral blocks ; coarse sand and pebbles ; stones.

Macrophiothrix longipeda (Lamarck)

Ophiura longipeda Lamarck, 1816 : 544.
Macrophiothrix longipeda : J. B. Balinsky, 1957 : 17 ; A. M. Clark, 1968, *Bull. Br. Mus. nat. Hist.* (Zool.) **16** : 300–302, figs 3m–o, 4p–r, 5p–r, 7l, m ; Macnae & Kalk, 1969 : 129. [Non *Ophiothrix longipeda* : H. L. Clark, 1923 : 340.]

D.d. up to 37 mm ; a.l. up to 625 mm ; d.d./a.l. 1/15–20.

A species of *Macrophiothrix* with the disc scales covered with cylindrical stumps, usually more than twice as long as broad but occasionally nearly granuliform, like the armament covering the radial shields ; arms very long ; dorsal arm plates not very broad, breadth : length *c.* 2 : 1, flat, trapezoidal with markedly acute latero-distal angles ; ventral arm plates somewhat broader than long, but not markedly so ; up to *c.* 10 arm spines, not clavate, though the middle ones usually somewhat truncated. Colour : disc and arms with scattered dark spots, especially along the proximal and distal edges of the dorsal arm plates.

TYPE LOCALITY. 'L'océan austral, près de l'Ile-de-France' (Mauritius).

SOUTHERN AFRICAN RECORD. 26/33/i. [The specimen from SE. Africa which I mentioned in 1968 was supposedly (but almost certainly erroneously) from Port Elizabeth. It was purchased from Mr Sowerby at the same time as other material from Mauritius (as well as the West Indies), which I think is significant.]

Ophiocnemis marmorata (Lamarck)

Ophiura marmorata Lamarck, 1816 : 543.
Ophiocnemis marmorata : H. L. Clark, 1923 : 341 ; Day, 1974 : 94.

D.d. up to *c.* 20 mm ; d.d./a.l. *c.* 1/6.

A species of Ophiotrichidae with the disc scaling restricted to narrow bands between the huge triangular bare radial shields, the bands only a little broader interradially than radially, armed on the dorsal side with slightly spaced coarse granules but these elongate at the ambitus becoming conical, the scaling ending abruptly just below the ambitus, the disc bare below ; arms markedly carinate, dorsal arm plates very broad, tent-like ; ventral arm plates much broader than long, with both proximal and distal edges concave, successive ones not contiguous ; up to five smooth tapering arm spines ; no tentacle scales on the conspicuously large pores. Colour marbled dark and light.

TYPE LOCALITY. Unknown.

SOUTHERN AFRICAN RECORDS. 29/31/s ; 23/35/s ; 6–57 metres. [The Natal square, based on sts NAD 8 and 63 (29°54'S 31°05'E and 29°22'S 31°36'E), provides the first positive record from South African waters.]

NATURE OF BOTTOM. Clay, muddy sand and shell (elsewhere from sand).

Genus *OPHIOGYMNA* Ljungman

See : Koehler, 1922 : 284.

A genus of Ophiotrichidae with the disc soft and puffy, scales and most of the radial shields covered and obscured by thick skin, scales bearing thorny stumps or spinelets; arms long and fairly flexible dorsoventrally, at least in the distal part, arm length : disc diameter often > 10 : 1; skin also covering or separating the dorsal arm plates, which may be somewhat fragmented; arm spines slender and opaque, fairly smooth except towards their tips; a tentacle scale only present proximally, if at all.

Ophiogymna capensis (Lütken)

Ophiothrix capensis Lütken, 1869, *K. dansk. Vidensk. Selsk. Skr.* **5** (8) : 59, 100; H. L. Clark, 1923 : 340.
Ophiothela capensis: Mortensen, 1933a : 340–341, figs 52b, 53b, pl. 19, fig. 26.

[Size of two syntypes (and only known specimens) not given.]

A species of *Ophiogymna* with the disc armed with slender spines between the large radial shields; dorsal arm plates more or less fragmented but obscured by the skin; arm spines up to 8, long and thin; no tentacle scales. Colour: a conspicuous dark line midradially between each pair of radial shields extending on to about the first five arm segments and ending abruptly, also transverse dark bands across every third to sixth arm segment dorsally, extending on to the uppermost spines.

Type locality. 'Cape'.

Nature of bottom. Epizoic on gorgonians.

Ophiogymna fulgens (Koehler)

Ophiothrix fulgens Koehler, 1905, *Siboga Exped.* **45b** : 107–109, pl. 10, figs 3–6.
Ophiogymna fulgens: Koehler, 1922 : 288–292, pl. 42, figs 1–8, pl. 43, figs 9, 10, pl. 44, fig. 8, pl. 60, fig. 6, pl. 103, fig. 8; Mortensen, 1933a : 338–340, figs 52a, 53a, pl. 19, fig. 25.
Placophiothrix phrixa H. L. Clark, 1939 : 88–90, figs 39, 40.

D.d. up to 12 mm.

A species of *Ophiogymna* with the disc armament rather variable, ranging from slender spines to almost granuliform stumps; dorsal arm plates mostly entire, slightly broader than long, trapezoidal but with the proximal edge concave and the distal convex, the lateral angles blunt, contoured with a median distal swelling but sunken proximally so not really carinate, successive ones not contiguous but separated narrowly by skin, as are the ventral arm plates; up to seven slender arm spines; a very small tentacle scale sometimes present on the proximal pores only.

Type locality. Paternoster Is, Lesser Sunda Is, Indonesia, 36 metres.

Southern African record. 29/31/d; 123 metres.

Genus *OPHIOTHELA* Verrill

See: Lyman, 1882 : 230.

A genus of Ophiotrichidae with the whole body covered with thick skin obscuring the underlying plates; disc coated with a very variable armament of granules, tubercles or short spinelets, rarely lacking altogether; arms relatively short, with a marked degree of dorsoventral flexure correlated with an epizoic habit (usually on gorgonians), appearing very convex dorsally since the arm spines are restricted to the ventral and lateral surfaces and the uppermost ones are reduced in size, dorsal arm plates lacking or rudimentary and restricted to a few basal segments, the arms often coated with granules above, at least proximally ; ventral arm plates sometimes evident through the skin ; arm spines mostly short, opaque and finely rugose ; a reduced tentacle scale or, more often, none.

Ophiothela danae Verrill

Ophiothela danae Verrill, 1869, *Proc. Boston Soc. nat. Hist.* **12** : 391 ; Matsumoto, 1917 : 230–232, fig. 67 ; Koehler, 1922 : 297–298, pl. 59, figs 1–3, pl. 103, fig. 1 ; Mortensen, 1933a : 342 ; A. M. Clark & Rowe, 1971 : 116, pl. 14, fig. 5 ; A. M. Clark, 1974 : 470.

? *Ophiothela dividua* von Martens, 1879, *Sber. Ges. naturf. Freunde Berlin* **1879** : 127–130, figs 1–4 ; H. L. Clark, 1923 : 343 ; J. B. Balinsky, 1957 : 22 ; Macnae & Kalk, 1969 : 129.

D.d. up to *c.* 7 mm, usually much less, 1–3 mm ; d.d./a.l. 1/3–5.

A fissiparous, commonly six-armed species of *Ophiothela* with the disc armament more often of rounded tubercles and granules than of spinelets, which are usually central, when present, occasionally more or less bare dorsally ; arms similarly covered dorsally with crowded or spaced granules and/or tubercles, sometimes reduced distally or even throughout, often displaying a segmental arrangement ; up to six arm spines, usually about four, blunt and distinctly rugose. Colour very variable, from black to white, usually patterned, southern African specimens described as having blue linear markings or reddish patches.

Type locality. Fiji Is.

Southern African records. 30/30/s ; 29/31/s, d ; 26/32/s ; 26/33/s ; 24/35/i ; 23/35/s [? 33/25 = 26/32] ; 0–220 metres.

Nature of bottom. Epizoic on gorgonians.

Ophiothela nuda (H. L. Clark)

Ophiopsammium nudum H. L. Clark, 1923 : 341–342.

Ophioteresis beauforti Engel, 1949 : 140–143, figs 1, 2.

Ophiothela beauforti : Macnae & Kalk, 1962 : 114, 117 ; J. B. Balinsky, 1957 : 22–24, pl. 4, fig. 16 ; Macnae & Kalk, 1969 : 129 ; A. M. Clark & Rowe, 1971 : 117.

Ophiothela nuda : A. M. Clark, 1974 : 469 ; Day, 1974 : 94.

D.d. up to *c.* 10 mm, d.d./a.l. 1/4–5.

A species of *Ophiothela* with normally five arms ; disc generally armed in the interradial ambital areas and below with rugose spinelets and dorsally with granules, tubercles or spinelets in various densities and combinations, sometimes lacking altogether ; arms dorsally with a similarly variable coating of granules ; up to eight finely rugose arm spines, usually about six. Colour dark brownish in life (Balinsky), variously patterned in preserved specimens.

Type locality. SE. of Tongaat River, Natal, 66 metres.

Southern African records. 30/30/s ; 29/31/s ; 26/33/i ; 24/34/s ; 23/35/s ; intertidal–66 metres.

Nature of bottom. Sand and shell ; muddy sand and shell ; shelly sand and rock ; fine grey sand and rock ; stones ; *Cymodocea* ; coral.

Genus *OPHIOTHRIX* Müller & Troschel

See : A. M. Clark, 1967, *Ann. Mag. nat. Hist.* **9** : 637.

A genus of Ophiotrichidae with the disc scaling more or less obscured by spines, spinelets or thorny stumps, often also with short stumps extending on to the radial shields ; arms mainly flexible horizontally, dorsoventral capability more limited ; dorsal arm plates rhombic or fan-shaped and not broadly contiguous in the subgenus *Ophiothrix* but hexagonal or trapezoidal and often more extensively contiguous in the subgenera *Keystonea*, *Placophiothrix* (unrepresented) and *Acanthophiothrix* (though in the last the plates are rarely much broader than long, if at all) ; arm spines usually long, glassy, more or less serrated and tapering ; tentacle scale single.

Ophiothrix (Acanthophiothrix) proteus Koehler

Ophiothrix proteus Koehler, 1905, *Siboga Exped.* **45b** : 100–101 ; 1922 : 260–261, pl. 36, figs 3, 4, pl. 101, fig. 3 ; Day, 1974 : 94.
Placophiothrix proteus : H. L. Clark, 1939 : 86 ; J. B. Balinsky, 1957 : 21 ; Macnae & Kalk, 1969 : 130.
Ophiothrix (*Acanthophiothrix*) *proteus* : A. M. Clark, 1974 : 465–466, fig. 11a, b.

D.d. up to *c.* 17 mm ; d.d./a.l. *c.* 1/10.

A species of *Ophiothrix* with the disc scales armed with trifid stumps and often also some scattered slender serrated spinelets or spines, the radial shields bare ; arms slender and relatively long, dorsal arm plates ranging from longer than broad in small specimens (d.d. < *c.* 7 mm) to broader than long in large ones (d.d. > *c.* 12 mm), hexagonal in shape, successive ones contiguous by the breadth of the median truncated part of the distal lobe, bearing a more or less sharply defined longitudinal crest ; ventral arm plates with the distal side straight or slightly convex ; arm spines very long and needle-like, up to six times the segment length, with many fine serrations. Colour brownish with a median yellow line along the arms (Balinsky) bordered with a pair of dark lines (green or purple – at least in preserved specimens).

Type locality. East Indies and/or Ceylon.

Southern African records. 30/30/s ; 29/31/s ; 26/33/s ; 25/34/d ; 23/35/s ; 'infralittoral' – 112 metres.

Nature of bottom. Mud ; green mud ; muddy sand and shell ; dark sandy mud ; stones ; *Cymodocea* beds.

Ophiothrix (Ophiothrix) aristulata Lyman

Ophiothrix aristulata Lyman, 1879, *Bull. Mus. comp. Zool. Harv.* **6** : 50–51, pl. 15, figs 421–424 ; 1882 : 223–224, pl. 21, figs 9–12 ; H. L. Clark, 1923 : 336–337 ; Mortensen, 1933a : 336–337 ; A. M. Clark, 1952 : 200. [Non *O. aristulata* : Bell, 1905b, nec Döderlein, 1910, nec Hertz, 1927a]

Placophiothrix aristulata : H. L. Clark, 1939 : 85.
Ophiothrix (*Ophiothrix*) *aristulata* : A. M. Clark, 1974 : 466–467, fig. 11c, d.

D.d. up to *c.* 16 mm ; d.d./a.l. *c.* 1/9.

A species of *Ophiothrix* with the disc scales armed with scattered slender serrated spines or spinelets (and sometimes also with trifid stumps around the ambitus, at least in smaller specimens (d.d. < *c.* 7 mm), radial shields bare ; arms fairly slender, dorsal arm plates ranging from slightly broader than long in small specimens to almost twice as broad as long in the largest ones, basically rhombic in shape with a rounded distal angle but becoming approximately hexagonal though still rounded distally, contiguous for only a third of their breadth at most, each plate medially flat proximally but swollen distally, lacking a continuous crest ; ventral arm plates broader than long with the distal side slightly convex ; arm spines up to ten, the longest ones about four to four and a half times the segment length. Colour of disc grey, arms pink with a pair of darker lines defining a light longitudinal median line.

TYPE LOCALITY. South of Cape Point, 275 metres.

SOUTHERN AFRICAN RECORDS. 30/15/d ; 31/15/d ; 31/16/d ; 32/16/d ; 32/17/d ; 34/17/d ; 34/18/d ; 35/18/d ; 36/21/d ; 200–420 metres.

NATURE OF BOTTOM. Stones ; 'crl' (? coral) and rock ; black specks ; grey sand and rock ; grey sand.

Ophiothrix (Ophiothrix) echinotecta Balinsky

Ophiothrix echinotecta J. B. Balinsky, 1957 : 16–17, fig. 6, pl. 3, figs 9, 10 ; Kalk, 1958 : 198 ; B. I. Balinsky *In* : Macnae & Kalk, 1969 : 99, 106, 129 ; A. M. Clark & Rowe, 1971 : 109.

D.d. up to 8 mm ; d.d./a.l. 1/5–8.

A species of *Ophiothrix* with the disc completely covered by multifid stumps (with at least five terminal points), less dense on the radial shields ; dorsal arm plates rhombic but broadest towards the distal end, the distal side often three-lobed, a small rugose granule present at the narrow junction point between many successive plates ; ventral arm plates broader than long, the distal side straight ; arm spines up to 10, the longest about three times the segment length. Colour unusually constant, dark or light green and white, dorsal arm plates transversely banded with a light chevron near the distal edge, sometimes brownish.

TYPE LOCALITY. Inhaca I., Delagoa Bay, intertidal.

SOUTHERN AFRICAN RECORDS. 26/33/i ; 24/35/i.

NATURE OF BOTTOM. In rock hollows under the echinoids *Echinometra* and *Stomopneustes*.

Ophiothrix (Ophiothrix) foveolata Marktanner-Turneretscher

Ophiothrix foveolata Marktanner-Turneretscher, 1887, *Annln naturh. Mus. Wien* **2** : 313, pl. 13, figs 32, 33 ; Koehler, 1922 : 238–239, pl. 47, figs 4–7, pl. 98, fig. 6 ; A. M. Clark & Rowe, 1971 : 110, pl. 15, fig. 3 ; Day, 1974 : 94.

Ophiothrix poecilodisca H. L. Clark, 1915, *Mem. Mus. comp. Zool. Harv.* **25** : 276, pl. 13, fig. 5 ; 1923 : 341.

Placophiothrix foveolata : J. B. Balinsky, 1957 : 20, pl. 4, fig. 15 ; Kalk, 1958 : 207, 214 : Macnae & Kalk, 1962 : 111, 114 ; B. I. Balinsky *In* : Macnae & Kalk, 1969 : 102, 106, 130.

D.d. up to 13 mm ; d.d./a.l. 1/4–5.

A species of *Ophiothrix* with the disc scales sparsely covered with slender spines, radial shields large and bare ; dorsal arm plates fan-shaped, not broadly contiguous ; ventral arm plates with the distal side concave ; up to eight rather smooth, glassy arm spines, the longest four or five times the segment length. Colour ranging from orange or light brown to violet, usually purplish, bright red in younger specimens, disc patterned with dark lines, especially along the adradial edge of the radial shields so as to emphasize their shape, also a conspicuous dark line across the distal edge of each arm segment, at least dorsally.

TYPE LOCALITY. Aru Is, East Indies.

SOUTHERN AFRICAN RECORDS. 26/32/i ; 26/33/i ; 24/34/s ; 23/35/s ; intertidal –50 metres.

NATURE OF BOTTOM. Coral reef, under dead coral blocks and in *Cymodocea* beds.

Ophiothrix (Ophiothrix) fragilis (Abildgaard *In* : O. F. Müller)

Asterias fragilis Abildgaard *In* : O. F. Müller, 1789, *Zoologica Danica*. Ed. 3. Hauniae. **3** : 28, pl. 98.
Ophiothrix fragilis : J. Müller & Troschel, 1842 : 110.

Forma *pentaphylla* :
Asterias pentaphylla : Pennant, 1777, *British Zoology*. Ed. 4. London. **4** : 64.*
Ophiothrix fragilis var. *pentaphyllum* : Koehler, 1908 : 635 ; 1923 : 9.
Ophiothrix aristulata : Döderlein, 1910 : 254, pl. 5, figs 4, 5 ; Hertz, 1927a : 35. [Non *O. aristulata* Lyman, 1879.]
Ophiothrix fragilis : H. L. Clark, 1923 : 337 ; Mortensen, 1933a : 338 ; Koehler, 1914 : 209–210 ; Madsen, 1971 : 213–214, fig. 36c ; A. M. Clark, 1974 : 467–469.

Forma *triglochis* :
Ophiothrix triglochis Müller & Troschel, 1842 : 114 ; Lyman, 1882 : 218 ; Koehler, 1904a, *Mém. Soc. zool. Fr.* **17** : 81–84, figs 41–45 ; Bell, 1905b : 259 ; Koehler, 1908 : 635 ; 1923 : 8 ; H. L. Clark, 1923 : 337–339 ; Mortensen, 1933a : 337–338 ; Stephenson, Stephenson & du Toit, 1937 ; 380 ; Bright, 1937 : 63 ; Stephenson, Stephenson & Bright, 1938 : 18 ; Eyre, 1939 : 304 ; Stephenson, 1944 : 317, 347 ; A. M. Clark, 1952 : 201 ; Day, Millard & Harrison, 1952 : 396 ; Macnae, 1957 : 362 ; Day, 1959 : 502, 544 ; Morgans, 1959 : 414, 422, 425 ; 1962 : 303, 308, 322 ; Day, Field & Penrith, 1970 : 80 ; Penrith & Kensley, 1970 : 201, 206, 208, 234.
Ophiothrix roseo-coerulans : Bell, 1905b : 258. [Non *O. roseocoerulans* Grube, 1868.]

D.d. up to *c.* 20 mm, usually *c.* 10–12 mm ; d.d./a.l. 1/5–7.

A very variable species of *Ophiothrix* with the disc either armed primarily with rugose spines and some trifid stumps obscuring the scaling but leaving bare the more or less large radial shields (forma *pentaphylla*) or with stumps alone and these extending on to and obscuring the radial shields (forma *triglochis*) or with different combinations or degrees of these characters ; arms fairly stout, dorsal arm plates

* Having proposed to the International Commission on Zoological Nomenclature in 1967 (*Bull. zool. Nom.* **24**(1) : 44–45) that *Asterias pentaphylla* Pennant, 1777 be suppressed in order to validate the more familiar *A. fragilis* Abildgaard for the type species of *Ophiothrix*, it would perhaps be more consistent if I were to abandon the usage of *pentaphylla* for the British and South African form of *Ophiothrix fragilis* with radial shields naked and disc scales bearing spines in favour of the long-forgotten name *rosula* Fleming, 1828. However, the Commission does not recognize infrasubspecific categories such as form or variety and the rules are not extended to cover them. Accordingly there seems no reason to introduce a further change at this level and we may yet have to revert to *pentaphylla* as the specific name should the Commission veto my proposal.

about half again as broad as long, successive ones only narrowly contiguous, shape rhombic, often with a pronounced median distal beak-like angle and usually distinctly carinate ; ventral arm plates with distal side concave ; arm spines up to 10, usually seven or eight. Colour almost limitlessly variable.

TYPE LOCALITY. Denmark (*O. fragilis*) ; Cornwall, England (forma *pentaphylla*), Port Natal (forma *triglochis*).

SOUTHERN AFRICAN RECORDS. Forma *pentaphylla* : 26/15/s ; 32/17/i ; 32/18/i, s; 33/17/i–d ; 34/18/s, d ; 34/18/FB/s ; 34/21/s ; 33/25/s ; 34/25/s ; 33/27/s ; 32/28/s ; 29/31/s ; intertidal–148 metres.

Forma *triglochis* : 28/16/– ; 30/17/i ; 31/17/i ; 33/17/s, d ; 33/18/i, s ; 34/18/i, s ; 34/18/FB/i, s ; 34/19/i, s ; 34/20/i ; 34/21/i, s ; 34/22/i, s ; 34/23/s ; 34/25/s, d ; 33/25/i, s ; 33/26/s ; 33/27/s ; 32/28/i, s ; 30/30/s ; 29/31/s, d ; 28/32/– ; intertidal –183 metres.

NATURE OF BOTTOM. Most often rock ; also various combinations and grades of sand and shell ; rarely green or black mud.

Ophiothrix (Ophiothrix) trilineata Lütken

Ophiothrix trilineata Lütken, 1869, *K. dansk. Vidensk. Selsk. Skr.* **5** (8) : 58–59, 100 ; H. L. Clark, 1923 : 341 ; A. M. Clark & Rowe, 1971 : 110–111, pl. 15, fig. 2.
Placophiothrix trilineata : J. B. Balinsky, 1957 : 21 ; Kalk, 1958 : 207, 216 ; Macnae & Kalk, 1969 : 130.

D.d. up to 10 mm ; d.d./a.l. 1/5–7.

A species of *Ophiothrix* with the disc scales sparsely armed with slender spines, radial shields large and bare ; dorsal arm plates rhombic in smaller specimens (d.d. < 7 mm) but the median distal lobe becomes truncated in larger specimens where the successive plates may be contiguous for as much as a third their total breadth ; ventral arm plates with distal side slightly concave ; arm spines up to seven, the longest about five times the segment length. Colour generally appearing dark purplish (Balinsky), arms dorsally with three longitudinal yellow lines separated by two narrow black lines.

TYPE LOCALITY. Samoa.

SOUTHERN AFRICAN RECORD. 26/33/i.

NATURE OF BOTTOM. Coral debris.

Ophiothrix (Keystonea) propinqua Lyman

Ophiothrix propinqua Lyman, 1861, *Proc. Boston Soc. nat. Hist.* **8** : 83 ; Koehler, 1922 : 256–257, pl. 38, figs 1, 2, pl. 101, fig. 4.
Ophiotrichoides propinqua : H. L. Clark, 1939 : 90–91 ; J. B. Balinsky, 1957 : 21–22 ; Kalk, 1958 : 216 ; B. I. Balinsky *In* : Macnae & Kalk, 1969 ; 106, 129.
Ophiothrix (*Keystonea*) *propinqua* : A. M. Clark, 1968, *Bull. Br. Mus. nat. Hist.* (Zool.) **16** : 283–284, fig. 2e ; A. M. Clark & Rowe, 1971 : 107, pl. 15, fig. 7.

D.d. up to *c.* 12 mm ; d.d./a.l. 1/9–11.

A species of *Ophiothrix* with the dorsal disc scales either quite bare or with some inconspicuous granuliform armament centrally, slightly obscured by thickening skin together with the bare radial shields in larger specimens, the ambital and ventral scales always with short stumps or spinelets ; arms relatively long (superficially resembling *Macrophiothrix* but for the smooth disc), dorsal arm plates broad fan-shaped but for the distal side which is flattened or trilobed, successive ones contiguous for more than a third of their breadth ; ventral arm plates broader than long, the distal side concave ; arm spines up to seven, the longer ones terminally truncated and somewhat opaque, about three times the segment length. Colour very variable (Balinsky) greenish, yellowish or brownish, occasionally bright red, variegated with white, arms more or less distinctly banded.

Type locality. Kingsmills (Gilbert) Is, Pacific.

Southern African record. 26/33/i, s.

Nature of bottom. Coral debris.

Family **AMPHIURIDAE** Ljungman

With subfamily **AMPHILEPIDINAE** Matsumoto

See : Mortensen, 1927 : 206 ; Spencer & Wright, 1966 : U100 ; A. M. Clark, 1970 : 1.

A family of Gnathophiurina with disc scaling usually well developed and distinct, often with a primary rosette of five radial plates around a central one, occasionally the disc armed with small spines and obscured by thickened skin or the scaling reduced to patches of very small scales adjoining the radial shields which are then bar-shaped rather than wedge-shaped ; apical jaw structure consisting of teeth which are usually broad rectangular but (Amphilepidinae) may be somewhat tapering, with a usually symmetrical pair of oral papillae superficial to them (or slightly offset in the Amphilepidinae), jaws otherwise armed with one, two or three superficial oral papillae distal to the infradental one, often also with an oral tentacle scale or additional papilla (sometimes two) inset into the oral slit on the side of the oral plate dorsal to the second oral tentacle ; arms usually moderately stout, not particularly long, l/d.d. usually less than 8/1, arm plates well developed ; arm spines opaque, usually mostly tapering and superficially appearing smooth ; one or two usually rounded tentacle scales, sometimes none.

Subfamily **AMPHILEPIDINAE**

Amphilepis scutata Mortensen

Amphilepis scutata Mortensen, 1933a : 372–373, fig. 76 ; A. M. Clark, 1974 : 464.

D.d. up to 5 mm ; d.d./a.l. *c.* 1/9.

A species of Amphiuridae with the disc scaling fairly coarse, the rosette, or at least the central plate, usually conspicuous, radial shields relatively large, their length about half the disc radius or more, completely separated though approximating distally, disc fully scaled below ; jaws thin (dorsoventrally) in comparison to the Amphiurinae, the teeth not very broad and the innermost oral papillae slightly offset laterally, not strictly infradental in position, in series with the scale of the first oral tentacle, which is usually very broad and superficial, not inset into the slit, the second oral tentacle also superficial and completely lacking any associated scales or papillae arising from near the edge of the adoral shields, as occur in *Amphiura* and *Amphioplus*, the oral formula (see A. M. Clark, 1970 : 5) being expressable as ,mt/o,o ; dorsal arm plates rounded triangular with the proximal angle truncated, not quite contiguous, ventral arm plates triangular with the distal side slightly convex ; only three arm spines, slender and pointed ; no tentacle scales.

Type locality. Off Durban, 410 metres.

Southern African records. 29/31/d; 200 (? 175) – 410 metres.

Nature of bottom. Sandy mud.

Subfamily **AMPHIURINAE**

Amphilycus scripta (Koehler)

Amphiura scripta Koehler, 1904a, *Mém. Soc. zool. Fr.* **17** : 70–71, figs 23, 24.
Amphilycus androphorus Mortensen, 1933c : *Vidensk. Meddr dansk naturh. Foren.* **93** : 185–188, figs 4–6 ; J. B. Balinsky, 1957 : 11 ; Macnae & Kalk, 1962 : 115, 118 ; B. I. Balinsky *In* : Macnae & Kalk, 1969 : 99, 106, 129.
Amphilycus scripta : A. M. Clark & Rowe, 1971 : 103, fig. 34a.

D.d. up to *c.* 5 mm ; d.d./a.l. *c.* 1/4.

A species of Amphiuridae which is sexually dimorphic, the male dwarfed and epizoic on the underside of the female, mouth to mouth, arms alternating ; disc scaling coarse, rosette normally conspicuous, radial shields relatively large, length about half the disc radius, almost completely contiguous, disc fully scaled below ; jaws modified in the female, sunken proximal to the oral shields which are rhombic but relatively very small, adorals larger and very broadly contiguous interradially, teeth broad and rounded but the infradental papillae usually more or less asymmetrical, no distal papilla superficial to the second oral tentacle but a continuous flange present along the edge of the oral plate which may represent a distal papilla fused with the oral tentacle scale so the oral formula could be expressed as m, $\overline{\text{tm}}$, ; dorsal arm plates broad rounded triangular, distal edge straight, barely, if at all, contiguous, even basally, ventral arm plates pentagonal with distal side convex ; up to five fairly long cylindrical spines on the proximal arm segments, near the arm tips three spines in each series modified into large proximally directed hooks ; one rounded tentacle scale. Colour pale greyish-brown.

Type locality. Oman.

Southern African records. 26/33/i ; 25/32/i.

Nature of bottom. Sand ; epizoic on underside of cake-urchins *Echinodiscus auritus* and *E. bisperforatus*.

Amphiodia sp. aff. ***A. microplax*** Burfield

See : Burfield, 1924, *Ann. Mag. nat. Hist.* (9) **13** : 146–148, figs 1, 2.
? *Amphiodia* sp. J. B. Balinsky, 1957 : 10.
Amphiodia sp. aff. *A. microplax* : A. M. Clark, 1974 : 461.

A species of Amphiuridae the disc of which is unknown ; oral shields longer than broad, jaws armed with three oral papillae each side in continuous series, the outermost not enlarged, no oral tentacle scale, oral formula m, mm, o–t ; three arm spines ; one tentacle scale.

SOUTHERN AFRICAN RECORDS. 29/31/s ; 38–57 metres. [? 26/33/i.]

NATURE OF BOTTOM. Coarse sand and shell ; coarse muddy sand and shell.

Genus ***AMPHIOPLUS*** Verrill

See : A. M. Clark, 1970 : 36.

A genus of Amphiuridae with the disc usually completely scaled but lacking in armament, original (as opposed to regenerated) discs often with a distinct rosette of primary plates, radial shields contiguous at their distal ends or for more or less of their length, rarely completely separated ; jaws armed with usually four oral papillae each side, sometimes slightly spaced or in a concave series incapable of closing the oral slit, and with a first oral tentacle scale present in the slit (subgenus *Amphioplus*), sometimes with only three papillae and an oral tentacle scale more or less in sequence with them (subgenus *Unioplus*) or with four superficial papillae in a straight row, the third enlarged, and no distinct oral tentacle scale (subgenus *Lymanella*) ; one or two tentacle scales, rarely absent or rudimentary.

Amphioplus (Amphioplus) pectinatus Mortensen

Amphioplus pectinatus Mortensen, 1933a : 367–368, fig. 72.
Amphioplus (*Amphioplus*) *pectinatus* : A. M. Clark, 1974 : 456–459, fig. 8.

D.d. up to *c.* 8 mm ; arms relatively long.

A species of *Amphioplus* with the rosette sometimes distinct, the disc fully scaled below, radial shields contiguous near or at their distal ends, the plate just distal to and below each radial shield bearing a spiniform 'comb' ; oral shields spearhead-shaped, truncated distally, adorals contiguous interradially to some extent, jaws armed with four oral papillae, with a space between the infradental one and the second revealing the oral tentacle scale which may be more or less in sequence with the papillae and also may be duplicated, the third oral papilla slightly enlarged, oral formula m,om$\overline{N,N}$m + t or m,omm,m + 2t ; dorsal arm plates rounded-triangular or almost elliptical, possibly sometimes hexagonal, contiguous medially, ventral arm plates pentagonal or squarish if the proximal angle is broadly truncated ; three to six tapering arm spines, usually up to four or five ; two tentacle scales.

TYPE LOCALITY. Off Durban, 410 metres.

SOUTHERN AFRICAN RECORDS. 29/31/s, d ; 77–410 metres.

NATURE OF BOTTOM. Mud ; green mud.

Amphioplus (Lymanella) furcatus Mortensen

? *Ophiophragmus gibbosus* Ljungman, 1867b, *Ofvers. K. VetenskAkad. Förh. Stockh.* **23** : 316.
Amphiura incana : Bell, 1905b : 258. [Non *A. incana* Lyman, 1879.]
Amphioplus furcatus Mortensen, 1933a : 370–372, fig. 75.
Amphioplus (Lymanella) furcatus : A. M. Clark, 1974 : 452–453.

D.d. up to *c.* 5 mm ; d.d./a.l. *c.* 1/8.

A species of *Amphioplus* with the rosette distinct on original discs, disc scaling fine, 11–13 scales on a line between the radial shields interradially in larger specimens (d.d. > 4 mm), the uppermost row of ventral scales along the ambitus with small spinose projections, radial shields almost fully contiguous, in length equal to about half the disc radius, length : breadth 2·3–2·6 : 1 ; oral shields spearhead-shaped, longer than broad, adorals triangular, just meeting interradially ; four oral papillae in a straight row, the third largest, formula m,mM (or m),m–t ; dorsal arm plates rounded triangular, barely contiguous, ventral arm plates broad pentagonal ; three slender pointed arm spines, about equal to the segment length ; two tentacle scales, the one on the ventral plate very large.

Type locality. Off Tugela River mouth, Natal, 46 metres.

Southern African records. 29/31/s ; 33–46 metres.

Nature of bottom. Black mud ; dark green mud.

Amphioplus (Lymanella) integer (Ljungman)

Amphipholis integra Ljungman, 1867b, *Ofvers. K. VetenskAkad. Förh. Stockh.* **23** : 313.
? *Ophiophragmus gibbosus* Ljungman, 1867b : 316.
Amphiura integra : Koehler, 1904a, *Mem. Soc. zool. Fr.* **17** : 65–66, figs 16, 17.
Amphioplus integer : H. L. Clark, 1923 : 330–331 ; Mortensen, 1933a : 368–370, figs 73, 74 ; Stephenson, Stephenson & du Toit, 1937 : 380 ; J. B. Balinsky, 1957 : 11 ; Macnae & Kalk, 1962 : 104, 107 ; B. I. Balinsky *In* : Macnae & Kalk, 1969 : 106, 129 ; Day, Field & Penrith, 1970 : 80 ; Day, 1974 : 54, 94.
Amphioplus hastatus : Day & Morgans, 1956 : 308 ; Day, 1974 : 94. [Both fide A. M. C., MS.] [Non *Amphipholis hastata* Ljungman, 1867.]
Amphioplus (Lymanella) integer : A. M. Clark & Rowe, 1971 : 102–103 ; A. M. Clark, 1974 : 453–455, fig. 6.

D.d. up to 5·5 mm ; d.d./a.l. *c.* 1/4.

A species of *Amphioplus* with the rosette distinct on original discs, the scaling moderately coarse, 7–11 scales on a line between the radial shields interradially, the ambitus smooth, radial shields almost fully contiguous, in length less than half the disc radius, proportions rather variable, length : breadth 1·6–2·5 : 1 (mean 2·2 : 1) in South African specimens, or 2·3–3·7 : 1 (mean 2·8 : 1) in those from Mozambique ; oral shields spearhead-shaped, longer than broad but with distal lobe short, adorals triangular, just meeting interradially, four oral papillae in a straight line, the third slightly enlarged, formula m,$\overline{\text{mN,N}}$m–t ; dorsal arm plates broad rounded triangular, narrowly contiguous, ventral arm plates broad pentagonal ; three fairly stout blunt arm spines, the upper one slightly flattened and the middle one easily exceeding the segment in length, especially in specimens from the False Bay area ; two large tentacle scales. Colour grey to dirty white (Balinsky).

TYPE LOCALITY. Port Natal (Durban).

SOUTHERN AFRICAN RECORDS. 32/16–17/s; 32/18/s; 33/18/s; 34/18/FB/i, s; 34/19/–; 34/20/–; 34/21/i, s; 34/22/s; 34/23/i, s; 33/25/s; 33/27/s; 32/28/i, s; 29/31/s; 26/33/i; 23/35/i; intertidal–62 metres.

NATURE OF BOTTOM. Rock; coarse sand and rock; rock and black mud; green sand and shell; black mud and rock.

Amphioplus (Unioplus) falcatus Mortensen

Amphioplus falcatus Mortensen, 1933a: 365–367, figs 70, 71, pl. 19, figs 18, 19.
Amphioplus (*Unioplus*) *falcatus*: A. M. Clark, 1974: 455–456, fig. 7.

D.d. up to 10 mm; d.d./a.l. *c.* 1/10.

A species of *Amphioplus* without a conspicuous rosette among the coarse scales, no spinose projections on the uppermost ventral scales; radial shields very long, in length more than half the disc radius, curved sickle- or banana-shaped and only their distal extremities just contiguous; oral shields triangular or with a small distal lobe, usually about as broad as long, adorals meeting broadly interradially, only three oral papillae usually present and an oral tentacle scale, the second from outermost papilla enlarged, formula m,oM,m+t (or m,om,m+t); dorsal arm plates fan-shaped or broad hexagonal with a slight median distal angle, ventral arm plates bell-shaped, broadest distally; three pointed arm spines, the middle one slightly hooked at the tip; one fairly elongate tentacle scale.

TYPE LOCALITY. Off Durban, 410 metres.

SOUTHERN AFRICAN RECORDS. 29/31/d; 138–410 metres.

NATURE OF BOTTOM. Usually sandy mud (once coarse sand and coral).

Amphioplus sp. aff. ***A. falcatus*** Mortensen*

Amphioplus sp. indet. A (aff. *A. falcatus*): A. M. Clark, 1974: 459–461, fig. 9.

D.d. up to *c.* 10 mm; arms relatively long.

A species of *Amphioplus* the disc of which is unknown; oral shields variable in shape, hexagonal, octagonal, pentagonal or circular, the inwardly abbreviated adorals widely separated from each other interradially by the extent of the broadly truncated end of the proximal lobe of the oral shields, four oral papillae of which the two middle ones are blunt spiniform or finger-shaped, oral formula m,mm,m+t; dorsal arm plates broad elliptical but for lateral angles, ventral arm plates squarish to hexagonal; four slightly flattened blunt-tipped arm spines for most of the arm, longer than the segments; no proper tentacle scales but one or two rudimentary accretions may be distinguishable on some pores; terminal arm segments relatively short and broad with the lowest of the three spines relatively long, twice the segment length.

* Of the southern African species of *Amphioplus*; in fact, the affinity is not close.

SOUTHERN AFRICAN RECORDS. 34/19/s ; 34/22/s ; 33/25/s ; 29/31/s ; 7–57 metres.

NATURE OF BOTTOM. Usually sand (once muddy sand and shell).

Genus *AMPHIPHOLIS* Ljungman

See : A. M. Clark, 1970 : 28.

A genus of Amphiuridae with the disc usually completely scaled, lacking armament, rarely the scaling reduced on the ventral side, rosette often distinct, radial shields usually contiguous for more than half their length ; jaws armed with three oral papillae each side in a continuous series, the outermost one extremely broad and opercular, no oral tentacle scale visible, formula $m,m\overline{N,N}-t$; usually two tentacle scales, sometimes one, rarely none.

Amphipholis similis Mortensen

Amphipholis similis Mortensen, 1933a : 363–364, fig. 69 ; A. M. Clark, 1974 : 450, fig. 5a.

D.d. up to only 3 mm ; d.d./al. *c.* 1/5.

A species of *Amphipholis* with the rosette distinct among the coarse disc scales, numerous scales ventrally, radial shields almost fully contiguous, in length about one-third the disc radius, fairly narrow, length : breadth *c.* 2 : 1 ; oral shields rhombic with rounded angles and distal lobe slightly shorter, about as long as broad, adorals meeting fairly widely interradially ; three tapering arm spines, all shorter than the segment ; two fairly large tentacle scales.

TYPE LOCALITY. Off Durban, 64 metres.

SOUTHERN AFRICAN RECORDS. 32/21/s ; 34/22/s ; 29/31/s, d ; 64–138 metres.

NATURE OF BOTTOM. Sand, shells ; sand and stone ; khaki sand.

Amphipholis squamata (Delle Chiaje)

Asterias squamata Delle Chiaje, 1828, *Memorie sulla storia e notomia degli animali senza vertebre del Regno di Napoli.* Naples. **3** : 74.
Amphiura kinbergi Ljungman, 1871, *Ofvers K. VetenskAkad. Förh. Stockh.* **28** : 646.
Amphiura squamata : Lyman, 1882 : 136.
Ophiactis minor Döderlein, 1910 : 253, pl. 5, fig. 3.
Amphipholis squamata : H. L. Clark, 1923 : 330 ; Mortensen, 1927 : 221–222, fig. 125 ; 1933a : 364–365 ; Stephenson, Stephenson & du Toit, 1937 : 380 ; Bright, 1937 : 63 ; Eyre, Broekhuysen & Crichton, 1938 : 110 ; Eyre, 1939 : 305 ; A. M. Clark, 1952 : 200 ; Day, Millard & Harrison, 1952 : 396 ; Macnae, 1957 : 362 ; J. B. Balinsky, 1957 : 10–11 ; Kalk, 1958 : 200, 207, 215, 237 ; Day, 1959 : 544 ; Grindley & Kensley, 1966 : 12 ; B. I. Balinsky *In* : Macnae & Kalk, 1969 : 106, 129 ; A. M. Clark, 1970 : 30–31 ; Day, Field & Penrith, 1970 : 81 ; Penrith & Kensley, 1970 : 208, 234 ; Day, 1974 : 94.
Amphipholis minor : Hertz, 1927a : 35.
Amphioplus squamata : Macnae & Kalk, 1962 : 111 [lapsus].

D.d. up to *c.* 5 mm ; d.d./a.l. 1/4–5.

A species of *Amphipholis* with the rosette only distinct among the moderately fine disc scales in young specimens (d.d. < *c.* 2 mm), numerous scales ventrally, radial shields almost fully contiguous, their length about one-third the disc radius, fairly narrow, length : breadth *c.* 2 : 1 ; oral shields rhombic, broader than long, adorals meeting fairly widely interradially ; dorsal arm plates rounded triangular, just contiguous except in small specimens, ventral arm plates pentagonal ; up to four short tapering arm spines ; two tentacle scales of moderate size. Colour greyish, a bright spot ringed by a darker area marking the distal ends of each pair of radial shields.

TYPE LOCALITY. Naples.

SOUTHERN AFRICAN RECORDS. 26/15/i ; 29/17/i ; 32/18/s ; 33/17/s, d ; 33/18/i, s ; 34/18/i, s, d ; 34/18/FB/i, s ; 34/19/i ; 34/21/i, s ; 34/22/i, s, d ; 34/23/i, s, d ; 34/25/d ; 33/25/i ; 33/27/s ; 32/28/i ; 31/29/s ; 30/31/s ; 29/31/d ; intertidal–172 metres.

NATURE OF BOTTOM. Rock ; rock, sand and shell ; sand with rocky patches ; coarse sand and rock ; coarse and fine shell ; coarse yellow-khaki sand ; smooth loose *Zonaria* ; dark mud and sand ; sand ; greenish-black mud ; khaki mud.

Amphipholis strata Mortensen

Amphipholis strata Mortensen, 1933a : 361–363, fig. 68, pl. 19, fig. 20 ; A. M. Clark, 1974 : 450–452, fig. 5b–d.

D.d. up to 8·5 mm ; d.d./a.l. 1/3–4.

A species of *Amphipholis* with the rosette distinct among the disc scales which are relatively coarse, especially the uppermost ventral ones which form a regular row of no more than nine squarish scales across each interradius, radial shields almost fully contiguous, in length about two-fifths the disc radius, broad D-shaped, length : breadth *c.* 1·5 : 1 ; oral shields rhombic, about as long as broad, adorals meeting fairly widely interradially ; dorsal arm plates rounded triangular or fan-shaped, contiguous, ventral arm plates broad pentagonal ; three arm spines, the middle one longest, cylindrical and blunt or even slightly clavate ; two very large tentacle scales. Disc pale, arms grey.

TYPE LOCALITY. Off Cape Point, 55 metres.

SOUTHERN AFRICAN RECORDS. 34/18/s ; 34/18/FB/s ; 34/22/s ; 34/25/s ; 29–85 metres.

NATURE OF BOTTOM. Sand ; coarse sand and rock ; sand, shells and *Phyllochaetopterus* ; rock and sand patches ; rock, sand and shell ; rock ; limestone.

Genus ***AMPHIURA*** Forbes

See : Matsumoto, 1917 : 194 ; A. M. Clark, 1970 : 7.

A genus of Amphiuridae with the disc usually fully scaled but without armament, sometimes the scaling reduced on the ventral side, rarely absent also dorsally except around the radial shields (subgenus *Fellaria*) which approximate and are often just contiguous distally and usually have l : br *c.* 3 : 1 ; jaws armed only with an infradental pair of papillae and one distal oral papilla each side, rarely two, arising from the oral plate where it meets the adoral shield, separated by a wide diastema in which the oral tentacle scale can be seen inset into the oral slit, formula usually m,$\overline{\text{on}}$,$\overline{\text{n}}$+t ; arms of moderate length, usually about five to seven times the d.d. except in species with very reduced disc scaling which tend to have relatively longer arms ; one or more of the middle arm spines often with specialized tips ; one or two, sometimes no tentacle scales.

I cannot account for the nomen nudum *Amphiura delagoa* (J.B.) Balinsky in Macnae & Kalk, 1962 : 104. The locality fits neither *Ophiactis delagoa* Balinsky nor *Amphiura inhacensis* Balinsky.

Amphiura (Amphiura) acutisquama A. M. Clark

Amphiura acutisquama A. M. Clark, 1952 : 200, 213–215, fig. 1.

D.d. *c.* 15 mm (only one specimen recorded) ; a.l. 100+ mm.

A species of *Amphiura* with the disc covered above and below with fine pointed scales, radial shields completely separated by several rows of scales, l : br *c.* 3 : 1 ; oral shields spearhead-shaped, nearly as broad as long, adorals not quite meeting interradially, distal oral papilla single, massive, usually with a double apex ; dorsal arm plates with a median distal angle, ventral arm plates truncated pentagonal ; up to five tapering pointed arm spines ; two tentacle scales of moderate size.

Type locality. West coast of South Africa, details unknown ; Cape Town University records give 32/16/d.

Amphiura (Amphiura) albella Mortensen

Amphiura albella Mortensen, 1933a : 359–361, fig. 67 ; A. M. Clark, 1974 : 444.

D.d. up to 6 mm ; d.d./a.l. *c.* 1/6.

A species of *Amphiura* with the disc fully scaled above, the primary rosette either distinct or not, the ventral interradii proximally bare ; oral shields variable, approximately spearhead-shaped with a more or less prolonged proximal angle but the distal lobe truncated, about as broad as long ; adorals meeting interradially, distal oral papilla single, broad rounded leaf-shaped ; dorsal arm plates broad fan-shaped, barely contiguous, ventral arm plates truncated pentagonal ; up to five tapering arm spines ; two tentacle scales of moderate size, at least on the proximal pores, distally there may be only one.

Type locality. Off Durban, 412 metres.

Southern African records. 30/31/vd ; 29/31/d ; 412–930 metres.

Nature of bottom. Soft mud and clay.

Amphiura (Amphiura) atlantica Ljungman

Amphiura atlantica Ljungman, 1867b, *Ofvers. K. VetenskAkad. Förh. Stockh.* **23** : 321 ; Mortensen, 1933b : 449–451, figs 17, 18 ; Madsen, 1971 : 181–182, fig. 15.

Amphiura dilatata Lyman, 1879, *Bull. Mus. comp. Zool. Harv.* **6** : 26, pl. 11, figs 314–316 ; 1882 : 135, pl. 29, figs 4–6 ; H. L. Clark, 1923 : 326–327.
Amphiura atlantica var. *dilatata* : Mortensen, 1933a : 351–353, figs 59, 60b ; A. M. Clark, 1952 : 200 ; Day, Field & Penrith, 1970 : 81.

D.d. up to 7 mm ; d.d./a.l. *c.* 1/5.

A species of *Amphiura* with the disc covered above with scales of moderate size, the primary rosette sometimes distinct, usually completely naked below, rarely with fine scaling ; oral shields variable in shape, the proximal lobe broadly rounded or flattened, distal lobe sometimes truncated, about as long as broad, adorals either meeting or just separated interradially, distal oral papilla single, spiniform ; dorsal arm plates fan-shaped or transverse oval, but usually with lateral angles, ventral arm plates relatively narrow truncated pentagonal ; up to six arm spines, the second from lowest more or less expanded axe-shaped at the tip ; a rudimentary tentacle scale sometimes present but usually totally lacking. Colour : arms uniformly orange.

Type locality. St Helena.

Southern African records. 31/17/d ; 32/16/d ; 32/17/d ; 33/18/d ; 34/17/d ; 34/18/d ; 34/18/FB/s ; 36/19/d ; 36/21/d ; 34/21/s ; 35/22/d ; 34/22/s ; 34/23/s, d ; 34/26/d ; 67–400 metres.

Nature of bottom. Usually khaki or green sand and/or mud ; in one case coarse sand and broken shell.

Amphiura (Amphiura) candida Ljungman

Amphiura candida Ljungman, 1867b, *Ofvers. K. VetenskAkad. Förh. Stockh.* **23** : 318–319 ; Mortensen, 1933a : 361 ; A. M. Clark, 1974 : 445, fig. 2. [Non *A. candida* : Koehler, 1904.]
Amphiura kalki : J. B. Balinsky, 1957 : 3–5, fig. 1, pl. 1, figs 1, 2 ; Macnae & Kalk, 1962 : 104, 114 ; B. I. Balinsky *In* : Macnae & Kalk, 1969 : 106 ; A. M. Clark *In* : Clark & Rowe, 1971 : 80, 97 ; Day, 1974 : 94.

D.d. up to 6 mm ; d.d./a.l. *c.* 1/5.

A species of *Amphiura* with the disc covered with fine, smooth scales on both sides, the primary rosette usually distinct ; oral shields rounded hexagonal, the proximal lobe truncated but the distal one more rounded, as long as broad or longer, adorals separated interradially by the width of the truncated end of the oral shield, distal oral papilla single, thick and blunt ; dorsal arm plates broad fan-shaped, ventral arm plates truncated pentagonal or rectangular, the distal edge straight ; up to eight arm spines all slightly exceeding the segment in length, the middle ones truncated with a very small distal hook ; two tentacle scales of moderate size. Colour : upper side of arms with bright orange-red midline.

Type locality. Mozambique.

Southern African records. 26/33/i, s ; 23/35/s ; intertidal–9 metres.

Nature of bottom. Sandy patches on mud flats with tubes of *Mesochaetopterus* ; sand ; muddy ; sand and shells.

Amphiura (Amphiura) capensis Ljungman

Amphiura capensis Ljungman, 1867b, *Ofvers. K. VetenskAkad. Förh. Stockh.* **23** : 320 ; Lyman, 1882 : 129, pl. 18, figs 14–16 ; Koehler, 1908 : 634 ; Döderlein, 1910 : 253–254, pl. 5, fig. 2 ; Koehler, 1914 : 190 ; H. L. Clark, 1923 : 327 ; Mortensen, 1933a : 348–350 ; Stephenson, Stephenson & du Toit, 1937 : 380 ; Bright, 1937 : 63, 76, 86, 87 ; Eyre, 1939 : 304 ; A. M. Clark, 1952 : 200 ; Day, 1959 : 544 ; Grindley & Kensley, 1966 : 13 ; Day, Field & Penrith, 1970 : 81 ; Penrith & Kensley, 1970 : 234 ; A. M. Clark, 1974 : 445–447.

? *Amphiura angularis* : H. L. Clark, 1923 : 327–328 ; Mortensen, 1933a : 354 ; J. B. Balinsky, 1957 : 5 ; Macnae & Kalk, 1969 : 129. [Non *A. angularis* Lyman, 1879.]

Amphiura adjecta Mortensen, 1933a : 355–357, fig. 62.

Amphiura compressa Mortensen, 1933a : 357–358, figs 63, 64.

D.d. up to 11 mm ; d.d./a.l. *c.* 1/6–7.

A species of *Amphiura* with the disc covered above with scales which appear relatively coarse in smaller specimens (d.d. < 5 mm), the primary rosette often distinct, the ventral scaling usually showing some degree of reduction proximally, occasionally even a patch of naked skin, radial shields markedly divergent inwardly, just contiguous distally, l : br *c.* 2·5 : 1 ; oral shields broad spearhead-shaped, as long as or longer than broad, the proximal and shorter distal lobes with rounded median angles, adorals only meeting interradially in the smallest specimens, usually separate, distal oral papilla single, conical or flattened, rarely spiniform ; dorsal arm plates rounded fan-shaped or transverse oval, ventral arm plates truncated pentagonal, hardly broader than long, even in the largest specimens ; up to seven arm spines, the upper ones more or less flattened and often spatulate or even axe-shaped at the tip, especially in larger specimens ; one rounded tentacle scale. Colour : disc greyish, sometimes nearly black, arms yellow or orange, sometimes with dark spots in indistinct bands.

TYPE LOCALITY. Near Port Natal (Durban) and the Cape of Good Hope.

SOUTHERN AFRICAN RECORDS. 26/15/i ; 27/15/s ; 28/16/s ; 29/16/i ; 29/17/i ; 30/17/i ; 31/17/i ; 31/18/i ; 32/17/i ; 32/18/s ; 33/17/s ; 33/18/i, s ; 34/18/i, s, d ; 34/18/FB/i, s ; 34/19/i ; 34/20/i ; 34/21/i ; 34/22/i, s ; 34/23/i ; 34/24/i ; 34/25/s, d ; 33/25/s ; 29/31/s ; (? 26/33) ; intertidal–180 metres.

NATURE OF BOTTOM. Under stones intertidally ; usually rock or rock and sand ; occasionally khaki sand and mud.

Amphiura (Amphiura) grandisquama natalensis Mortensen

Amphiura grandisquama var. *natalensis* Mortensen, 1933a : 353–354, fig. 60a.

Amphiura grandisquama natalensis : A. M. Clark, 1974 : 447–448, fig. 3.

D.d. not known to exceed 4 mm.

A subspecies of *Amphiura* with the disc fully scaled, coarsely so above ; oral shields triangular or rhombic, a little broader than long, adorals nearly or quite contiguous interradially, distal oral papilla single, unremarkable ; dorsal arm plates rounded fan-shaped, ventral arm plates truncated pentagonal ; up to five arm spines, the lowest one becoming markedly longer than the rest, up to three times the segment length, sometimes clavate or slightly curved ; tentacle scale single, relatively large, completely covering the pore.

Type locality. Off Durban, 412 metres.

Southern African records. 29/31/d ; 370–412 metres.

Nature of bottom. Green sand and mud.

Amphiura (Amphiura) incana Lyman

Amphiura incana Lyman, 1879 : 20, pl. 11, figs 285–287 ; 1882 : 128, pl. 33, figs 5–7 ; H. L. Clark, 1923 : 328–329 ; Hertz, 1927a : 34, pl. 7, fig. 1 ; Mortensen, 1933a : 351, fig. 60c ; A. M. Clark, 1952 : 200 ; Morgans, 1962 : 308, 309, 310, 312, 313, 315, 322 ; Day, Field & Penrith, 1970 : 81 ; Madsen, 1971 : 173–177, figs 8–10. [Non *A. incana* : Bell, 1905b, = *Amphioplus furcatus*.]

D.d. up to 8 mm ; d.d./a.l. *c.* 1/10.

A species of *Amphiura* with the disc covered above and below by coarse scales ; oral shields very variable, ranging from rhombic to nearly circular, usually sunken in the middle, about as long as broad, adorals separated or just meeting interradially, distal oral papilla single, short and very broad except in small specimens, d.d. $<$ 4 mm ; dorsal arm plates rounded fan-shaped or transverse oval, ventral arm plates pentagonal, the distal edge distinctly concave ; up to eight short thick blunt arm spines, shorter than the segment, somewhat flattened ; tentacle scales two, small or moderate in size. Colour : disc grey, arms with a pink, orange or red longitudinal stripe.

Type locality. Simon's Bay, 18–36 metres.

Southern African records. 32/18/s ; 34/18/FB/s ; 34/21/s ; 34/23/s, d ; 34/25/s ; 33/25/s ; 33/27/s ; 30/31/s ; 10–110 metres.

Nature of bottom. Usually sand and/or rock, sometimes green or black mud.

Amphiura (Amphiura) inhacensis J. B. Balinsky

Amphiura inhacensis J. B. Balinsky, 1957 : 5–7, fig. 2, pl. 4, figs 13, 14 ; Macnae & Kalk, 1969 : 129.

D.d. up to 5 mm.

A species of *Amphiura* with the disc covered above and below with fine scales, delicate on the upper side medially ; oral shields broad spearhead-shaped, broader than long, the proximal angle rounded, adorals widely separated interradially, distal oral papilla single, peg-like, blunt ; dorsal arm plates truncated rounded fan-shaped, ventral arm plates rounded pentagonal ; up to five, sometimes six, bluntly pointed arm spines, the lowest longest and slightly exceeding the segment ; tentacle scale single, rounded.

Type locality. Inhaca I., Delagoa Bay (26/33), reef.

Amphiura (Amphiura) linearis Mortensen

Amphiura linearis Mortensen, 1933a : 354–355, fig. 61.

D.d. up to 3·5 mm (only two specimens recorded).

A species of *Amphiura** with the disc covered above and below with fine scales, radial shields very long and narrow, l : br 5+ : 1, almost parallel and widely separated ; oral shields broad spearhead-shaped with the proximal lobe rounded and longer than the distal, as broad as long, adorals just meeting interradially, distal oral papilla single, pointed ; dorsal arm plates narrow fan-shaped, longer than broad, ventral arm plates narrow truncated pentagonal ; up to six flattened arm spines, the lowest longest and markedly exceeding the segment in length ; tentacle scale single, moderately large.

Type locality. Off Durban (30/31), 90–165 metres.

Amphiura (Amphiura) simonsi A. M. Clark

Amphiura simonsi A. M. Clark, 1952 : 215–217, fig. 2 ; Morgans, 1962 : 322 ; A. M. Clark, 1974 : 448–449, fig. 4.

D.d. up to *c.* 7 mm ; d.d./a.l. probably *c.* 1/10.

A species of *Amphiura* with the disc covered above with fine scales but (probably) naked below, easily lost, the radial shields relatively small, l : br 2·4–3·0 : 1 ; oral shields relatively small, very broad rounded triangular, the distal side flattened (except for the madreporite), adorals relatively large, with broad distal lobes between the oral shield and the first lateral arm plate each side and broadly contiguous interradially, two† short conical or blunted distal oral papillae each side, the outer one probably not a true papilla but a calcified extension of the rim of the second oral tentacle pore ; dorsal arm plates deep fan-shaped, broadest near the proximal end, ventral arm plates blunt rectangular ; up to five arm spines, the lowest one flattened or cylindrical but tapering to a blunt point, the second developing a conspicuous, distally curved and sometimes barbed, hyaline hook beyond the base of the arm, the spines above it more or less flattened and paddle-shaped ; tentacle scale single, fairly small, often absent from several of the basal pores. Colour : arms pale orange, barred.

Type locality. Simon's Bay, 27–28 metres.

Southern African records. 34/18/FB/s ; 34/21/s ; 34/25/d ; 33/25/s ; 29/31/s ; 27–110 metres.

Nature of bottom. Usually sand and shell, shingle or gravel ; rarely mud.

* As mentioned in 1974 (p. 475) *Amphiura linearis* should be compared with *Ophiopsila bispinosa*; the small size and broken arms of the type material would explain the absence of the sword-like inner tentacle scale characteristic of *Ophiopsila*.

† On further consideration, the oral frame from False Bay which in 1974 I designated '? *Amphiura* sp. indet. C' is probably referable to *A. simonsi* in spite of the single distal oral papilla; the loss of all but the arm bases accounts for the lack of the hooked spines.

Amphiura (Amphiura) uncinata Koehler

Amphiura uncinata Koehler, 1904b, *Siboga Exped.* **45a** : 76–77, pl. 14, figs 3, 4 ; 1922 : 160, pl. 65, figs 6–8, pl. 96, fig. 4 ; Mortensen, 1933a : 358–359, figs 65, 66 ; H. L. Clark, 1939 : 58.

D.d. up to 10 mm ; arms very long, probably d.d./a.l. 1/10+.

A species of *Amphiura* with the disc fully scaled above but with only scattered scales in the skin of the proximal parts of the ventral interradii ; oral shields spear-head-shaped, longer than broad, the proximal lobe longer than the distal, adorals meeting interradially, distal oral papilla single, short but broad, semicircular or conical ; dorsal arm plates broad fan-shaped, ventral arm plates broad pentagonal ; arm spines six proximally, the middle ones with a hyaline, distally curved, hooked tip, even on the first free arm segments ; tentacle scales two, of moderate size.

TYPE LOCALITY. East of Java, East Indies, 250–330 metres.

SOUTHERN AFRICAN RECORDS. 29/31/d, 350*–412 metres.

NATURE OF BOTTOM. Sandy mud.

Amphiura spp.

Amphiura spp. A. M. Clark, 1974 : 449–450.

At least two further species, probably of *Amphiura*, occur in South African waters, as mentioned in my 1974 paper (pp. 449–450). These can be distinguished as follows :

Amphiura sp. A. Disc probably naked below ; radial shields relatively long, distally contiguous ; oral shields rhombic, adorals just separated interradially, distal oral papillae two each side and spiniform, formula m,om,m+t ; dorsal arm plates almost circular ; arm spines four, stout and blunt ; tentacle scale single, poorly calcified. Locality : 34/25, 108 metres.

Amphiura sp. B. Disc unknown ; oral shields pentagonal, proximal lobe angular, the two distal angles rounded, adorals broadly contiguous interradially, distal oral papillae two each side, rounded, the inner one larger, formula m,$o\overline{n},\overline{n}$m+t ; dorsal arm plates broad oval ; arm spines three, slender, acute ; tentacle scale single. Locality : 29/31, 118 metres.

As noted above (p. 157) sp. C of 1974 is probably conspecific with *Amphiura simonsi*.

Amphiura (Fellaria) africana (J. B. Balinsky)†

Ophionephthys africana J. B. Balinsky, 1957 : 7–9, fig. 3, pl. 1, figs 3, 4 ; Macnae & Kalk, 1962 : 107 ; B. I. Balinsky *In* : Macnae & Kalk, 1969 : 101, 106, 129.

Amphiura (*Fellaria*) *africana* : A. M. Clark, 1970 : 18 ; A. M. Clark & Rowe, 1971 : 80, 95.

D.d. up to 10 mm ; arms very long, d.d./a.l. probably 1/10+.

* This depth is that of 'Anton Bruun' st. 390–E, where an additional specimen was collected: 29°42′S 31°38′E.

† As noted in 1971, I think that this nominal species will prove to be synonymous with *A.* (*Fellaria*) *heptacantha* (Mortensen, 1940) from the northern Gulf of Oman.

A species of *Amphiura* (subgenus *Fellaria*) with the disc very fragile, completely naked except for a few scales around the bar-like, almost parallel, narrowly separated radial shields, l : br *c.* 5 : 1 ; oral shields very broad oval, twice as broad as long, adorals widely separated interradially, distal oral papilla single, thick and rounded ; dorsal arm plates narrow rectangular but with rounded distal angles, barely contiguous, if at all, ventral arm plates similarly narrow, longer than broad ; arm spines up to seven, the lowest exceeding the segment in length, the rest shorter, all flattened and expanded and almost digitate at the tip ; tentacle scales lacking. Colour : dull brown, the ventral arm plates darker.

TYPE LOCALITY. Inhaca I., Delagoa Bay (26/33), infralittoral fringe.

NATURE OF BOTTOM. Blue-grey mud.

Ophiocentrus dilatatus (Koehler, 1905)

Ophiocnida dilatata Koehler, 1905, *Siboga Exped.* **45b** : 30–31, pl. 12, figs 2–4.
Ophiocentrus dilatatus : J. B. Balinsky, 1957 : 7 ; Macnae & Kalk, 1962 : 115 ; 1969 : 129 ; A. M. Clark & Rowe, 1971 : 80, 94.

D.d. up to 15 mm ; arms long, d.d./a.l. probably 1/10+.

A species of Amphiuridae with the disc large and puffy, the scaling on both sides obscured by thick skin bearing scattered spinelets, radial shields relatively small, contiguous distally, l : br *c.* 2·5 : 1 ; oral shields triangular with rounded angles or broad oval, broader than long, adorals separate or only narrowly contiguous interradially, distal oral papilla single, broad, flaring towards its free edge ; dorsal arm plates broad oval but the distal edge almost straight, ventral arm plates rounded rectangular, barely contiguous ; up to seven spines on the broadest part of the arms beyond the base, their tips broad, rounded and somewhat rugose ; tentacle scales lacking. Colour : disc violet and orange, arms buff with mauve lines and patches.

TYPE LOCALITY. Aru Is, East Indies, 13 metres.

SOUTHERN AFRICAN RECORD. 26/33/i.

NATURE OF BOTTOM. Sand.

Ophionephthys lowelli A. M. Clark

Ophionephthys lowelli A. M. Clark, 1974 : 462–464, fig. 10.

D.d. up to *c.* 8 mm ; arms very long, d.d./a.l. probably 1/10+.

A species of Amphiuridae with the disc fragile, covered above with extremely fine, probably continuous scaling except towards the periphery where it stops short at a tangential row of large scales extending interradially from the middle of the abradial edge of each radial shield, ambital and ventral areas of disc naked, radial shields relatively large, in length more than half the disc radius, l : br nearly 4 : 1, contiguous distally for nearly half their length ; oral shields rhombic or inverted triangular if the proximal angle is truncated, as long as broad or broader, adorals more or less widely separated interradially, two spiniform, rugose-tipped oral papillae both on the oral plate and in series with the infradental papilla, formula m,mm,o+t; dorsal arm plates bucket-shaped, widest proximally, about as broad as long, ventral arm plates similar in shape; arm spines up to five proximally, somewhat flattened and blunt-tipped, the lowest sometimes exceeding the segment in length; tentacle scale single, moderate, longer than broad.

TYPE LOCALITY. NE. of East London, 55 metres.

SOUTHERN AFRICAN RECORDS. 33/27/s; 32/28/s; 51–55 metres.

NATURE OF BOTTOM. Brown sand and shell.

Paracrocnida sacensis* (J. B. Balinsky)

Ophiophragmus sacensis J. B. Balinsky, 1957 : 9–10, fig. 4, pl. 2, figs 5, 6; Macnae & Kalk, 1962 : 107; B. I. Balinsky *In*: Macnae & Kalk, 1969 : 101, 106, 129.
Paracrocnida sacensis: A. M. Clark & Rowe, 1971 : 82, 100.

D.d. up to 6 mm; d.d./a.l. 1/7–8.

A species of Amphiuridae with the disc covered above with fairly large coarse scales changing abruptly at the ambitus to much smaller thick scales which cover the ventral side, giving an irregular outline, rosette conspicuous, compact, radial shields short and broad, almost completely contiguous; oral shields rhombic, proximal and distal angles about equal, adorals almost meeting inwardly, oral papillae totalling four each side, the third the largest (Balinsky describes three but his figure shows a fourth papilla arising from the first lateral arm plate or partly from the adoral shield); dorsal arm plates kidney-shaped but with laterodistal angles, the distal edge markedly concave medially; ventral arm plates broad rectangular, also concave distally but convex proximally, not contiguous; arm spines seven, the middle four broad and flattened; tentacle scales two, relatively large. Colour brownish-grey.

TYPE LOCALITY. Inhaca, Delagoa Bay (26/33), infralittoral fringe.

NATURE OF BOTTOM. Blue-grey mud.

Family **OPHIACTIDAE** Matsumoto

See: Mortensen, 1927 : 198; Spencer & Wright, 1966 : U100.

A family of Gnathophiurina with disc scaling well developed, often showing a primary rosette in the centre, scales often armed with scattered spinelets, radial shields triangular, contiguous or approximating distally, size varying from conspicuous to little larger than the adjacent scales; jaws armed apically with broad rounded or rectangular teeth and a single broad blunt superficial oral papilla, rarely absent, separated by a diastema (in which an elongated oral tentacle scale within the slit can be seen) from the one, sometimes two (or rarely more) distal

* As noted in 1971, I think that this nominal species may prove to be synonymous with *P. persica* Mortensen, 1940, from the Persian Gulf and Red Sea (if that can itself be distinguished from *P. sinensis* (A. H. Clark, 1917) from the China Sea) if the number of arm spines can exceed seven in larger specimens.

oral papillae each side ; arms usually moderately stout and not particularly long, a.l./d.d. rarely more than 8/1, arms occasionally numbering six in fissiparous species which divide in half across the disc and regenerate ; arm spines usually fairly stout, opaque, superficially smooth and tapering or sometimes rugose and blunt ; tentacle scale normally single, large and rounded.

Genus *OPHIACTIS* Lütken

See : Mortensen, 1927 : 199.

Diagnosis as for the family, but not more than two distal oral papillae.

Ophiactis abyssicola (M. Sars)

Amphiura abyssicola M. Sars, 1861, *Oversigt af norges Echinodermer*. Christiania. p. 18, pl. 2, figs 7–12.

Ophiactis abyssicola : H. L. Clark, 1923 : 334–335 ; Mortensen, 1927 : 202–203, fig. 114 ; 1933a : 347.

D.d. up to 9 mm ; d.d./a.l. *c.* 1/6.

A species of *Ophiactis* with five arms ; disc covered above and below with scales, the upper ones usually large and small intermixed and armed with relatively large scattered spinelets, radial shields conspicuous, separated by a single row of large scales ; oral shields broad rhombic, adorals broadly contiguous interradially, two rounded distal oral papillae each side ; dorsal arm plates fan-shaped, not broadly contiguous, ventral arm plates bell-shaped, broadest distally ; arm spines four proximally, slender.

TYPE LOCALITY. Norway.

SOUTHERN AFRICAN RECORDS. 31/16/d ; 33/17/d ; 34/17/vd ; 167–1646 (? 1830) metres.

NATURE OF BOTTOM. Dark green sand ; green sand and black specks ; fine sand ; green mud ; grey mud.

Ophiactis carnea Ljungman

Ophiactis carnea Ljungman, 1867b, *Ofvers. K. VetenskAkad. Förh. Stockh.* **23** : 324–325 ; Lyman, 1882 : 120 ; H. L. Clark, 1923 : 332–333, pl. 20, figs 3, 4 ; Mortensen, 1933a : 342–345, figs 54–56 ; Stephenson, Stephenson & du Toit, 1937 : 380 ; Eyre & Stephenson, 1938 : 39 ; A. M. Clark, 1952 : 199 ; J. B. Balinsky, 1957 : 11–12 ; Kalk, 1958 : 197, 200, 215, 237 ; Morgans, 1959 : 414, 422 ; 1962 : 303 ; Macnae & Kalk, 1962 : 114 ; B. I. Balinsky *In* : Macnae & Kalk, 1969 : 106, 129 ; Day, Field & Penrith, 1970 : 81.

Ophiactis africana capensis Hertz, 1927a : 29.

D.d. up to 6 mm ; d.d./a.l. 1/5–6.

A species of *Ophiactis* with usually five arms ; disc covered above and below with scales of fairly even size, spinelets usually restricted to the ambital area, sometimes lacking, radial shields small to moderate, less than half the disc radius in length, just contiguous distally ; oral shields broad rhombic, adorals broadly contiguous interradially. distal oral papilla single, rounded, infradental papilla often lacking, exposing the lowest tooth ; dorsal arm plates ranging from broad oval to rounded fan-shaped but broadly contiguous, ventral arm plates octagonal or broadly truncated pentagonal but with the two distal angles rounded off to some extent ; arm spines up to five, mostly tapering though the middle ones may be blunt. Colour reddish-brown and whitish, some specimens probably greenish in life.

Type locality. Port Natal (Durban).

Southern African records. 33/18/i, s ; 34/18/s ; 34/18/FB/i, s ; 34/19/i, s ; 34/20/i ; 34/21/s ; 34/22/i, s ; 34/23/i, s ; 33/25/s ; 34/25/s, d ; 33/26/s ; 33/27/s ; 33/28/s ; 32/28/s ; 30/30/s ; 29/31/i, s, d ; 29/32/s ; 28/32/– ; 26/33/i ; 24/34/s ; intertidal–220 metres.

Nature of bottom. Usually sand or shell of some kind, sometimes rock or coral.

Ophiactis delagoa J. B. Balinsky

Ophiactis delagoa J. B. Balinsky, 1957 : 12–13, fig. 5, pl. 2, figs 7, 8 ; Kalk, 1958 : 216, 237 ; Macnae & Kalk, 1969 : 129.

D.d. up to 5 mm ; d.d./a.l. 1/3–4.

A species of *Ophiactis* with five arms ; disc covered with scales above and below, those of the upper side fairly coarse, spinelets only present in the ambital areas, if at all, radial shields moderate in size, not contiguous ; oral shields broad spearhead-shaped, adorals widely separated interradially, distal oral papilla single ; dorsal arm plates fan-shaped but distal edge medially straight, contiguous for about half their breadth, ventral arm plates truncated pentagonal, distal edge slightly convex ; arm spines up to five, tapering to blunt tips. Colour olive-green to brownish, a dark spot near the distal end of each radial shield, arms banded transversely.

Type locality. Inhaca, Delagoa Bay, 26/33, 9 metres.

Nature of bottom. Coral debris.

Ophiactis hemiteles H. L. Clark

Ophiactis hemiteles H. L. Clark, 1915, *Mem. Mus. comp. Zool. Harv.* **75** : 262, pl. 10, figs 7, 8 ; J. B. Balinsky, 1957 : 13 ; Macnae & Kalk, 1969 : 129.

D.d. up to 5 mm ; arms long, d.d./a.l. 1/8+.

A species of *Ophiactis* with five arms ; disc covered above with fine scales but below with skin showing a few scattered scales each armed with a small spinelet, radial shields moderate in size, length less than half the disc radius, distally contiguous ; oral shields broad oval, adorals contiguous interradially, distal oral papillae two each side ; dorsal arm plates oval, broadly contiguous, ventral arm plates rounded octagonal ; arm spines up to eight, small, all sharp or the upper distal ones flattened at the tips. Colour olive-greenish on white background.

Type locality. Torres Strait, northern Australia, reef.

South African record. 26/33/i.*

Nature of bottom. Coral reef.

* See note about identity on p. 105.

Ophiactis modesta Brock

Ophiactis modesta Brock, 1888, *Z. wiss. Zool.* **47** : 482 ; J. B. Balinsky, 1957 : 14 ; Kalk, 1958 : 200, 237 ; Macnae & Kalk, 1962 : 111 ; 1969 : 129 ; A. M. Clark & Rowe, 1971 : 105.
Ophiactis ? modesta : Day, 1974 : 94.

D.d. up to 8 mm ; d.d./a.l. 1/7–8.

A fissiparous species of *Ophiactis* with usually six arms, sometimes five, rarely seven ; disc scaled above and below, usually with some spinelets in the ambital areas, radial shields of moderate size, less than half the disc radius ; oral shields rhombic or oval, longer than broad, adorals separate interradially, distal oral papilla single ; dorsal arm plates oval, becoming very broad in larger specimens (d.d. $>$ 6 mm) br : l up to 4 : 1, broadly contiguous, ventral arm plates truncated pentagonal, the distal angles rounded ; arm spines up to five, smooth, tapering to blunt tips. Colour bluish-green.

TYPE LOCALITY. Amboina, East Indies.

SOUTHERN AFRICAN RECORDS. 26/33/i ; 24/34/s ; 23/35/s ; intertidal–22 metres.

NATURE OF BOTTOM. Reef or flats ; shell, sand and rock.

Ophiactis nidarosiensis Mortensen

Ophiactis nidarosiensis Mortensen, 1920, *Vidensk. Meddr dansk naturh. Foren.* **72** : 60, fig. 5 ; 1927 : 200, fig. 111 ; 1933a : 346–347, fig. 58a.

D.d. 2·5 mm (in the type material) ; d.d./a.l. 1/5–6.

A fissiparous species of *Ophiactis* with usually six arms ; sometimes five, rarely seven ; disc covered with scales, coarser above than below, radial shields about half the disc radius in length ; oral shields rhombic, adorals contiguous interradially, distal oral papillae usually two each side, sometimes only one ; dorsal arm plates narrow fan-shaped, not broader than long, ventral arm plates pentagonal, the distal end convex ; arm spines four proximally, tapering and pointed. [The relatively large size of the radial shields and the narrowness of the dorsal arm plates are at least partly attributable to the small size.]

TYPE LOCALITY. Norway, *c.* 200 metres.

SOUTHERN AFRICAN RECORD. No details.

Ophiactis plana Lyman

Ophiactis plana Lyman, 1869, *Bull. Mus. comp. Zool. Harv.* **1** : 330–331 ; H. L. Clark, 1923 : 333 ; Mortensen, 1933a : 345–346, fig. 57 ; H. L. Clark, 1939 : 76–77 ; A. M. Clark, 1974 : 464–465.
? *Ophiactis plana* : Day & Morgans, 1956 : 308.
? *Ophiactis lymani* Ljungman, 1871, *Ofvers. K. VetenskAkad. Förh. Stockh.* **28** : 629 ; Mortensen, 1933b : 442–449, figs 15a–d, 16d, e ; J. B. Balinsky, 1957 : 14.
? *Ophiactis parva* Mortensen, 1926, *Trans. zool. Soc. Lond.* **22** : 123–124, fig. 12 ; 1940 : 70–71.

D.d. up to *c.* 4 mm but rarely exceeding 3 mm ; d.d./a.l. 1/3–4.

A fissiparous species of *Ophiactis* with usually six arms ; disc covered with relatively coarse scales above and below, with few, if any, spinelets, radial shields relatively small, distinctly less than half the disc radius ; oral shields rhombic or rounded triangular or almost circular, as broad as or broader than long, adorals either contiguous or separate, distal oral papilla single ; dorsal arm plates fan-shaped, barely contiguous proximally, separated distally, ventral arm plates truncated pentagonal, the distal edge straight or slightly concave medially ; arm spines up to four, smooth and tapering to blunt tips.

Type locality. Florida Strait, *c.* 200 metres.

Southern African records. 34/18/d ; 35/18/d ; 35/19/d ; 35/22/d ; 34/23/d ; 34/25/s ; 33/26/d ; 33/28/d ; 30/30/d ; 30/31/d ; 29/31/s, d ; 57–238 metres.*

Nature of bottom. Coarse khaki sand and stones ; khaki sand ; rocks on sand ; rock ; shell and rock ; coral ; mud.

Ophiactis savignyi (Müller & Troschel)

Ophiolepis savignyi Müller & Troschel, 1842 : 95.

Ophiactis savignyi : Mortensen, 1933a : 348 ; Day & Morgans, 1956 : 308 ; J. B. Balinsky, 1957 : 14 ; Kalk, 1958 : 197, 200, 207, 213, 214, 237 ; Macnae & Kalk, 1962 : 114 ; B. I. Balinsky *In* : Macnae & Kalk, 1969 : 99, 101, 104, 106, 129 ; Madsen, 1971 : 207–208, fig. 33 ; Day, 1974 : 94.

D.d. up to 6 mm ; d.d./a.l. *c.* 1/7.

A fissiparous species of *Ophiactis* with usually six arms, sometimes five, occasionally seven ; disc covered above and below with fairly coarse scales bearing scattered spinelets, radial shields relatively large and conspicuous, more than half the disc radius in length ; oral shields usually rhombic, adorals contiguous interradially or sometimes only approximating, distal oral papillae two each side ; dorsal arm plates oval or rounded rectangular, less than twice as broad as long, the middle of the distal edge often with a small median projection emphasized by a spot each side of it, ventral arm plates octagonal ; arm spines up to six, very short, stout-based, the middle ones abruptly truncated and markedly rugose at the tip. Colour bright green with white and darker markings, a white patch at the distal end of each radial shield, the rest of the plate dark.

Type locality. Red Sea.

Southern African records. 30/30/i ; 29/31/i, s ; 26/33/i ; 24/35/i ; 23/35/i, s; intertidal–64 metres.

Nature of bottom. In crevices of sponges, coral fragments, coralline algae, etc.

Suborder LAEMOPHIURINA Matsumoto

Ophiuroids with a moderately large disc covered with thin skin in which continuous scaling is more or less evident but may be obscured by armament of spinelets,

* Re-examination of the specimen from Durban, LWST, named by me *Ophiactis plana* (see Day & Morgans, 1956, and A. M. C., 1974) together with the specimens from NAD 15N and 81R (A. M. C., 1974 : 464), all of which have d.d. 1·5 mm or less, throws doubt on the wisdom of giving them a positive identification. Possibly they represent young specimens of *O. modesta* which may also show more fan-shaped than oval dorsal arm plates, this being a juvenile feature.

thorny stumps or granules, when this is particularly dense, flat, wedge-shaped or bar-like radial shields usually distinct but sometimes very reduced or similarly obscured, either separated or distally approximating or contiguous, their distal articulations with the genital plates rarely distinct externally ;* oral shields well developed on all the jaw angles, adorals large, oral papillae papilliform or spiniform, continuous ; arms simple, the disc formed around or slightly above their bases, more flexible horizontally than vertically, usually slender and covered with only thin skin through which the arm plates are distinct, dorsal as well as ventral arm plates entire but relatively small, usually only the proximal successive ones contiguous, most being separated by the large* lateral arm plates, which are conspicuous both above and below and bear conspicuous series of laterally projecting spines, only the lowest of which may be modified into a hook distally ; tentacle pores usually small* but armed with at least one tentacle scale.

Family **OPHIACANTHIDAE** Perrier

See : Matsumoto, 1917 : 92 ; Mortensen, 1927 : 182 ; Spencer & Wright, 1966 : U98.

A family of Laemophiurina with the disc usually bearing some kind of armament, often concealing the scaling ; jaws armed apically with a vertical series of teeth and usually a single infradental papilla, occasionally several, with a continuous series of oral papillae each side often numbering three ; arms more flexible horizontally than vertically, except in a few epizoic species, often more or less constricted at the joints or even moniliform (like a string of beads) ; arm spines usually long, the proximal upper ones nearly always exceeding the segment in length ; tentacle pores more often small than large, usually armed with a single scale.

Amphilimna cribriformis A. M. Clark

Amphilimna cribriformis A. M. Clark, 1974 : 442–444, fig. 1.

D.d. up to 6·5 mm ; arms long, d.d./a.l. probably 1/12–15.

A species of Ophiacanthidae with the disc not constricted interradially, covered above and below with scales which are indistinct under the skin and bear scattered needle-like spines, radial shields long and narrow, l : br *c.* 5 : 1, probably fully contiguous in life but gaping apart distally in preserved specimens, a pair of stout-based distally directed spines on the genital plate below and distal to each radial shield ; oral shields rounded triangular, about as long as broad, adorals widely separated interradially, oral papillae very irregular, short and rounded adapically but the more distal three or four each side spiniform ; dorsal arm plates delicate, semitransparent, oval, broader than long and contiguous, at least proximally, basal ventral arm plates truncated pentagonal (their proximal limits indistinct) broadest distal to the tentacle pores, the following plates with the median part of the distal side slightly concave ; arm spines up to six, flattened and tapering, not exceeding the segment in length, on each of the segments bordering the genital slits all but the

* *Amphilimna*, whose position in the family Ophiacanthidae is debatable, is exceptional with regard to these characters.

lowest spine markedly flattened and fused together to form a curved flange each side; tentacle pores large, those of the first 10–12 arm segments with two scales, the one on the ventral arm plate spiniform but soon reduced and lost, the small papilliform second scale of the proximal segments on the lateral arm plate becoming enlarged into a flat oval scale capable of covering the pore.

TYPE LOCALITY. NE. of Durban, 118 metres.

SOUTHERN AFRICAN RECORDS. 29/31/s, d; 86–200 metres.

NATURE OF BOTTOM. Sandy mud.

Anamphiura valida H. L. Clark

Anamphiura valida H. L. Clark, 1939: 70–72, figs 26A, 27; A. M. Clark, 1974: 478–479, fig. 16.

D.d. up to 5 mm; d.d./a.l. 1/2·2–2·5.

A species of Ophiacanthidae (see below) with the disc covered above and below by coarse convex scales bearing scattered pointed spinelets, the primary rosette conspicuous, the radial shields short and broad, proximally contiguous but distally gaping apart, a pair of conical spines projecting from the genital plate distal to each radial shield; oral shields broad rhombic with rounded angles, adorals broadly contiguous inwardly, apex of each jaw with two or sometimes three often irregular infradental papillae flanked each side by a pointed first oral tentacle scale and separated by a diastema from the two or sometimes three papillae each side distal to the second oral tentacle pore on the edge of the adoral shield; dorsal arm plates oval or almost circular, thick, convex and just contiguous; ventral arm plates constricted in their proximal half but with prolonged lateral angles distal to the tentacle pores, successive ones contiguous, the distal edge with a very obtuse median angle; arm spines up to six, conical and pointed, relatively short, the lowest and longest about equal to the segment in length, the upper ones shorter, on the first arm segment the two lowest spines each side are unmodified but the upper ones are markedly flattened and fused together to form a curved flange each side, bordering the genital slit; tentacle pores large with a rounded scale (occasionally double on the first segment) on the lateral arm plate, opposed on the first three to six pores by a papilliform scale based on the ventral arm plate towards the distal end of the pore.

Affinities. In addition to the points of resemblance between *Anamphiura valida* and *Amphilimna* which I detailed in 1974, Dr L. P. Thomas (during a visit to the British Museum) noticed that *Anamphiura* also has some basal arm spines modified into a flange opposite the genital slits, which I had overlooked. (Owing to the few arm segments underlying the disc in this small species, the flange is limited to the first arm segment only and does not involve the two lowest spines each side, whereas in *Amphilimna* only one normal spine occurs on the basal segments. The spines of the second segment are all free in *Anamphiura* but the uppermost ones may be flattened and spatulate.) This genus must therefore be arrayed with *Amphilimna* which Thomas has referred to the heterogeneous family Ophiacanthidae, on the basis of its jaw structure. If the more compact disc and relatively short arms of *Anamphiura valida* are explained by the small size of the known specimens, then it may be congeneric with *Amphilimna olivacea*, as Thomas (1975) suggests.

TYPE LOCALITY. Off Zanzibar, 238–293 metres.

SOUTHERN AFRICAN RECORD. 29/31/d ; 350 metres.

Genus *OPHIACANTHA* Müller & Troschel

See : Mortensen, 1927 : 188.

A genus of Ophiacanthidae with the disc not constricted interradially, covered with rugose stumps or granules or spinelets, or a combination, obscuring the scales, radial shields inconspicuous, separated, bar-like but only their distal ends visible externally ; oral shields usually broad rhombic, rarely narrower than long, adorals relatively large and usually broadly contiguous inwardly, oral papillae usually comprising a single infradental papilla with about three smaller ones each side, sometimes more, the distalmost papilla sometimes enlarged ; dorsal arm plates small, fan-shaped or triangular with the distal side convex and becoming rhombic, ventral arm plates often pentagonal or bell-shaped with the distal side usually convex, only the basal plates of both series, if any, contiguous ; arm spines mostly tapering and pointed, the upper proximal ones usually very long, sometimes serrated ; tentacle pores small, tentacle scales usually single, small papilliform.

Ophiacantha baccata Mortensen

Ophiacantha baccata Mortensen, 1933a : 319–322, figs 40, 41.

D.d. up to 5 mm ; d.d./a.l. *c.* 1/6.

A species of *Ophiacantha* with the disc armed above and below with small, primarily trifid, thorny stumps, only the distal ends of the radial shields naked, if anything ; oral shields broad rhombic, sunken in the middle, adorals with a narrow distal lobe and broadly contiguous interradially, usually three oral papillae each side of a broad infradental papilla, the outermost one markedly broader or sometimes* replaced by two narrow papillae ; arms moniliform, very constricted between the widely separated fan-shaped or rhombic dorsal arm plates, which are markedly swollen, ventral arm plates pentagonal, the two laterodistal angles rounded and the distal side between them often slightly concave, all the arm plates smooth ; arm spines up to seven proximally, long slender and rugose ; tentacle scale small single, elongated and rugose.

TYPE LOCALITY. Off Durban, 232 metres.

SOUTHERN AFRICAN RECORDS. 35/22/d ; 33/27/s ; 29/31/d ; 27/32/d ; 24/35/d ; 88–440 metres.

NATURE OF BOTTOM. Hard sand ; coarse khaki sand and stones ; shelly sand.

* The presence of four similar oral papillae each side occurs in three specimens taken on the Still Bay transect at 35°22′S 22°31′E, 200 metres.

Ophiacantha nerthepsila H. L. Clark

Ophiacantha nerthepsila H. L. Clark, 1923 : 319, fig. 1, pl. 19, figs 3, 4 ; Mortensen, 1933a : 316–317, fig. 37 ; Day, Field & Penrith, 1970 : 80.
Ophiacantha Barracoutae Koehler, 1923 : 3–5, figs 1–3.

D.d. up to 7 mm ; d.d./a.l. 1/3·5–4·0.

A species of *Ophiacantha* with the disc covered above with fine granules concealing the scales, often with scattered spinelets in addition, the granules sometimes extending on to the distal part of the ventral side but the lower scales mostly bare, radial shields widely separated by the width of the arm, only a small triangular part of each exposed ; oral shields broad rhombic, sunken medially, adorals without a distal lobe but broadly contiguous interradially, usually three oral papillae each side of the enlarged infradental one, the distalmost papilla the broadest ; dorsal arm plates triangular with the distal side convex, sometimes bell-shaped if the two proximal sides are sinuous, ventral arm plates pentagonal with the two laterodistal angles rounded and the distal side straight, texture of arm plates smooth ; arm spines up to seven, long and smooth ; tentacle scale single, blunt spiniform.

Type locality. Between Port Elizabeth and East London, 42 metres.

Southern African records. 33/18/s ; 34/18/s ; 34/18/FB/s ; 34/19/s ; 34/20/s ; 34/21/s ; 34/23/s, d ; 34/25/s ; 33/26/s ; 33/27/s ; 32/28/s ; 22–110 metres.

Nature of bottom. Usually rock ; isolated records of : coarse sand and shell ; sand, mud and rock ; rock and khaki sand ; sand ; limestone.

Ophiacantha scutigera Mortensen

Ophiacantha scutigera Mortensen, 1933a : 317–319, figs 38, 39, pl. 19, figs 6, 7.

D.d. up to 5 mm ; d.d./a.l. 1/6.

A species of *Ophiacantha* with the disc covered more or less densely with low rounded rugose stumps or granules and scattered slender spinelets obscuring the scales except between the radial shields which are small but distinct, separated by less than their own breadth or much less than the arm breadth ; oral shields rhombic or broad pentagonal, adorals without a distal lobe but broadly contiguous interradially, oral papillae three (sometimes four) each side of the broader infradental papilla, the outermost papilla sometimes slightly enlarged ; dorsal arm plates triangular with the distal side convex, ventral arm plates fan-shaped, ventral and lateral arm plates textured with transverse concentric striations ; arm spines up to ten proximally, the proximal upper ones long and smooth, the lower ones slightly serrated ; tentacle scale single, rounded.

Type locality. Off Durban (29/31 and 30/31), 165–366 metres.

Ophiacantha striolata Mortensen

Ophiacantha striolata Mortensen, 1933a : 322–324, figs 42, 43.

D.d. 3 mm ; d.d./a.l. 1/4.

A species of *Ophiacantha* with the disc covered above and below with small, primarily trifid, thorny stumps, only the widely separated ends of the radial shields visible ; oral shields almost triangular, the distal lobe very slight, adorals with a narrow distal lobe and broadly contiguous interradially, oral papillae three each side, similar, narrow and pointed ; dorsal arm plates small, triangular, with the distal side convex, ventral arm plates fan-shaped, all arm plates textured with concentric transverse striations ; arm spines up to eight proximally, all deeply serrated ; tentacle scale single, pointed and furrowed.

TYPE LOCALITY. Off Durban, 412 metres.

SOUTHERN AFRICAN RECORDS. 33/28/s ; 29/31/d ; 84–412 metres. [? 34/18/158 metres–juvenile.]

NATURE OF BOTTOM. Fine khaki sand and gravel.

Genus ***OPHIOMITRELLA*** Verrill

See : Mortensen, 1927 : 185.

A genus of Ophiacanthidae with the disc not constricted interradially, covered with usually large scales bearing spaced short thick stumps or granules, radial shields broad rounded, approximating or sometimes partially contiguous ; oral shields usually broader than long, adorals relatively large and broadly contiguous interradially, usually three papilliform oral papillae each side of an infradental one ; dorsal arm plates small, rhombic or fan-shaped, not contiguous, ventral arm plates fan- or bell-shaped, not contiguous ; arm spines long and slender ; tentacle pores small, armed with a single scale.

Ophiomitrella corynephora H. L. Clark

Ophiomitrella corynephora H. L. Clark, 1923 : 322–324, fig. 2, pl. 19, figs 5, 6 ; Mortensen, 1933a : 331–333, fig. 48 ; A. M. Clark, 1952 : 199, 212 ; 1974 : 441.

D.d. up to 10 mm ; d.d./a.l. 1/2·1–3·9.

A species of *Ophiomitrella* with some disc scales bearing a cylindrical, rugose-tipped stump, radial shields separated by less than their own breadth ; oral shields rhombic, adorals without a distal lobe, broadly contiguous interradially, oral papillae usually three each side (also a calcified elevation distal to the outermost one arising from the inner abradial edge of the first ventral arm plate adjoining the oral tentacle pore), jaws sunken ; dorsal arm plates rhombic, consecutive ones widely separated, ventral arm plates with the proximal angle very obtuse, except on the basal plates and a rounded distal angle ; arm spines up to seven proximally, finely serrated, the upper ones long and pointed, easily exceeding the segment in length, the two lowest short and blunt ; tentacle scale single, fairly long and tapering. Colour white.

TYPE LOCALITY. West of Cape Peninsula, 420 metres.

SOUTHERN AFRICAN RECORDS. 30/15/d ; 31/16/d ; 33/17/s ; 34/17/d ; 34/18/d ; 35/22/d ; 34/23/d ; 34/24/d ; 34/25/d ; 79–420 metres.

NATURE OF BOTTOM. Rock ; rock and fine sand ; coarse khaki sand and stones ; rocks and khaki sand ; green sand ; dark green mud ; on alcyonarian.

Ophiomitrella hamata Mortensen

Ophiomitrella hamata Mortensen, 1933a : 333–335, figs 50, 51, pl. 19, fig. 12.

D.d. 4 mm ; d.d./a.l. 1/3.

A species of *Ophiomitrella* with most disc scales bearing a very short rugose-tipped stump, radial shields just contiguous distally ; oral shields broad rhombic, adorals without a distal lobe, broadly contiguous interradially, oral papillae usually three each side, fairly thick, similar, jaws not sunken ; dorsal arm plates rhombic, not contiguous, ventral arm plates with the proximal angle very obtuse after the basal plates and a rounded distal lobe, becoming flattened medially or even slightly concave on the plates in the middle of the arm ; arm spines up to five proximally, serrated, not exceeding the segment in length ; tentacle scale small, pointed.

TYPE LOCALITY. Off Durban (29/31), 412 metres ; on a gorgonian.

Ophioplinthaca sexradia Mortensen

Ophioplinthaca sexradia Mortensen, 1933a : 326–327, fig. 45.

D.d. 2·5 mm ; d.d./a.l. 1/4–5.

A species of Ophiacanthidae with six arms, the disc deeply notched interradially, covered with a few coarse scales sparsely armed with abbreviated stumps or granules, radial shields large, more than half the disc radius in length and distally contiguous ; oral shields rhombic except that the distal angle is very rounded, broader than long, adorals large but short, lacking a distal lobe but very broadly contiguous interradially, three papilliform oral papillae each side of an almost spiniform infradental papilla, the distalmost one each side broadened ; dorsal arm plates rounded triangular, widely separated, ventral arm plates pentagonal but distally convex ; arm spines four proximally, thick, tapering and much shorter than the segment ; tentacle pores small, with a single small papilliform scale.

TYPE LOCALITY. Near East London (33/27), 44 metres.

Ophiothamnus remotus Lyman

Ophiothamnus remotus Lyman, 1878, *Bull. Mus. comp. Zool. Harv.* **5** : 149–150, pl. 8, figs 201–203 ; 1882 : 212–213, pl. 14, figs 1–3 ; Bell, 1905b : 258 ; H. L. Clark, 1923 : 324–325 ; Mortensen, 1933a : 327–330, figs 46, 47a.

Ophiothamnus remotus var. *cordatus* Mortensen, 1933a : 330–331, fig. 47b.

D.d. up to 3·5 mm ; d.d./a.l. 1/3–4.

A species of Ophiacanthidae with the disc notched interradially and covered above and below with only a few scales, some bearing a tapering spine, radial shields large triangular but with the two inner sides convex, more than half the disc radius in length and contiguous for more than half their length ; oral shields small triangular with the distal side slightly convex, as long as broad or slightly longer (forma *remotus*) or heart-shaped with the distal side slightly concave and broader than long (forma *cordatus*), adorals very large, widely contiguous interradially, one pointed papilliform infradental oral papilla and three lateral papillae each side, the outermost one very broad and opercular ; dorsal arm plates rounded triangular, broader than long, not contiguous even proximally, ventral arm plates pentagonal, not contiguous ; arm spines seven proximally, slender smooth and pointed, only the two uppermost exceeding the segment in length ; tentacle pores small, scale single, small, pointed. Colour orange.

TYPE LOCALITY. Agulhas Bank, 275 metres.

SOUTHERN AFRICAN RECORDS. 31/16/d ; 34/17/d ; 34/18/d ; 35/18/d ; 35/22/d ; 35/23/d ; 34/24/d ; 34/25/– ; 33/27/s ; 33/28/d ; 32/28/s ; 30/31/d ; 29/31/d ; 27/32/d ; 88–457 metres.

NATURE OF BOTTOM. Rock ; coarse khaki sand and stones ; coarse khaki sand ; hard sand ; fine khaki sand and gravel ; green mud and sand ; black sandy mud.

Ophiotreta durbanensis (Mortensen)*

Ophiacantha (*Ophiotreta*) *durbanensis* Mortensen, 1933a : 324–325, fig. 4, pl. 19, figs 13–15.

D.d. 7 mm (only two specimens known) ; d.d./a.l. *c.* 1/6.

A species of Ophiacanthidae with the disc not constricted interradially, closely covered above by a coat of fine granules interspersed with scattered spinelets, granules extending on to the median distal parts of the ventral interradii but leaving the scales exposed near the genital slits, radial shields showing as a small oval naked patch, widely separated ; oral shields spearhead-shaped, longer than broad, the surface of the short distal lobe slightly sunken, adorals with a distal lobe lateral to the oral shield, approximating or just contiguous interradially, oral papillae numerous, five or six each side besides an oral tentacle scale at the distal end of the series mounted on the abradial inner edge of the first ventral arm plate ; dorsal arm plates fan-shaped or triangular with the distal side convex, lateral angles acute, just contiguous, ventral arm plates relatively broad fan-shaped, the proximal angle truncated on the basal segments and the lateral angles interrupted by the tentacle pore and scales ; arm spines up to five, the uppermost longest and tapering, equalling two arm segments in length, middle ones truncated and finely serrated ; tentacle pores small, with two papilliform scales aligned parallel, the inner one absent on the distal arm segments.

TYPE LOCALITY. Off Durban, 412 metres.

SOUTHERN AFRICAN RECORDS. 29/31/d ; 27/32/d ; 400–450 metres.

Suborder CHILOPHIURINA Matsumoto

Ophiuroids with a moderately large disc covered with thin skin over more or less well-developed continuous scaling, usually only obscured if dense granulation is present (Ophiocomidae, Ophiodermatidae and Ophioleucidae), more often no armament, radial shields flat and oval or triangular, often conspicuous, separate or more or less contiguous, their distal articulations with the genital plates concealed externally ; oral shields well developed on all the jaw angles, adorals usually well developed, oral papillae in continuous series ; arms simple, often fairly stout, the

* See footnote on p. 121.

disc either formed around their bases or above, flexible horizontally, dorsal, ventral and lateral arm plates usually all well developed; arm spines erect or appressed, none becoming hooked distally; tentacle pores moderately large, at least proximally, but sometimes rapidly reduced or even lost after the first few segments, armed with usually one or two scales on the proximal segments, sometimes more.

Family **OPHIOCOMIDAE** Ljungman

See: Matsumoto, 1917: 340; Mortensen, 1927: 177; Spencer & Wright, 1966: U97.

A family of Chilophiurina with the disc overlying the arm bases, disc scales covered with a dense coat of granules, sometimes mixed with or replaced by spinelets, which may be spaced, or with naked skin more or less obscuring the scales, which in *Ophiopsila* are extremely fine or even rudimentary, radial shields not distinct in adults, except in *Ophiopsila* where they appear as long and narrow, more or less widely separated; oral and adoral shields well developed, though the adorals are rarely contiguous interradially except in *Ophiopsila*, naked, as are the oral plates, teeth broad, quadrangular, the more superficial ones replaced by a cluster of rounded tooth papillae similar in size to the three to five oral papillae each side, which are usually contiguous with them except in *Ophiopsila* where there is a more or less marked diastema revealing the first oral tentacle scale inset into the slit (in *Ophiopsila* the number of tooth papillae is small, three to seven in the species included and small specimens, d.d. < 5 mm, may have only two papillae resembling the paired infradental oral papillae of amphiurids), the second oral tentacle scale more or less in series with the oral papillae at the distal end, though often overlaid to some extent; arms more or less long and stout, often widening out beyond the base, the plates all well developed, successive dorsal and ventral ones contiguous for most of the arm, except in *Ophiopsila*; arm spines erect, often long or stout, or both, tentacle pores moderate in size, usually with two parallel distally directed oval scales, sometimes only one, the inner scale elongated and sword-like in *Ophiopsila*.

Genus ***OPHIOCOMA*** L. Agassiz

See: H. L. Clark, 1921, *Pap. Dep. mar. Biol. Carnegie Instn Wash.* no. 10: 120; Devaney, 1970, *Smithson. Contr. Zool.* no. 51: 9.

A genus of Ophiocomidae including many species which reach a large size, d.d. > 20 mm, the disc covered above with a continuous coat of granules extending on to at least the distal parts of the ventral interradii, if not completely concealing all the scales as well as the radial shields (except in juveniles, d.d. < 5 mm, where the granule coating is more or less imperfect), sometimes the ambital granules elongated into short blunt papillae; oral shields large, oval, hexagonal or pentagonal if the distal lobe is angular, the proximal end being usually more or less truncated to a degree matching the interradial separation of the adoral shields, jaw angles showing a similar range of form in most species, tooth papillae numerous, the superficial ones in series with the three to five oral papillae each side; dorsal arm plates broad, fan-shaped, oval or hexagonal, ventral plates almost square, the proximal angle being usually very obtuse, but the distal angles broadly rounded; arm spines stout, three to seven, sometimes alternating three and four on successive segments or on opposite sides of the same segment, the lower ones sometimes flattened spatulate, the upper ones cylindrical or cigar-shaped, exceptionally the uppermost one on alternate sides up to five times as long as the segment (in *O. pusilla* a few of the second from uppermost spines may be clavate); tentacle scales usually two, sometimes one, oval.

Ophiocoma erinaceus Müller & Troschel

Ophiocoma erinaceus Müller & Troschel, 1842 : 98; H. L. Clark, 1921, *Pap. Dep. mar. Biol. Carnegie Instn Wash.* no. 10 : 127; 1923 : 348; J. B. Balinsky, 1957 : 25–26; Kalk, 1958 : 207, 216, 237; Macnae & Kalk, 1969 : 130; A. M. Clark & Rowe, 1971 : 86, 119, pl. 17, figs 5, 6.

D.d. up to 28 mm; d.d./a.l. 1/4–5, rarely more.

A species of *Ophiocoma* with the disc covered above with relatively coarse spherical granules, 9–16/mm² or three or four to a linear mm, only extending on to the distal parts of the ventral interradii; oral shields variable, circular to hexagonal; dorsal arm plates fan-shaped with the distal side more or less flattened; arm spines rarely more than five, progressively longer upwards, often alternating three and four on successive series of the same side of the arm or opposite series of the same segment, the alternate uppermost spines very stout and often long, exceptionally up to five times the segment length, usually cigar-shaped; tentacle scales two proximally, sometimes falling to one on a number of more distal segments. Colour densely black (or very dark grey) all over, the tube feet black or red.

Type locality. 'Red Sea; Indian Ocean.'

Southern African record. 26/33/i.

Nature of bottom. Between branches of coral.

Ophiocoma pica Müller & Troschel

Ophiocoma pica Müller & Troschel, 1842 : 101; H. L. Clark, 1921, *Pap. Dep. mar. Biol. Carnegie Instn Wash.* no. 10 : 127–128, pl. 13, fig. 8; J. B. Balinsky, 1957 : 26; Kalk, 1958 : 237; Macnae & Kalk, 1969 : 130; A. M. Clark & Rowe, 1971 : 86, 118, pl. 18, fig. 4.

D.d. up to 25 mm; d.d./a.l. 1/2–4.

A species of *Ophiocoma* with the disc covered above with moderately fine granules, 50–70/mm², only extending on to the distal parts of the ventral interradii; oral shields usually oval; arm spines up to six proximally, relatively long and slender, especially distally, the second from uppermost spine usually the longest; tentacle scales two. Colour dark brown or black with radiating golden lines on the disc and often transverse segmental ones on the arms.

Type locality. Unknown.

Southern African records. 28/32/–; 26/33/i.

Nature of bottom. In branches of corals, as for *O. erinaceus*.

Ophiocoma pusilla (Brock)

Ophiomastix pusilla Brock, 1888, *Z. wiss. Zool.* **47** : 499–500.
Ophiocoma pusilla : H. L. Clark, 1921, *Pap. Dep. mar. Biol. Carnegie Instn Wash.* no. 10 : 131 ; Devaney, 1970, *Smithson. Contr. Zool.* no. 51 : 25–28, figs 26, 29 ; A. M. Clark & Rowe, 1971 : 86, 118.
Ophiocoma insularia : J. B. Balinsky, 1957 : 26 ; Kalk, 1958 : 207, 216, 237 ; Macnae & Kalk, 1969 : 130. [Non *O. insularia* Lyman.]

D.d. up to 10 mm ; d.d./a.l. *c.* 1/5.

A small species of *Ophiocoma* with the disc covered above with moderately fine granules only extending on to the distal parts of the ventral interradii ; arm spines up to six but rarely more than five, relatively long and slender, the second from uppermost on and about the eighth free segment (for not more than three successive segments) often appearing clavate, especially in wet specimens, the skin near the tip much thickened though the underlying spine is only slightly expanded ; tentacle scales usually two proximally, one distally. Colour : disc chocolate brown, often with large round darker spots circumscribed with light brown, the spots sometimes so frequent that the light rings run together into a reticulum, arms banded dark and light brown, the distal edge of each dorsal arm plate dark.

TYPE LOCALITY. Amboina, Kei Is.

SOUTHERN AFRICAN RECORD. 26/33/i.

NATURE OF BOTTOM. Dead or sometimes live coral heads.

Ophiocoma scolopendrina (Lamarck)

Ophiura scolopendrina Lamarck, 1816, **2** : 544.
Ophiocoma scolopendrina : H. L. Clark, 1921, *Pap. Dep. mar. Biol. Carnegie Instn Wash.* no. 10 : 125–126 [non pl. 13, fig. 9 = *O. pusilla*] ; 1923 : 348 ; J. B. Balinsky, 1957 : 25 ; Kalk, 1958 : 205, 207, 237 ; Macnae & Kalk, 1962 : 110 ; B. I. Balinsky *In* : Macnae & Kalk, 1969 : 99–101, 106, 130 ; A. M. Clark & Rowe, 1971 : 86, 119, pl. 17, figs 3, 4. [Non *O. scolopendrina* (part) Lyman, 1882 : 170 (specimens supposedly from Simon's Bay) followed by Day, Field & Penrith, 1970 ; in 1957 I renamed these specimens *O. erinaceus* but in addition the locality is almost cetainly erroneous, as with some other 'Challenger' material said to be from 10–20 fathoms in Simon's Bay ; see p. 55.]

D.d. up to 32 mm ; d.d./a.l. 1/4·5–7·5.

Diagnosis as for *O. erinaceus* except that the uppermost arm spine rarely exceeds three times the segment length, the tentacle scales are more often two throughout the arm and the colour is variegated shades of grey or sometimes brown above, though some large specimens may be uniformly black above but always distinctly lighter on at least the proximal part of the under side and the tube feet are pale. Other differences are in habits and habitat.

TYPE LOCALITY. 'Isle de France' (Mauritius).

SOUTHERN AFRICAN RECORDS. 30/30/i ; 29/31/? i ; 26/33/i ; 24/35/i.

NATURE OF BOTTOM. Algal carpet on flat beach, under rocks or loose coral pieces or in crevices, or on *Cymodocea* beds.

Ophiocoma valenciae Müller & Troschel

Ophiocoma valenciae Müller & Troschel, 1842 : 102 ; H. L. Clark, 1923 : 349 ; Mortensen, 1933a : 375 ; Eyre & Stephenson, 1938 : 38, 43 ; Stephenson, 1944 : 305, 347 ; J. B. Balinsky, 1957 : 27 ; Kalk, 1958 : 200, 207, 237 ; Macnae & Kalk, 1962 : 110 ; B. I. Balinsky *In* : Macnae & Kalk, 1969 : 101, 106, 130 ; A. M. Clark & Rowe, 1971 : 86, 119, pl. 18, fig. 1.

D.d. up to 23 mm ; d.d./a.l. 1/5·25–8·5, mean 1/7·0.

A species of *Ophiocoma* with the disc covered above and below with moderately fine granules which become distinctly elongated near the ambitus and sometimes leave bare a narrow band of scales near the genital slits ; dorsal arm plates broad oval ; arm spines up to six proximally, the uppermost ones shorter than the middle ones ; tentacle scale single. Colour : disc brownish, arms dark greenish with darker bands.

TYPE LOCALITY. Red Sea.

SOUTHERN AFRICAN RECORDS. 31/29/i ; 30/30/i ; 29/31/i ; 26/33/i ; 24/35/i.

NATURE OF BOTTOM. On rocks or reef or among worm tubes, or under weed, as for *O. scolopendrina*.

Ophiocomella sexradia (Duncan)

Ophiocnida sexradia Duncan, 1887, *J. Linn. Soc.* Zool. **21** : 92.
Ophiocoma parva H. L. Clark, 1915, *Mem. Mus. comp. Zool. Harv.* **25** : 292, pl. 14, figs 8, 9 ; J. B. Balinsky, 1957 : 27 ; Kalk, 1958 : 207, 216, 237 ; B. I. Balinsky *In* : Macnae & Kalk, 1969 : 104, 106, 130.
Ophiocomella sexradia : A. M. Clark & Rowe, 1971 : 86, 118.

D.d. up to 6 mm but rarely > 4 mm ; d.d./a.l. *c.* 1/4 ; six arms.

A small fissiparous species of Ophiocomidae, the disc armed at first with short spaced blunt spinelets which become relatively shorter and more nearly granuliform and crowded in larger specimens, radial shields indistinct ; oral shields rather variable in shape, rhombic, spearhead-shaped or hexagonal, adorals not contiguous interradially, tooth papillae usually four to six, in series with the usually three oval oral papillae each side ; dorsal arm plates deep fan-shaped, only about as broad as long (to be expected at this small size), ventral arm plates with a more or less truncated proximal angle and the distal end rounded ; arm spines up to four, sometimes five proximally, tapering to blunt or truncated tips ; tentacle scale single, except on the pores of the first segment which may have two, oval. Colour : arms banded brown or red.

TYPE LOCALITY. Mergui Archipelago, Burma.

SOUTHERN AFRICAN RECORD. 26/33/i.

NATURE OF BOTTOM. Coral debris and live coral.

Genus ***OPHIOMASTIX*** Müller & Troschel

See : H. L. Clark, 1921, *Pap. Dep. mar. Biol. Carnegie Instn Wash.* no. 10 : 133.

A genus of Ophiocomidae with the disc armed with spaced spinelets, sometimes few or totally lacking in the species included, though some other species have crowded spinelets or elongated papilliform tubercles, radial shields obscured in adults ; jaw angles rather variable, as in *Ophiocoma* the oral shields usually large and adorals small and widely separated interradially, or the plates may be obscured by thickened skin ; arm plates usually as in *Ophiocoma* ; arm spines with the uppermost one of some series, even of alternate series but more often sporadic ones, markedly elongated in comparison with the adjacent spines and often abruptly clavate or even digitate at the tip ; tentacle scales one or two, oval.

Ophiomastix variabilis Koehler

Ophiomastix variabilis Koehler, 1905, *Siboga Exped.* **45b** : 69–71, pl. 6, fig. 16, pl. 16, figs 3, 4 ; A. M. Clark & Rowe, 1971 : 86–87, 121 ; Devaney, 1974, *Micronesica* **10** : 171–174.
Ophiomastix bispinosa H. L. Clark, 1917, *Bull. Mus. comp. Zool. Harv.* **61** : 442–443, pl. 2, figs 1, 2 ; A. M. Clark & Rowe, 1971 : 86–87, 120, 121, fig. 39.
Ophiomastix notabilis H. L. Clark, 1938, *Mem. Mus. comp. Zool. Harv.* **55** : 337–338, fig. 27; J. B. Balinsky, 1957 : 27 ; Macnae & Kalk, 1969 : 130.

D.d. up to 15 mm ; d.d./a.l. 1/7–9.

A species of *Ophiomastix* with the disc covered by thick skin with scattered spinelets ; limits of oral and adoral shields obscured by thick skin, about three oral papillae each side, the distalmost one often distinctly broader than the rest ; arms covered with thick skin obscuring the plates, dorsal arm plates broad fan-shaped or hexagonal, contiguous, ventral plates rectangular, or pentagonal if the proximal angle is developed, apparently not contiguous ; arm spines up to four in the largest specimens but usually three and two alternating for at least part of the arm, the uppermost spine of the second to fifth consecutive series abruptly elongated, up to three times the segment length, and sometimes clavate or cigar-shaped, otherwise slightly tapering but with the tip broad ; tentacle scale single, rounded. Colour black with narrow white bands across the arms at intervals of four to six segments, ventral arm plates brownish, some arm spines banded.

TYPE LOCALITY. Paumotu Is (? Tuamotus), E. of Tahiti.

SOUTHERN AFRICAN RECORD. 26/33/i.

NATURE OF BOTTOM. Reef.

Ophiomastix venosa Peters

Ophiomastix venosa Peters, 1851 : 464–465 ; Koehler, 1904a, *Mém. Soc. zool. Fr.* **17** : 73–74, figs 28, 29 ; H. L. Clark, 1923 : 349 ; J. B. Balinsky, 1957 : 27–28 ; Kalk, 1958 : 237 ; Macnae & Kalk, 1969 : 130.
Ophiomastus [sic] *venosa* : Macnae & Kalk, 1962 : 111.

D.d. up to 35 mm ; d.d./a.l. 1/3·5–4·5 (up to 6 according to Peters).

A species of *Ophiomastix* with the disc usually completely naked of spines, sometimes a few scattered ones ; oral shields variable in form, their limits not obscured ; dorsal arm plates broad fan-shaped or hexagonal, ventral arm plates octagonal or truncated pentagonal, contiguous ; arm spines two or three, the uppermost of every second or third series markedly elongated, often clavate and sometimes cloven or digitate at the tip ; tentacle scales two proximally then one. Colour light greyish brown, disc patterned with meandering black lines, usually two opposite the base of each arm.

TYPE LOCALITY. Mozambique between 12 and 15°S.

SOUTHERN AFRICAN RECORD. 26/33/i.

NATURE OF BOTTOM. In algal carpet on flat beach, as for *Ophiocoma scolopendrina*.

Genus *OPHIOPSILA* Forbes

See : Mortensen, 1927 : 179 ; Madsen, 1971 : 221.

A genus of Ophiocomidae reaching a moderate size, d.d. rarely > 15 mm, with very fine or barely perceptible disc scales more or less obscured in the skin, radial shields bar-like, distinct for much of their length ; oral shields rhombic with rounded angles or spearhead-shaped, or the proximal lobe may be truncated, tooth papillae relatively few in the species represented, only three to seven at d.d. < 10 mm, oral papillae usually only two or three each side, rounded or spiniform, and separated from the apical tooth papillae by a slight diastema in which the inset first oral tentacle scale can be seen ; dorsal arm plates deep fan-shaped or hexagonal, ventral arm plates pentagonal, the distal angles rounded, sometimes not contiguous beyond the arm bases ; arm spines numerous, usually about 10 on the broadest part of the arm in specimens with d.d. 8–10 mm, flattened, the lowest the longest, the middle ones shortest ; tentacle scales two, the inner one on the ventral arm plate becoming very long and sword-shaped after the basal segments and (in preserved specimens at least) aligned obliquely distally across the ventral plate also crossing the corresponding scale, outer scale papilliform or short spiniform.

Ophiopsila bispinosa A. M. Clark

Ophiopsila bispinosa A. M. Clark, 1974 : 472–475, fig. 13.

D.d. up to *c.* 10 mm ; arms very long, d.d./a.l. 1/10+.

A species of *Ophiopsila* with sometimes six arms, the disc scaling rudimentary and transparent except around the long, narrow, bar-like, widely separated radial shields ; oral shields spearhead-shaped, longer than broad, the adorals narrowly contiguous or approximating interradially, only three to five tooth papillae, the two most superficial ones usually forming a symmetrical pair, three or sometimes two, spiniform oral papillae each side and an inconspicuous distal second oral tentacle scale more or less inset into the slit ; dorsal arm plates deep fan-shaped or hexagonal, becoming broader than long on the broadest part of the arms of larger specimens, d.d. > 8 mm, but mostly relatively narrow, ventral arm plates similar in proportions but pentagonal with the laterodistal angles produced on the basal segments, then broadly truncated, the distal edge between them distinctly concave, successive plates becoming separated ; arm spines up to 10, flattened and tapering, the lowest slightly curved and up to almost twice the segment length ; tentacle scales two, the outer one also becoming spiniform after the basal segments but not more than half as long as the inner one.

Type locality. Off the Tugela River mouth, Natal (29/31), 77–150 metres.

Nature of bottom. Mud ; green mud ; sandy mud ; coarse sand and coral.

Ophiopsila seminuda A. M. Clark

Ophiopsila seminuda A. M. Clark, 1952 : 200, 218–219, fig. 3 ; Day, Field & Penrith, 1970 : 81 ; A. M. Clark, 1974 : 470–472, fig. 12.

D.d. up to 10 mm ; d.d./a.l. 1/8+.

A species of *Ophiopsila* with the disc scales fine but continuous above, more tenuous but still distinct below, in larger well-preserved specimens at least, radial shields more or less tapering proximally owing to the overlying scales, widely separated ; oral shields rhombic or hexagonal, usually broader than long but sometimes as long as broad if the distal lobe is prolonged, adorals usually separate interradially but sometimes contiguous, up to seven tooth papillae on each jaw and usually two broad flat blunt oral papillae each side, with a more or less evident smaller second oral tentacle scale distally ; dorsal arm plates hexagonal or oval, in larger specimens, d.d. $>$ 8 mm, becoming distinctly broader than long on the broadest part of the arm beyond the base, ventral arm plates truncated pentagonal, the basal ones with the laterodistal angles produced, the distal edge barely concave ; arm spines up to 10, all spatulate with broad rounded tips, the lowest about half again as long as the segment ; tentacle scales two, the outer one remaining relatively short and blunt. Colour : disc dark, arms white.

Type locality. False Bay, 27–28 metres.

Southern African records. 34/18/FB/s ; 35/22/d ; 34/25/d ; 33/25/s ; 33/27/s ; 22–180 metres.

Nature of bottom. Coarse or shelly sand and rock ; sand and shell ; khaki sand and brown shell ; limestone and sand ; shell ; coarse sand ; dark mud and sand.

Family **OPHIONEREIDIDAE** Ljungman

See : Matsumoto, 1917 : 315 (Ophiochitonidae) ; Spencer & Wright, 1966 : U97.

A family of Chilophiurina with the disc overlying the arm bases, disc scales distinct, though occasionally very fine, rarely with any armament, radial shields small or moderate in size, more or less widely separate ; oral shields usually spear-head-shaped (i.e. with the distal lobe blunter than the proximal and the broadest part distal to the middle of the length) though sometimes described as rhombic, often variable in proportions from broader than long to longer than broad, adorals only narrowly contiguous, if at all, in adult specimens, naked as are the oral plates ; teeth broad quadrangular or with a blunt angled apex, oral papillae numbering four or more each side, continuous distally with or overlapping the scale of the second oral tentacle pore which is prolonged up into the slit, the adapical oral papilla each side often partially infradental but never a cluster of infradental (tooth) papillae ; arms more or less long and stout, widening beyond the base, arm plates well developed, successive dorsal and ventral ones mostly contiguous ; arm spines well developed, erect ; tentacle pores moderate in size, usually with one large rounded scale, sometimes a smaller second one partially underlying it.

Genus *OPHIONEREIS* Lütken

See : A. M. Clark, 1953, *Proc. zool. Soc. Lond.* **123** : 65–92.

Diagnosis as for the family. Additionally the presence of a pair of symmetrical supplementary dorsal arm plates on at least the proximal free arm segments may be noted.

Ophionereis australis (H. L. Clark)

Ophiochiton australis H. L. Clark, 1923 : 345–347, fig. 3.
Ophionereis australis : Mortensen, 1933a : 374–375, fig. 77 ; J. B. Balinsky, 1957 : 24 ; Kalk, 1958 : 237 ; Macnae & Kalk, 1969 : 130.

D.d. up to 11 mm ; d.d./a.l. 1/7–9.

A species of *Ophionereis* with fairly coarse disc scales, genital papillae bordering at least the proximal ends of the slits ; oral shields spearhead-shaped or rhombic, longer than broad, adorals separate or barely contiguous ; dorsal arm plates broad rectangular or hexagonal but with a slight median distal angle in larger specimens, d.d. > 5 mm (but fan-shaped in smaller ones), the proximal plates bordered on each side by an inconspicuous triangular supplementary dorsal arm plate, even the largest ones not running the full length of the segment and in small specimens only present on a few proximal segments, ventral arm plates bell-shaped, broadest distal to the tentacle scales and with the distal side convex ; arm spines three, thick and blunt. Colour brownish, radial shields white with brown edges.

Type locality. SE. of the Tugela River mouth, 86 metres.

Southern African records. 30/31/s, d ; 29/31/s, d ; 26/33/i ; 24/35/d ; intertidal–205 metres.

Nature of bottom. Broken shell ; coarse sand and rock ; gravelly with masses of compacted worm tubes.

Ophionereis dubia (Müller & Troschel)

Ophiolepis dubia Müller & Troschel, 1842 : 94.
Ophionereis dubia : H. L. Clark, 1923 : 343–344 ; Mortensen, 1933a : 374 ; Stephenson, Stephenson & du Toit, 1937 : 380 ; A. M. Clark, 1953, *Proc. zool. Soc. Lond.* **123** : 83–88, figs 9, 10, pl. 2, figs 1, 2 ; Morgans, 1962 : 308, 315 ; Day, Field & Penrith, 1970 : 81.

D.d. up to 9 mm but rarely more than 7 mm ; d.d./a.l. 1/6–8.

A species of *Ophionereis* with the disc scaling very fine, no genital papillae ; oral shields spearhead-shaped or rhombic, usually slightly broader than long, adorals separate or barely contiguous ; dorsal arm plates triangular with the median distal angle rounded, about as broad as loug, bordered on each side by a supplementary dorsal arm plate running the full length of at least the proximal segments, ventral arm plates squarish proximally but becoming pentagonal with a proximal lobe, the median part of the distal side straight or sometimes slightly concave; arm spines three, tapering to blunt points. Colour of disc pale with a reticulate pattern including a transverse bar or a Y opposite the base of each arm across the radial shields, arms with narrow dark bands at intervals of several segments.

TYPE LOCALITY. Egypt, probably Koseir (Qoseir).

SOUTHERN AFRICAN RECORDS. 34/18/FB/s; 34/19/i; 34/20/i; 34/21/i; 32/28/s; 30/31/d; 29/31/s; intertidal–165 metres.

NATURE OF BOTTOM. Sand and shell; broken shell; broken shell and stones; discoloured shell.

Note: See also under *O. vivipara*.

Ophionereis porrecta Lyman

Ophionereis porrecta Lyman, 1960, *Proc. Boston Soc. nat. Hist.* **7**: 260; H. L. Clark, 1923: 344–345; Mortensen, 1933a: 373–374; Stephenson, Stephenson & Bright, 1938: 18; J. B. Balinsky, 1957: 24; Kalk, 1958: 207, 237; Macnae & Kalk, 1969: 130.

D.d. up to 15 mm; d.d./a.l. *c.* 1/7–8.

A species of *Ophionereis* with the disc scaling distinct, appearing moderately coarse in smaller specimens, d.d. < 8 mm, but relatively finer, except peripherally, in larger ones, genital papillae bordering the slits; oral shields oval or pear-shaped, usually slightly longer than broad, adorals separate; dorsal arm plates becoming distinctly broader than long, trapezoidal, the middle part of the distal side flattened, with a well-developed supplementary plate each side distinct for most of the arm, the proximal ones running the full length of the segment, ventral arm plates squarish or pentagonal with rounded distal angles and the edge between them straight or slightly concave; arm spines three, the middle one distinctly the largest, tapering to blunt tips. Colour variegated light and darker brown.

TYPE LOCALITY. Unknown.

SOUTHERN AFRICAN RECORDS. 34/20/i; 33/25/i, s; 33/27/s; 32/28/i, s; 29/31/s; 30/31/d; 26/33/i; intertidal–165 metres.

NATURE OF BOTTOM. Usually rock or combinations of stones, coarse sand and shell.

Ophionereis vivipara Mortensen

Ophionereis vivipara Mortensen, 1933c, *Vidensk. Meddr dansk naturh. Foren.* **93**: 191–192, fig. 7; J. B. Balinsky, 1957: 24; Kalk, 1958: 200, 207, 237; Macnae & Kalk, 1969: 130.

D.d. 3 mm; d.d./a.l. 1/5.

A species of *Ophionereis* with disc scales so reduced as to be barely perceptible in the skin when wet, no genital papillae; oral shields rounded triangular,* adorals broadly contiguous inwardly;* dorsal arm plates rhombic, longer than broad,* only narrowly contiguous, supplementary dorsal arm plates only half the segment length,* ventral arm plates narrow pentagonal,* the short distal side convex; arm spines three, not exceeding the segment in length.

Note: The characters marked * are almost certainly immature features correlated with the small size of the type material. A badly broken Survey specimen from 22 metres off southern Mozambique (24°46′S 34°50′E) and four better ones from near East London (33°09′S 28°02′E) in 84 metres, which I have identified only as '*Ophionereis* sp. aff. *O. dubia*', are possibly referable to *O. vivipara*. They have d.d. 5–7 mm. Their disc scaling is more tenuous than usual in *O. dubia* and probably more comparable to that of *O. vivipara*. In spirit there is no sign of the brown patch on the disc which Balinsky thinks is characteristic of *O. vivipara* but there are narrow brown bands at intervals of three to six segments on the arms (though *O. dubia* also has similar bands). Their dorsal arm plates are broadest at or near the proximal end and taper distally, somewhat as in *O. dubia*, but are more often truncated with a short straight median distal side than rounded, unlike *O. dubia*. On the distal quarter of the arms the younger plates are of the juvenile form with a prolonged proximal lobe so that they are widest medially and the supplementary plates only extend for the distal half of the segment. However, these distal main dorsal arm plates are more hexagonal than rhombic, the distal lobe still being truncated. It remains for more material to show the affinities of these specimens.

Type locality. Cannoniers Point, Mauritius.

Southern African record. 26/33/i.

Nature of bottom. In algal carpet.

Family **OPHIODERMATIDAE** Ljungman

See: Matsumoto, 1917: 307; Spencer & Wright, 1966: U97.

A family of Chilophiurina with the disc surrounding the arm bases and covered above and below with a dense coat of granules (sometimes also with spinelets), often completely concealing the widely spaced radial shields or leaving only a small patch exposed; oral shields rather variable in proportions, usually spearhead-shaped but sometimes truncated oval, especially when accompanied by a distal supplementary shield which completes the curvature (*Ophiarachnella*), usually naked together with the lateral part of each adoral shield between the oral shield and the first ventral arm plate each side, the rest of the jaw covered with granulation coarser than that on the disc, sometimes (e.g. *Cryptopelta*) this coarse granulation extending over the rest of the adorals and the oral shield as well, though tending to leave bare the madreporite, teeth not very broad (in the included species), overlain by the innermost oral papilla from each side but never a cluster, the numerous oral papillae forming a linear series with the second oral tentacle scale (inclining up into the slit) at the distal end; arms well developed, often rectangular in cross-section, broadest basally, appearing rather snake-like, successive dorsal and ventral arm plates mostly broadly contiguous; arm spines short, rarely as long as the segment and usually only half as long, aligned distally and more or less appressed to the arm (in the included species), peg-like or conical, the lower ones often flattened and spatulate; tentacle pores small to moderate, usually with two flat oval scales aligned parallel to the arm axis, the shorter outer one overlying the base of the lowest arm spine.

Cryptopelta aster (Lyman)

Ophiopeza aster Lyman, 1879, *Bull. Mus. comp. Zool. Harv.* **6** : 50, pl. 14, figs 395–397 ; 1882 : 12, pl. 21, figs 16–18.
Cryptopelta aster : H. L. Clark, 1923 : 350–351 ; Mortensen, 1933a : 376–379, figs 78a, 79a, 80a, d, pl. 19, fig. 21.

D.d. up to 13 mm ; d.d./a.l. 1/2 in small specimens, usually 1/3·0–3·5 in larger ones.

A species of Ophiodermatidae with the disc completely covered above and below by a uniform coat of granules extending on to the first arm segments and also covering the entire adoral shields as well as the oral plates and oral shields, though sometimes leaving bare the madreporite, the granulation on the jaws coarser ; two genital slits in each interradius ; usually eight or nine oral papillae in each series, besides the second oral tentacle scale, the proximal ones narrower and more pointed ; dorsal arm plates hexagonal, broadest near the middle, broader than long in larger specimens, d.d. $>$ 10 mm, ventral arm plates bell-shaped, convex distally ; arm spines up to seven, rarely eight, less than half the segment length ; tentacle scale single. Colour ranging from cream through orange to brick-red on the disc, sometimes orange and red mottled, the arms banded.

Type locality. South from Cape Point, 274 metres.

Southern African records. 34/17/d ; 34/18/d ; 35/18/d ; 35/21/d ; 34/25/s ; 33/27/s ; 29/31/d ; 75–420 metres.

Nature of bottom. Usually rock ; sometimes broken shell and rock or sand, shell and rock.

Ophiarachnella capensis (Bell)

Pectinura capensis Bell, 1888, *Proc. zool. Soc. Lond.* **1888** : 282, pl. 16, figs 3, 4.
Ophiarachnella capensis : H. L. Clark, 1923 : 351 ; Mortensen, 1933a : 380–381, fig. 82 ; Stephenson, Stephenson & du Toit, 1937 : 380 ; Stephenson, Stephenson & Bright, 1938 : 18 ; Stephenson, 1944 : 347 ; Day, 1959 : 544 ; Day, Field & Penrith, 1970 : 81 ; A. M. Clark, 1974 : 480–481.

D.d. up to 15 mm (19 mm on Vema Seamount, west from South Africa) ; d.d./a.l. *c.* 1/5.

A species of Ophiodermatidae with the disc covered above and below by uniform granules leaving bare an oval or pear-shaped area on each radial shield ; two genital slits in each interradius ; oral shields with a separate distal supplementary oral shield both bare, also the lateral part of each adoral shield, the rest of the jaw covered with granules coarser than on the disc, usually seven or eight oral papillae in each series, besides the second oral tentacle scale, the proximal ones narrower and pointed ; dorsal arm plates hexagonal, broadest near their distal ends, ventral arm plates bell-shaped, convex distally ; arm spines up to seven, very short and conical, even the lowest less than half the segment length ; two parallel tentacle scales. Colour of preserved specimens yellowish with an irregular dark spot in the middle of the disc and the arms banded.

Type locality. 'Cape of Good Hope' (Dr Andrew Smith).

SOUTHERN AFRICAN RECORDS. 34/18/i ; 34/18/FB/i ; 34/19/i ; 34/20/i ; 34/21/i ; 34/24/i ; 33/25/s ; 33/26/i ; 33/28/s ; 29/31/s ; intertidal–91 metres.

NATURE OF BOTTOM. Rock and black specks ; sand and shell ; under stones at low water.

Ophiochasma nitida Hertz

Fig. 276a, b [see Appendix, p. 260]

Ophiochasma nitida Hertz, 1927a : 116–117, pl. 9, figs 13, 14 ; Mortensen, 1933a : 216.

D.d. up to 12 mm (only three specimens known) ; d.d./a.l. *c.* 1/6.

A species of Ophiodermatidae with the disc granulation probably completely covering the scales in life (partly rubbed off in the known specimens) but leaving almost completely bare the very large, convex, oval radial shields, which are separated, even interradially, by not more than their own breadth ; two genital slits in each interradius ; oral shields hexagonal or shield-shaped with a proximal angle and a very short distal lobe, longer than broad, a regular, almost semicircular disc plate like a supplementary oral shield present distal to each main oral shield but in life probably covered more or less completely with granules (in one specimen a small rounded convex area near the proximal side may have been bare in life), lateral parts of adoral shields bare but their proximal and distal ends and the oral plates and distal edges of the oral shields covered with granules not appreciably larger than those on the disc, oral papillae numbering seven or eight (rarely nine) each side, in series with the second oral tentacle scale, the distal papillae (when seven) markedly broader ; dorsal arm plates broad hexagonal, their surface convex, ventral arm plates broad octagonal, or the three distal sides forming a continuous curve ; up to 10 short arm spines, appressed or slightly projecting, tapering to points, the lower ones blunter, none more than half as long as the segment ; two tentacle scales.

TYPE LOCALITY. Agulhas Bank (35/21), 102 metres (Hertz's depth of 86 metres is incorrect for this station).

Ophioderma wahlbergi Müller & Troschel

Fig 276c, d [see Appendix, p. 262]

Ophioderma wahlbergi Müller & Troschel, 1842 : 87 ; Bell, 1905b : 255 ; H. L. Clark, 1923 : 353.
Ophioderma leonis Döderlein, 1910 : 252–253, pl. 5, fig. 1 ; H. L. Clark, 1923 : 351–352 ; Mortensen, 1933a : 381–382 [non fig. 83] ; Day, 1959 : 502, 544 ; Morgans, 1959 : 424, 425 ; Grindley & Kensley, 1966 : 12 ; Day, Field & Penrith, 1970 : 81.

See Appendix for discussion of synonymy (p. 262).

D.d. up to 32 mm ; d.d./al. usually *c.* 1/3 but individual arms may be shorter giving a ratio nearer 1/2 (Döderlein).

A species of Ophiodermatidae with the disc granulation complete or in larger specimens, d.d. > 15 mm, leaving parts of some or all of the radial shields bare, rarely some other disc scales partly exposed ; genital slits divided into two short consecutive ones on each side of each ventral interradius ; oral shields rounded triangular with the distal side more or less convex, sometimes so much so as to make the shape rhombic, usually slightly broader than long, lateral parts of adoral shields bare but their proximal ends and the oral plates covered with granules slightly coarser than those on the disc, oral papillae numbering five or six each side, in series with the second oral tentacle scale, the distal papillae broader, no supplementary oral shields ; arm spines seven or eight basally (nine on a few segments at d.d. > 30 mm only) ; tentacle scales two. Colour of preserved specimens grey, brown or black above, paler below, sometimes with an abrupt rather than gradual change.

TYPE LOCALITY. Supposedly Port Natal (Durban) but it is incredible that such a conspicuous shallow-water ophiuroid should have escaped notice there in recent years. In the absence of other records from east of Cape Agulhas it seems most likely that Wahlberg collected it in the vicinity of Cape Town, where he stopped *en route* to Natal. I am obliged to Dr Andersson for this information.

SOUTHERN AFRICAN RECORDS. 26/15/? i or s ; 33/18/i, s ; 34/18/s ; 34/18/FB/i, s ; 34/19/i ; intertidal–75 metres.

NATURE OF BOTTOM. Shell, stones, rock ; fine green sand and shell.

Ophiopeza fallax Peters

Ophiopeza fallax Peters, 1851 : 465–466 ; A. M. Clark & Rowe, 1971 : 127, pl. 21, figs 5, 6.
Ophiopezella decorata Mortensen, 1933a : 379–380, fig. 81, pl. 19, fig. 24 ; J. B. Balinsky, 1957 : 28 ; Kalk, 1958 : 238 ; B. I. Balinsky *In* : Macnae & Kalk, 1969 : 106, 130.

D.d. up to 15 mm ; d.d./a.l. 1/3·0–3·5.

A species of Ophiodermatidae with a coat of extremely fine indented granules covering it usually completely, including the radial shields but sometimes leaving one or more of these or a few peripheral scales partially bare ; two genital slits in each interradius ; oral shields and the lateral part of each adoral bare but the rest of the jaw covered by granules, if a supplementary oral shield is developed distal to the main one it is normally covered by the disc granulation, usually eight, sometimes nine oral papillae in each series, besides the second oral tentacle scale ; dorsal arm plates fan-shaped or sometimes hexagonal, rather variable, their distal edge medially convex, straight or slightly concave, even irregularly notched (sometimes all these on consecutive segments of the same specimen), ventral arm plates bell-shaped, convex distally ; up to 13 short, mostly conical arm spines, even the lowest hardly more than half the segment length ; two tentacle scales. Colour mottled or variegated brown and grey on the disc, arms with dark brown bands dorsally.

TYPE LOCALITY. Querimba, northern Mozambique.

SOUTHERN AFRICAN RECORDS. 29/31/i ; 26/33/i ; 25/32/i ; 24/35/i.

NATURE OF BOTTOM. Reef.

Family **OPHIOLEUCIDAE** Matsumoto

See : Matsumoto, 1917 : 303 ; Spencer & Wright, 1966 : U96.

A family of Chilophiurina with the disc usually distinctly overlying the arm bases and partially or completely covered with a coat of granules, often also concealing the radial shields, otherwise the scales naked or skin-covered ; oral and adoral shields well developed, sometimes covered with granules which may also obscure the oral plates, teeth not very broad, mostly longer than broad and blunt papilliform, oral papillae more or less numerous, the series continuous distally with one or more scales of the second oral tentacle pore, apically one to three papillae below the lowest tooth but no cluster of papillae ; arms long and slender, often carinate (at least in the species represented), at least the proximal consecutive dorsal and ventral arm plates broadly contiguous ; arm spines very variable in size and number, usually aligned distally and inconspicuous but sometimes longer and flaring ; tentacle pores variable in size, sometimes enlarged proximally but becoming small on the arms, with one to several scales, usually several on the pores of the first segment.

Ophiernus vallincola Lyman

Ophiernus vallincola Lyman, 1878, *Bull. Mus. comp. Zool. Harv.* **5** : 122–123, pl. 6, figs 170–172 ; 1882 : 32–33, pl. 24, figs 16–18, pl. 38, figs 6–9 ; H. L. Clark, 1923 : 365 ; 1939 : 134.

D.d. up to 21 mm ; d.d./a.l. 1/5·5–7·0.

A species of Ophioleucidae with scales developed in the peripheral part of the disc between the radial shields and bearing scattered (easily lost) granules, the ventral interradii similarly covered but the whole central part of the upper side covered with skin, the bare radial shields almost circular, each pair only narrowly separated (contrary to Lyman's descriptions but as shown in his figures) ; oral shields rounded triangular, bare like the adorals, which may or may not meet interradially, and the oral plates, oral papillae five or six each side (including a small, conical infradental one offset from the rest) in series with three or four scales half encircling the second oral tentacle pore, of which the two more proximal and superficial are papilliform and may also count as oral papillae but the two distalmost, arising from the sunken first ventral arm plate, are squamiform ; arms slightly keeled above, proximal dorsal arm plates broader than long, the distal edge almost straight, proximal ventral arm plates bell-shaped and contiguous but becoming reduced and separated for the distal half of the arm ; arm spines three, slender and pointed, on the basal segments arising from the middle of the lateral arm plates but shifting to the distal edge on segments beyond the disc, mostly only half as long as the segment but longer on some of the proximal free segments, almost equal to the segment length ; tentacle scales mostly two, papilliform and on the proximal side of the smaller pores on the arms but often three with one on the lateral side of the enlarged proximal pores. Colour 'madder blue and dragon's blood red' on the disc, arms red, the distal ends of the segments paler (H. L. Clark, 1939 ; Arabian Sea specimens).

Type locality. Azores, 1830 metres.

Southern African records. 34/17/vd ; 840–1628 metres.

Nature of bottom. Green mud.

Ophiocirce inutilis Koehler

Ophiocirce inutilis Koehler, 1904b, *Siboga Exped.* **45a** : 13–14, pl. 3, figs 4, 5 ; H. L. Clark, 1939 : 131–132 ; A. M. Clark, 1974 : 476–477, fig. 14.

D.d. up to 10 mm ; d.d./a.l. 1/7·0–8·5.

A species of Ophioleucidae with the disc completely covered (including the radial shields) by very fine granules above and below, uniform except on the disc margin radially where they are slightly enlarged; coarser granules extending on to the lateral arm plates below the disc and over the entire jaws, obscuring the oral and adoral shields and the oral plates, though sometimes leaving a patch of an oral shield naked, seven or eight oral papillae, the distalmost one much broadened, the second oral tentacle scale sometimes superficial and in sequence with the papillae but often inset into the slit; arms slightly keeled above, dorsal arm plates with the distal edge almost straight or with an obtuse angle, proximal ventral arm plates markedly wider distal to the tentacle pores than proximally, their distal edge slightly concave medially; arm spines only two, the longest ones only just over half the segment length, lateral arm plates with a fingerprint-like texture; tentacle scale single, large and rounded on most pores but a few proximal pores may have a smaller second scale behind the first and the pores of the first segment have four to six scales encircling all but their proximal sides.

Type locality. Near Timor, East Indies, 112 metres.

Southern African record. 24/35/d; 132 metres.

Ophiopallas paradoxa Koehler

Ophiopallas paradoxa Koehler, 1904b, *Siboga Exped.* **45a**: 12–13, pl. 3, figs 1–3; 1922: 436–437, pl. 79, figs 1, 2; A. M. Clark, 1974: 477–478, fig. 15.
Ophiopallas paradoxa altera Hertz, 1927b: 110, pl. 9, fig. 5.

D.d. up to 8 mm; d.d./a.l. *c.* 1/12.

A species of Ophioleucidae with the disc closely and completely covered (including the radial shields) by moderately coarse granules, giving way to papillae along the edges of the long genital slits, which extend up around the sides of the arm bases; oral shields very large, bare, with a rounded proximal angle, lateral constrictions for the genital slits and a very broad distal lobe, broadest distally and longer than broad altogether, adorals separate proximally, a few coarse granules scattered on the oral plates, four to six oral papillae, the outermost one usually opercular, in series with a second oral tentacle scale; dorsal arm plates with the distal edge almost straight or slightly notched medially, proximal ventral arm plates as broad proximal to the pores as distally; arm spines up to eight, the upper ones on the proximal free segments long and evenly tapering, exceeding the segment in length but the lower spines progressively reduced to only half the segment length; tentacle scale single and very large and oval on the pores after those of the first segment which have two scales.

Type locality. Celebes–Moluccas area, East Indies, 204–450 metres.

Southern African records. 24/35/d; 27/32/d; 347 – 376 (?384) metres.

Family **OPHIURIDAE** Lyman

See: Matsumoto, 1917: 233 (Ophiolepididae); Mortensen, 1927: 228 (Ophiolepidae); Spencer & Wright, 1966: U93 (Ophiuridae).

A family of Chilophiurina with the disc surrounding the arm bases, covered above and below with naked scales (rarely a few inconspicuous spinelets or tubercles), primary rosette often conspicuous as are the radial shields, scaling sometimes obscured by thick skin, a comb of papillae or spinelets often present distal to each radial shield continuous ventrally with papillae bordering the genital slits ; oral and adoral shields variously developed (ill-defined in the aberrant *Astrophiura permira* where the arm bases are integrated into the disc to form a flat pentagonal body with diminutive arms at the angles) ; teeth narrow, pointed or rounded, no tooth papillae but usually a single infradental papilla and a continuous series of pointed or rounded oral papillae each side leading to the single or multiple scales of the second oral tentacle pore according to whether this is inset into the oral slit (e.g. in *Ophiolepis*) or superficial on the ventral face (*Ophiura*) ; arms moderate in length or short, usually widest basally and tapering, flattened cylindrical, not at all moniliform, the proximal ends of the segments not constricted, dorsal and ventral arm plates usually small, only the basal ones, if any, contiguous with each other, occasionally both series of plates lost altogether beyond the arm base, the rest of the arm covered by the large lateral arm plates, spines often short and appressed but sometimes the upper ones at least longer, even needle-like and erect ; tentacle pores either small throughout and armed with two semicircular scales or, more often, large on the disc with multiple papilliform or squamiform scales, reducing on the arms or even lost altogether.

Amphiophiura trifolium Hertz

Amphiophiura trifolium Hertz, 1927b : 78–79, pl. 6, figs 14, 15 ; H. L. Clark, 1939 : 108–109 ; A. M. Clark, 1974 : 476.

D.d. up to 11 mm ; d.d./a.l. *c.* 1/3–4.

A species of Ophiuridae with a thick disc covered with rather ill-defined scales and plates, the central plate usually and the primary radials and interradials often enlarged, radial shields broad rounded triangular, usually just contiguous distally, less than half the disc radius in length, arm combs well developed with coarse papillae becoming broader ventrally, overlying a secondary comb of finer blunt papillae ; oral shields with an extremely large broadly rounded distal lobe, the much narrower proximal angle constricted opposite the ends of the genital slits, forming what has been called a trefoil shape, adoral shields narrow, contiguous proximally, oral papillae five or six each side of an apical one, none very well defined ; proximal dorsal arm plates narrow bell-shaped, longer than broad, about four proximal ones contiguous, the rest separate, ventral arm plates squatter bell-shaped, broad distal to the tentacle pore but markedly tapering proximally, none contiguous ; arm spines three or two, very small, up to a third the segment length, distally the middle spine (or upper one of two) becoming hook-like.

TYPE LOCALITY. Off southern Somalia, 1289 and 1644 metres.

SOUTHERN AFRICAN RECORDS. 27/34/vd ; 24/36/vd ; 1140–1335 metres.

NATURE OF BOTTOM. Hard sand and rock ; globigerina ooze.

Astrophiura permira Sladen

Astrophiura permira Sladen, 1879, *Ann. Mag. nat. Hist.* **5** : 401–415, pl. 20 ; Hertz, 1927b : 83–85, pl. 7, figs 4, 5 ; Mortensen, 1933a : 394–396, figs 90, 91.
Astrophiura cavellae Koehler, 1915 : 1, figs 1–6.

D.d. (expanded disc) up to 14 mm, the free arm length probably not exceeding this measurement, more likely half of it (arms very breakable).

A species of Ophiuridae with the disc enlarged into a flattened pentagon (often arched so as to be concave below, apparently for clinging limpet-like to the substrate) by the incorporation of a number of proximal arm segments of which the lateral plates are highly modified, wing-like, extending laterodistally, the innermost meeting that of the adjacent arms, all stopping short on a line at right angles to the interradius where their spines form a continuous horizontal fringe to the enlarged 'disc', the plates in the centre of the upper side fairly regular and symmetrical but the superficial ventral plates very thin, polygonal and forming a translucent mosaic ; oral shields, except the distinct madreporite, only definable as the innermost plate of the mosaic in each interradius, the adorals also indistinct, the jaws dominated by the broadly contiguous stout oral plates, three or four oral papillae each side of an apical one ; the proximal dorsal and ventral arm plates incorporated in the disc rectangular, markedly reduced and lost at the bases of the free parts of the arms, most of which are made up by the lateral arm plates ; arm spines three or less and very inconspicuous on the free arms but three or four and distinctly larger on each lateral plate of the 'disc' ; tentacle pores much enlarged on the 'disc', with one or two scales, pores lacking on the free arms.

Type locality. Madagascar.

Southern African records. 34/18/d ; 35/22/d ; 27/32/d ; 200–376 metres. [The maximum depth from a new record at 27°33′S 32°44′E.]

Nature of bottom. Coarse khaki sand and stones ; grey sand, rock and 'coral'. [Matsumoto found his Japanese *Astrophiura kawamurai*, which Mortensen thinks conspecific with *A. permira*, attached to a stone.]

Dictenophiura anoidea H. L. Clark*

Dictenophiura anoidea H. L. Clark, 1923 : 361–363, pl. 19, figs 1, 2 ; Mortensen, 1933a : 388–390, fig. 86 ; 1936 : 339 ; Morgans, 1962 : 303, 315, 322 ; Day, Field & Penrith, 1970 : 81.
Ophiura carnea : Hertz, 1927b : 69. [Non *O. carnea* M. Sars, 1857.]

D.d. up to 10 mm ; d.d./a.l. *c.* 1/2·5.

A species of Ophiuridae with the disc flat but thick at the edge, rising vertically from the base of each arm, plates rather coarse, the primaries usually distinct, radial shields partly contiguous, with a swollen median indented or medially furrowed plate distal to each pair bearing a fine secondary arm comb each side opposing the papilliform primary comb ; oral shields with a sharp proximal angle and a longer distal lobe tapering only very slightly to a broadly rounded distal end, adorals

* Madsen (1971) reduces *Dictenophiura* to the rank of a subgenus of *Ophiura*, I think with some justification.

contiguous inwardly, about three oral papillae each side of an apical one, the distalmost the broadest, second oral tentacle pore superficial, with about three squamiform scales each side ; dorsal arm plates fan-shaped, the proximal ones overlapping, ventral arm plates triangular but distally convex, not contiguous ; arm spines three, tapering, the two lower ones up to half the segment length, the upper one slightly spaced and longer but still less than the segment length, even at its longest ; tentacle scales three on the first arm pore, falling to one on the third and subsequent pores. Colour orange with orange bands on the arms.

TYPE LOCALITY. False Bay, 40 metres.

SOUTHERN AFRICAN RECORDS. 32/17/d ; 34/18/s, d ; 34/18/FB/s ; 34/20/s ; 34/21/s ; 34/23/s, d ; 34/24/d ; 34/25/s, d ; 33/25/s ; 33/27/s ; 32/28/s ; 38–181 metres.

NATURE OF BOTTOM. Usually sand (coarse or fine) and shell, occasionally rock and once green mud.

Ophiocten latens Koehler

Ophiocten latens Koehler, 1906, *Mém. Soc. zool. France* **19** : 13, pl. 1, figs 9, 10 ; Mortensen, 1933a : 392–393.
Ophiocten pacificum : H. L. Clark, 1923 : 364. [Non *O. pacificum* Lütken & Mortensen, 1899.]

D.d. up to 12 mm ; arms usually badly broken.

A species of Ophiuridae with the flattened disc covered with some spaced, enlarged symmetrical circular plates, including a primary rosette, all encircled by multiple series of small scales, radial shields widely separated by almost their own breadth, the disc hardly at all indented midradially distal to each pair and sometimes bearing a series of (easily lost) comb papillae but no papillae on the basal dorsal arm plates ; oral shields pentagonal, the proximal angle very obtuse, much broader than long, adorals short, contiguous inwardly, about five indistinct oral papillae each side of an apical papilla, second oral pore superficial, with three squamiform scales ; dorsal arm plates trapezoidal, proximal ones broadly contiguous, the distal edge only slightly convex, ventral arm plates very short with a slight median proximal angle and a deeply convex distal lobe, none contiguous ; arm spines three, slender and pointed, not exceeding the segment in length ; tentacle scales two, small and pointed.

TYPE LOCALITY. NE. of the Azores, 4060 metres.

SOUTHERN AFRICAN RECORDS. 33/17/vd ; 34/17/vd ; 1646(? 1463)–2177 metres.

NATURE OF BOTTOM. Green mud ; grey mud.

Ophiolepis cincta Müller & Troschel

Ophiolepis cincta Müller & Troschel, 1842 : 90 ; Mortensen, 1933a : 382–383 ; J. B. Balinsky, 1957 : 28 ; Kalk, 1958 : 207, 216, 238 ; B. I. Balinsky *In* : Macnae & Kalk, 1969 : 106, 130 ; A. M. Clark & Rowe, 1971 : 90, 129.

D.d. up to 18 mm (rarely > 14 mm) ; d.d./a.l. usually 1/3·5–4·0.

A species of Ophiuridae with the disc covered with scales of two sizes, large proximally truncated ones encircled by single series of much smaller scales, the radial shields not appreciably larger than the large scales ; oral shields as long as or longer than broad, with a proximal angle between two slightly concave proximal sides and a distal lobe with parallel sides and a rounded distal end, adorals broad and fairly short, broadly contiguous inwardly, about four rounded oral papillae each side, the outermost broadest and in series with a single curved oral tentacle scale, the second oral pore being inset into the slit, not superficial ; dorsal arm plates broad rounded, like the larger disc scales bordered laterally and distally by a series of much smaller scales, ventral arm plates truncated pentagonal, slightly broader distal to the tentacle pores ; arm spines three or four, short, conical ; tentacle scales two, slightly elongate semicircular, together making an ovate shape. Colour brownish on the disc with irregular whitish patches, white bands on the arms, the white areas finely marbled with greyish.

Type locality. Red Sea.

Southern African records. 29/31/i ; 26/32/i ; 26/33/i.

Nature of bottom. On coral, coral debris or under stones.

Ophiomisidium pulchellum (Wyville Thomson)

Ophiomusium pulchellum Wyville Thomson, 1877, *The Voyage of the 'Challenger' : The Atlantic.* London. **2** : 67–69, figs 18, 19 ; Lyman, 1882 : 96–98, pl. 3, figs 1–3.

Ophiomisidium pulchellum : H. L. Clark, 1923 : 356–357 ; Hertz, 1927b : 82 ; A. M. Clark, 1974 : 476.

D.d. up to 5 mm ; d.d./a.l. 1/1·5–2·0.

A species of Ophiuridae with the upper side of the disc covered by only the large rosette, the radial shields with a small radial plate distal to each pair and two plates in each interradius, the outer one forming the more or less vertical high margin, at least the primary rosette plates and in the largest specimens all the others too with a median knob-like tubercle (two on the marginal plates), ventrally a narrow median flattened plate projecting beyond the margin in the space between the adjacent arm spines of the extended second lateral arm plates ; oral shields small, pentagonal or rhombic, as long as broad or longer, adorals larger, broadly contiguous inwardly, obviously homologous with the lateral arm plates, oral papillae fused together each side of the apex, second 'oral' tentacle pore with the adjacent first ventral arm plate completely superficial and unconnected with the oral slit, its scale single and rounded ; arms very short, at d.d. 4 mm consisting of only about 15 segments (two within the disc), tapering markedly near the disc but more attenuated distally ; dorsal arm plates broad triangular but very small, widely separated even proximally, markedly convex in the largest specimens with a high median distal tubercle, like those on the disc, the ventral arm plates bell-shaped on the proximal segments with pores, usually only the first two or three just contiguous, on the segments without pores the ventral plates very reduced and widely separated, the second lateral arm plate markedly broadened below the disc so as to form much of the ventral edge of the disc, its three or sometimes four spines enlarged, flattened and aligned horizontally, other segments with two or three short blunt spines, rarely four on the first free segment, rapidly becoming reduced ; up to five pairs of tentacle pores including the potentially 'oral' pair in each radius (the South African paratype with d.d. 5 mm has a sixth pore on one side only of two fifth arm segments but at d.d. 2·5 mm there are only a total of three pairs of pores in each radius), each pore with a single scale.

TYPE LOCALITY. SW. of the Canary Islands, 3063 metres.

SOUTHERN AFRICAN RECORDS. 34/18/s ; 35/18/d ; 35/21/d ; 35/22/d ; 70–275 metres.

NATURE OF BOTTOM. 'Rough' ; sand ; coarse khaki sand and stones.

Ophiomusium lymani Wyville Thomson

Ophiomusium lymani Wyville Thomson, 1873, *The Depths of the Sea.* London. pp. 172–175, figs 32, 33 ; H. L. Clark, 1923 : 364 ; Mortensen, 1933a : 394.
Ophiomusa lymani : Hertz, 1927b : 103–105 ; H. L. Clark, 1939 : 128.

D.d. up to 48 mm ; d.d./a.l. 1/4–6.

A species of Ophiuridae with the thick disc covered with large and small plates among which the primary rosette is often distinct, the large plates of the upper side, including the long radial shields (length about half the disc radius) studded with more or less numerous little tubercles, more densely so in larger specimens, d.d. > 30 mm ; oral shields triangular, bordered distally by a broader pentagonal plate covering much of the interradius, adorals large, broadly contiguous inwardly, oral papillae five or six each side forming a close series but no evident second oral tentacle scale at the distal end, the pore being concealed within the oral slit which is demarcated distally by the small pentagonal first ventral arm plate ; arms slender but stiff, dorsal arm plates very small, triangular with the distal edge convex, all widely separate, ventral arm plates only present on the first two arm segments, to which the pores are also restricted, pentagonal but excavated for the tentacle scales, lateral arm plates covering most of the arms ; arm spines extremely small, usually six to eight but up to 13 in the largest specimens ; tentacle scales single, oval, only two (very occasionally three) pairs developed in each radius. Colour yellowish-brown.

TYPE LOCALITY. Faeroe Channel, 'warm area'.

SOUTHERN AFRICAN RECORDS. 33/16/vd ; 34/17/vd ; 1380–1830 metres.

NATURE OF BOTTOM. Green mud.

Genus ***OPHIURA*** Lamarck

See : Matsumoto, 1917 : 266.

A genus of Ophiuridae with the disc flat (in most preserved specimens at least), covered with mainly small scales or plates but the primary rosette usually distinct, radial shields not broadly contiguous, if at all, an arm comb usually distinct distal to each one between which the disc is more or less indented, disc scales occasionally obscured by thin skin or bearing scattered inconspicuous spinelets ; oral shields usually well developed but the distal lobe (in the species represented) not markedly enlarged, oral papillae more or less numerous, second oral tentacle pore superficial, with at least two scales but more often several each side ; arms of moderate size or stout, dorsal arm plates usually well developed, at least the proximal ones more or less broadly contiguous except in small specimens, ventral arm plates small, few if any contiguous ; arm spines ranging from small, peg-like and appressed to long and needle-like (especially the upper ones) and somewhat flaring but usually not conspicuous ; tentacle pores present for most or all of the arms but sometimes very small beyond the disc with only one or no scales, the basal ones with at least two scales.

Ophiura (Ophiura) affinis simulans (Mortensen) ***new comb.***

Ophiocten amitinum : H. L. Clark, 1923 : 363–364 ; Mortensen, 1933a : 390–391, fig. 88a ; Morgans, 1962 : 317, 322.

Ophiocten amitinum var. *microplax* Mortensen, 1933a : 391–392, fig. 88b.

Ophiocten amitinum var. *simulans* Mortensen, 1936 : 337, fig. 48b ; Day, Field & Penrith, 1970 : 81.

D.d. up to 7·5 mm ; d.d./a.l. *c.* 1/3.

Nomenclature. Following Mortensen's comment (1936) that the South African ophiuroid he then called *Ophiocten amitinum* var. *simulans* may be indistinguishable from the north European *Ophiura affinis*, I made a comparison between them and also the holotype and paratypes of *Ophiocten amitinum* Lyman from Kerguelen. This substantiated Mortensen's observation and at the same time revealed several minor but consistent differences between *Ophiocten amitinum* and *simulans* :

Ophiocten amitinum	*simulans*
Radial shields well separated by a band of scales leaving no distinct notch.	Radial shields approximating distally and only narrowly separated, notch distinct.
Comb papillae squat, even the longer ones not more than twice as long as broad.	Longer comb papillae almost spiniform.
Arms convex above.	Arms distinctly carinate.
Uppermost arm spine at its longest more than twice the segment length, markedly thicker than the next spine.	Uppermost arm spine only slightly exceeding the segment length, if at all, not appreciably thicker than the next spine.
Oral shields not longer than broad, smooth.	Oral shields longer than broad, often half again as long, their surface textured into folds.
Tentacle scales on arm pores longer than broad, tapering to a point.	Tentacle scales not longer than broad, evenly rounded.

Cumulatively, I think that these differences amount to at least a specific distinction between *Ophiocten amitinum* from Kerguelen and the South African species. The differences between the latter and *Ophiura affinis* are much less, the only ones I could find being the slightly smaller arm spines of the proximal segments with even less difference between the uppermost and middle ones and the less tapering upper comb papillae in European specimens. The distinction between *Ophiura* and *Ophiocten* is clearly very slight, with *Ophiura affinis* bridging the gap between them.

Diagnosis. A species of *Ophiura* with the flattened disc covered with some spaced large symmetrical circular plates, including the rosette, all encircled by much smaller scales (as in *Ophiocten*), radial shields approximating distally and only narrowly separated, if at all, by a wedge of scales, the edge of the disc slightly indented midradially distal to them so that the two arm combs are discrete, some smaller papillae also present in the indentation on the basal dorsal arm plates of larger specimens, d.d. > 5 mm, sometimes forming a V-shaped series so as to make a secondary arm comb ; oral shields longer than broad, the distal lobe only slightly tapering to a broadly rounded end, the surface rather corrugated, adorals meeting inwardly, ribbon-like, usually only three oral papillae each side of the apical one, the distalmost one very broad, the oral tentacle pore slightly set back, with only one scale each side ; dorsal arm plates trapezoidal, the proximal ones broadly contiguous, distinctly carinate but in profile forming a flat top to the arm, ventral arm plates approximately rhombic, none contiguous, no hollows between the basal ones ; arm spines three, slender and pointed, the two upper ones not markedly dissimilar and about equal to or slightly longer than the segment ; two tentacle scales on the first two arm pores, then one, broad and rounded, not longer than broad, the tentacles and pores distinct for most of the arms. Colour in one case brown on the disc and arms orange, in another the disc pale orange, white below.

Mortensen's forma *microplax* from the shore near Port Elizabeth is based on small specimens up to only 2·5 mm d.d., which appear to differ from offshore specimens of the same size in having relatively smaller radial shields.

Type locality. False Bay, 35 metres.

Southern African records. 32/17/d ; 33/17/s–d ; 34/18/d ; 34/18/FB/s ; 34/21/s ; 34/23/d ; 34/25/s–d ; 33/25/i ; 33/26/s, d ; 33/27/d ; intertidal–208 metres.

Nature of bottom. Usually sand, fine, coarse, yellow to green or mixed with shell, occasionally muddy.

Ophiura (Ophiura) flagellata (Lyman)

Ophioglypha flagellata Lyman, 1878, *Bull. Mus. comp. Zool. Harv.* **5** : 69, pl. 2, figs 49–51 ; 1882 : 42, pl. 4, figs 16–18.

Ophiura flagellata : Matsumoto, 1917 : 273–274 ; Koehler, 1922 : 375–377, pl. 85, figs 1, 6, 7, pl. 86, figs 1–4, 10 ; H. L. Clark, 1923 : 359–360 ; Mortensen, 1933a : 383–384.

D.d. up to 26 mm ; no intact arm measurements recorded.

A species of *Ophiura* with the disc flat, the scales obscured by skin though a central plate may be distinguishable in some dried specimens, radial shields very short and widely separated, arm combs well developed, their spinelets short and conical or sometimes elongate, the combs widely separated mid-radially ; oral shields rounded pentagonal or constricted medially opposite the genital slits and the distal lobe somewhat narrowed, as long as or longer than broad, adorals not touching inwardly, about five conical oral papillae each side of the apical one, the outermost not broadened, in series with the multiple abradial scales of the oral pores which open into the edge of the oral slit ; dorsal arm plates very broad hexagonal, broadly contiguous, proximal ventral arm plates almost rhombic, just contiguous, no hollows between them ; arm spines three, similar and close together, flat with spatulate tips and about equal to the segment in length ; tentacle scales numerous, still three or four well out along the arms.

TYPE LOCALITY. SE. of Japan, 620 metres.

SOUTHERN AFRICAN RECORDS. 33/16/vd ; 34/17/vd ; 840–1830 metres.

NATURE OF BOTTOM. Sandy mud ; green mud ; grey mud.

Ophiura (Ophiura) kinbergi Ljungman

Ophiura kinbergi Ljungman, 1867a, *Ofvers. K. VetenskAkad. Förh. Stockh.* **1866** : 166 ; Matsumoto, 1917 : 271–272, fig. 73.

D.d. up to 10 mm ; d.d./a.l. 1/3·5–4·0.

A species of *Ophiura* with the disc covered with thin scales of two sizes, larger rounded symmetrical ones, including the primary rosette, encircled by much smaller ones, radial shields conspicuous, longer than broad, approximating distally but separate, arm combs conspicuous, formed of needle-like spiniform papillae ; oral shields pentagonal but more or less constricted medially opposite the genital slits, longer than broad, adorals narrowly touching inwardly, about four pointed oral papillae each side of the apical one, about three scales each side of the oral pore ; arms convex above, dorsal arm plates trapezoidal with the distal edge somewhat convex, the proximal ones broadly contiguous, ventral arm plates all very small and widely separated, the lateral plates partially gaping between the basal three to five ventral plates and sunken to form a conspicuous hollow on each segment ; arm spines three, tapering, similar and evenly spaced, on the proximal segments about equal in length to the segment ; tentacle scales three or four on the first few segments, reducing to one broad rounded scale on the free arms.

TYPE LOCALITY. 'Sidney, New Holland.'

SOUTHERN AFRICAN RECORD. 29/31/43 metres. EXTENSION OF RANGE FROM SOUTHERN ARABIA [St. NAD 86, 29°10'S 31°51'E].

NATURE OF BOTTOM. Sand.

Ophiura (Ophiura) trimeni Bell

Ophiura trimeni Bell, 1905b : 257–258, pl. 1, figs 3, 4 ; H. L. Clark, 1923 : 360–361 ; Mortensen, 1933a : 384–385, fig. 84 ; A. M. Clark, 1974 : 475–476.

Gymnophiura novembris Hertz, 1927b : 72–73, pl. 6, figs 9, 10 ; Mortensen, 1933a : 393–394, fig. 89.

D.d. up to 10 mm ; d.d./a.l. *c.* 1/5.

A species of *Ophiura* with the disc scales obscured by skin, though the rosette or at least the central plate is usually distinguishable among the smaller scales, some of which bear inconspicuous scattered spinelets, radial shields long and narrow, their length equal to about half the disc radius, diverging inwardly, approximating distally, arm combs small and inconspicuous, the few small papillae at the upper end often lost in preservation ; oral shields broad pentagonal, distally flattened and constricted medially, broader than long or sometimes as broad as long, adorals meeting broadly inwardly, three to five small pointed oral papillae each side of the apical one continuous with the numerous blunter abradial scales of the oral tentacle pores which open into the edge of the slit ; dorsal arm plates relatively long and narrow, only a few of the proximalmost ones contiguous, the first two or three plates with the distal edge medially concave but the shape rapidly changing to elongate bell-shaped with a prolonged median distal blunt angle, ventral arm plates small, only the first two or three just contiguous, not separated by hollows ; arm spines three, almost needle-like, the uppermost one abruptly stouter and longer, exceeding the segment in length and sometimes equal to two segments ; tentacle scales numerous on the basal segments but reduced to one small finger-like papilla on the proximal third of the free arms where the lateral plates are notched to accommodate the pores. Colour red and white or orange.

Type locality. West of Cape Town, 285–420 metres. (Although two of Bell's four samples are labelled as syntypes, it is not possible to narrow down the type locality since they have station numbers, 110 and 115, which cannot be matched with the reference numbers in Bell's report.)

Southern African records. 32/16/d ; 32/17/d ; 33/16/d ; 33/17/d ; 34/17/d ; 34/18/d ; 35/18/d, vd ; 36/21/d ; 183–564 metres.

Nature of bottom. Dark green or green sand ; green sand and black specks ; 'rough' bottom ; green sand and mud ; dark green mud ; a small specimen from a sponge.

Ophiura (Ophiuroglypha) costata costata (Lyman)

Ophioglypha costata Lyman, 1878, *Bull. Mus. comp. Zool. Harv.* **5** : 76–77, pl. 4, figs 92–94 ; 1882 : 50, pl. 5, figs 1–3.
Ophiozona capensis Bell, 1905b : 256–257, pl. 1, figs 1, 2.
Ophiura costata : H. L. Clark, 1923 : 357–358 ; A. M. Clark, 1952 : 201.
Ophiuroglypha capensis : Hertz, 1927b : 90–91, pl. 7, fig. 10.
Ophiura (Ophiuroglypha) costata : Mortensen, 1933a : 385–386, fig. 85a, d.

D.d. up to 25 mm ; d.d./a.l. 1/2·2–3·0.

A subspecies of *Ophiura* with the disc scales somewhat irregular, mostly coarse, the rosette usually distinct, a broad subambital plate in each interradius sometimes bulging but otherwise the contours of the disc fairly smooth, radial shields oval, broadest medially, more than half as broad as long, the length equal to about two-fifths of the disc radius, separated except in small specimens, d.d. $<$ 5 mm, by a single series of broad midradial scales continuous with the basal dorsal arm plates, each arm comb of flat squarish squamiform papillae not curving into the radial notch, the two combs of each radius separated by more than half the breadth of the first free arm segment ; oral shields pentagonal, as long as or longer than broad, except in small specimens where they are relatively broader, the two distal angles rounded, adorals meeting broadly inwardly, oral plates sunken distally so that a ring-shaped depression encircles the jaws, four or five rectangular ill-defined oral papillae each side of the apical one, the distalmost broader and overlapped by the proximalmost scales of each side of the second oral pore, which are not in series with it ; arms stout basally, markedly tapering, almost flat above in profile, proximal dorsal arm plates truncated fan-shaped, becoming triangular and separated, ventral arm plates mostly rhombic and separate, no hollows between the basal ones ; arm spines three, very small peg-like, the uppermost sometimes slightly spaced on the earlier free segments ; tentacle pores of the first three (sometimes four) arm segments large with several scales each side but the pores of the next segment abruptly reduced (the main subgeneric distinction) indistinct and with a single small papilliform scale placed close to the lowest arm spine. Colour light reddish or white with orange between the disc plates.

Type locality. Agulhas Bank, 275 metres.

Southern African records. 28/14/d ; 29/14/d ; 30/15/d ; 30/16/d ; 31/16/d ; 33/17/s ; 33/18/s, d ; 34/17/d ; 34/18/d ; 35/18/d ; 35/21/d ; 35/22/d ; 34/25/d ; 79–420 metres. Bell's record from 800 to 900 fathoms, 40 miles approximately SW. of Cape Point must be treated with suspicion ; the sample (ref. no. 16928a) cannot be found among the South African Ophiuridae in the British Museum collection.

Nature of bottom. Rock ; polyzoa and rock ; coarse khaki sand and stones ; sand and gravel ; khaki sand ; sand ; black sandy mud ; dark green mud.

Ophiura (Ophiuroglypha) costata tumida Mortensen*

Ophiura (Ophiuroglypha) tumida Mortensen, 1933a : 387–388, fig. 85b, c, pl. 19, figs 22–23.

D.d. up to 15 mm ; d.d./a.l. 1/2·2–3·0.

Larger specimens, d.d. > 9 mm, differ from *O. costata costata* in having the disc plates markedly convex, especially the broad subambital plate in each interradius, the arm combs approximating closer mid-radially (separated by only about two-fifths of the breadth of the first free arm segment as opposed to about three-fifths in most specimens of *O. costata costata*), the ring-shaped depression around the jaws less well marked and the dorsal arm plates convex so as to give a scalloped profile to the arms.

Type locality. Off Durban, 232 metres.

Southern African records. 29/31/d, 124–440 metres. The maximum depth is derived from a new record of 20 specimens from 29°45′S 31°40′E, the largest with d.d. 10 mm, some of which are markedly convergent to *O. costata costata*, having the arm combs fairly well separated midradially. Only the largest specimen has the disc plates markedly tumid, while the upper profile of the arms is only slightly undulating in all the specimens.

* Here reduced to subspecific rank, its differences from *O. costata* being only of degree.

Ophiura (Ophiuroglypha) irrorata (Lyman)

Ophioglypha irrorata Lyman, 1878, *Bull. Mus. comp. Zool. Harv.* **5** : 73–74, pl. 4, figs 106–108 ; 1882 : 47–48, pl. 5, figs 7–9.
Ophiura irrorata : Matsumoto, 1917 : 277–278 ; H. L. Clark, 1923 : 358–359 ; 1939 : 109.
Ophiuroglypha irrotata (sic) : Hertz, 1927b : 86–87.
Ophiura (*Ophiuroglypha*) *irrorata* : Mortensen, 1933a : 388.

D.d. up to 33 mm ; d.d./a.l. *c.* 1/5.

A species of *Ophiura* with the disc covered with numerous small, ill-defined scales among which the primary rosette is often distinct, radial shields conspicuous, diverging and markedly tapering proximally, approximating but separated distally by a distinct notch at the base of the arm, comb papillae usually short and rounded, combs well separated midradially ; oral shields rather variable in shape, shorter or longer than broad, adorals narrow, meeting inwardly, six to eight usually blunt oral papillae each side of the apical one, in sequence with the abradial series of scales of the oral tentacle pores ; dorsal arm plates trapezoidal, many proximal ones contiguous in larger specimens, d.d. > 20 mm, arm profile almost flat above, ventral arm plates proximally bell-shaped, becoming triangular with the distal edge convex, usually all separate, no hollows between the basal plates ; arm spines three, small and appressed, the upper one spaced from the other two which are placed towards the ventral side though in small specimens all three are equally spaced, distally the middle arm spine becoming hooked ; tentacle scales numerous on several proximal pores but then reduced to a single scale, from about the tenth arm segment the pores barely distinct.

Type locality. Far to the south of Cape Agulhas (36°48′S 19°24′E), 3475 metres.

Southern African records. 33/16/vd ; 34/17/vd ; 36/19/vd ; 1645–3475 metres.

Nature of bottom. Globigerina ooze ; grey ooze ; green mud.

Class *ECHINOIDEA*

INTRODUCTION. For a detailed study of the recent taxa of this class the monograph of Mortensen (1928–1951) should be consulted. However, the outline classification used here follows the consensus of opinion among palaeontologists which was adopted by Durham and others in the *Treatise on Invertebrate Paleontology* (1966). This scheme has the merit of approximating the Stomechinidae and Arbaciidae with the Echinoida and Temnopleuroida rather than with such families as the Diadematidae.

The following terms are used in the descriptions of echinoids and may need clarification :

Ambitus – the broadest part of the test through which the horizontal diameter of regular species or length and breadth of irregular ones are measured.

Apical system – the central (or nearly central) group of plates on the upper side of the test. This consists of (a) *genital* plates (each with a pore in mature specimens), interradial in position and numbering five in regular echinoids but fewer in most irregular species, (b) five radial *ocular* plates adjacent to which the double series of new interambulacral and ambulacral plates respectively are developed during growth, both these surrounding (c) a circular or oval group of periproctal plates in regular species. One genital plate is perforated by hydropores and is the *madreporite*. In regular echinoids the apical system encircles the periproct and takes two basic forms ; when *dicyclic* the genital plates form a continuous ring separating the smaller oculars from the periproct, these being termed *exsert*, whereas in monocyclic species the plates of the two series alternate and the oculars are *insert* ; intermediate conditions frequently occur. In irregular species the apical system adopts a degree of bilateral symmetry and the posterior genital plate is fused or lost, the number of genital pores being usually less than five.

Corona, or test – the usually rigid main part of the skeleton formed of 10 pairs of vertical series of plates, five ambulacral and five interambulacral. The five radial ambulacral paired series are perforated by normally paired pores for each tube foot, a pair on each plate when the plates are simple (Cidaroida and most irregular species), though more often the plates are compound and more or less completely fused with several pore-pairs but mostly a single primary tubercle and spine ; the commonest compound plates are *trigeminate* where three plates are fused, but *polyporous* species with four or more pairs of pores per plate are not rare.

Fasciole – in heart urchins, bilaterally symmetrical bands of many fine ciliated spinelets (for creating water currents), their basal tubercles noticeable on bare tests, usually characteristic in shape and occurrence. A *subanal* fasciole forms a loop below the periproct ventro-posteriorly. An *internal* fasciole encircles the adapical part of the frontal ambulacrum and the apical system, excluding the paired petals, whereas a *peripetalous* fasciole also encompasses the paired petals. From the latter a *latero-anal* fasciole may extend backwards and dip below the periproct, or conversely a subanal fasciole may have an upward *anal* branch ending each side of the periproct.

Gill slits – ten notches in the lowest interambulacral plates bordering the peristome, accommodating the gills in regular species except Cidaroids.

Labrum – the convex interradial lip posterior to the semilunar peristome in heart-urchins backed by the usually richly spined *sternum.*

Lunules – in certain Clypeasteroids, dorsoventral slots through the test.

Pedicellariae – minute stalked grasping organs, each with usually three calcified valves, found on various parts of the surface; even the largest barely visible with the naked eye. The most important taxonomically are the *globiferous* ones, with associated poison glands, tubular valves and one or more terminal injecting teeth, and the *tridentate* with relatively large, parallel-sided or petaloid valves. Other common forms, especially in regular echinoids, include strong bulbous *ophicephalous* pedicellariae and smaller short-valved *triphyllous.*

Periproct – the circular or oval extensible area around the anus embedded with small plates, among which an enlarged *suranal* plate may be distinct. In regular species it is encircled by the apical system but in irregular echinoids it is posterior.

Peristome – the extensible area between the mouth and the corona; in regular species usually with only a single pair of *buccal* ambulacral plates in each radius, though often accompanied by secondary platelets, but in echinothuriids with numerous imbricating series of ambulacral plates and in cidaroids both ambulacral and interambulacral series.

Petals – modified aboral parts of the ambulacra of most irregular echinoids delimited by enlarged tube-foot pores, usually ending abruptly well short of the ambitus.

Tubercles – the knobs on the test for articulation of the spines. They may be *perforate* with a central hole or *imperforate* and the raised boss around the larger tubercles may be ornamented with small bumps (*crenulate*) or smooth. The muscles are attached to the smooth *areole* around the tubercle. The areoles are very conspicuous around the primary spines of cidaroids.

The measurements generally used for regular echinoids are: h.d. – the horizontal diameter and v.d. – the vertical diameter (both best measured with calipers); also used are the diameters of the apical system and peristome and often the length of the longest spines. For irregular species the length, breadth, and in heart-urchins height, are usually taken; sometimes the length of the petaloid area from the tip of the anterior petal to a line taken between the tips of the posterior paired petals may also be measured. These values are normally expressed as percentages of the h.d. or length. The figures given under each species are derived from specimens in the upper half of the size range (in terms of h.d. or l.) unless otherwise stated. (Higher values are usually obtained for the peristome and apical system diameters in immature specimens.)

The number of species included is 58 but all records of *Diadema setosum* may be mistakes for *D. savignyi.*

DISTRIBUTION TABLE FOR ECHINOIDEA

	Luderitz Bay area	Lambert's Bay area	Cape Town area	False Bay	Cape Agulhas area	Mossel Bay–Knysna area	Port Elizabeth area	Durban area	Lourenço Marques area	Other localities
CIDARIDAE										
Acanthocidaris maculicollis (de Meijere)	..	..	..	..	..	..	..	?	..	East Indies, S. Japan
Eucidaris metularia (Lamarck)	..	..	..	..	..	..	..	..	i	Tropical Indo-West Pacific
Goniocidaris sp.	..	..	..	..	..	d	..	..	..	
Histocidaris sp. aff. *H. elegans* (A. Agassiz)			'South African Seas'							S. Japan–SE. Australia
Kionocidaris striata Mortensen	..	..	..	..	..	..	..	d	..	
Prionocidaris pistillaris (Lamarck)	..	..	..	..	..	..	i	s	i	East Africa–Mauritius and Seychelles
Stereocidaris capensis Döderlein	..	..	..	..	..	vd	vd	..	..	
Stereocidaris excavata Mortensen	..	..	..	..	..	d	d	..	..	
Stereocidaris squamosa Mortensen	..	..	..	..	..	..	..	s d	..	Tanzania, Saya da Malha Bank
ECHINOTHURIIDAE										
Araeosoma paucispinum H. L. Clark	..	..	..	..	..	..	..	d	..	
Hygrosoma petersi (A. Agassiz)	..	..	d vd	..	..	..	..	(d vd ?)	..	N. Atlantic, West Indies
Phormosoma bursarium A. Agassiz	..	..	..	..	..	..	..	d vd	..	Indo-West Pacific
Phormosoma placenta africana Mortensen	..	vd	vd	..	..	..	..	..	..	

Sperosoma biseriatum Döderlein	..	..	..	..	..	..	..	vd	..	East Africa, Arabian Sea
PEDINIDAE										
Caenopedina capensis H. L. Clark	..	..	vd	..	..	..	..	..	..	
ASPIDODIADEMATIDAE										
Aspidodiadema africanum Mortensen	..	..	..	..	..	..	vd	..	..	
DIADEMATIDAE										
Astropyga radiata (Leske)	..	..	..	..	..	..	..	..	i	Tropical Indo-West Pacific
Chaetodiadema africanum H. L. Clark	..	..	..	..	..	..	..	d	..	
Diadema savignyi Michelin	..	..	..	..	..	..	..	i	..	Tropical Indo-West Pacific
Diadema setosum (Leske)	..	..	..	..	..	..	..	i	i	Tropical Indo-West Pacific
Echinothrix calamaris (Pallas)	..	..	..	..	..	..	..	i	i	Tropical Indo-West Pacific
SALENIIDAE										
Salenia phoinissa Agassiz & Clark	..	..	..	..	..	d	..	..	..	
STOMECHINIDAE										
Stomopneustes variolaris (Lamarck)	..	..	..	..	..	..	..	i	i	Tropical Indo-West Pacific
ARBACIIDAE										
Coelopleurus interruptus Döderlein	..	..	..	..	..	d	s	..	..	
Coelopleurus maillardi (Michelin)	..	..	..	..	..	..	d	–	–	Indian Ocean, East Indies and Philippines
TEMNOPLEURIDAE										
Lamprechinus nitidus Döderlein	..	..	..	..	..	vd	..	..	..	
Orechinus monolini (A. Agassiz)	..	..	..	..	..	vd	..	..	..	East Indies, Hawaiian and Kermadec Is.

Distribution Table for Echinoidea (*cont.*)

	Luderitz Bay area	Lambert's Bay area	Cape Town area	False Bay	Cape Agulhas area	Mossel Bay–Knysna area	Port Elizabeth area	Durban area	Lourenço Marques area	Other localities
Salmaciella dussumieri erythracis (H. L. Clark)	..	..	..	..	..	..	..	s	..	East Africa, Seychelles
Salmacis bicolor L. Agassiz	..	..	..	..	..	..	..	i s	i	Tropical Indo-West Pacific
Temnopleurus reevesi (Gray)	..	..	..	..	..	d	..	s	..	East Indies, Philippines, S. China Sea and Japan
Temnopleurus toreumaticus (Leske)	..	..	..	..	..	..	..	..	i s	Tropical Indo-West Pacific
TOXOPNEUSTIDAE										
Toxopneustes pileolus (Lamarck)	..	..	..	..	..	..	..	..	i	Tropical Indo-West Pacific
Tripneustes gratilla (Linnaeus)	..	..	..	..	..	..	i	i s	i s	Tropical Indo-West Pacific
ECHINIDAE										
Dermechinus horridus (A. Agassiz)	..	d	d	..	..	vd	..	..	..	Chile and SE. Australia
Echinus gilchristi Bell	d	d	s d	s	..	s d vd	s d	..	..	
Parechinus angulosus (Leske)	i	i s	i s	i s	..	i s	i s	i	..	
Parechinus angulosus forma *annulatus* Mortensen	..	..	..	i s	..	..	..	..	..	
Parechinus angulosus forma *pallidus* H. L. Clark	..	d	s	s	..	..	..	..	..	
Polyechinus agulhensis (Döderlein)	..	vd	vd	..	..	vd	..	..	..	

ECHINOMETRIDAE										
Echinometra mathaei (de Blainville)	..	..	..	..	..	..	..	i	i	Tropical Indo-West Pacific
Echinostrephus molaris (de Blainville)	..	..	..	..	..	..	..	i	..	Tropical Indo-West Pacific
CLYPEASTERIDAE										
Clypeaster eurychorius H. L. Clark	..	..	..	..	..	..	..	d	..	Tanzania, ? Mauritius
Clypeaster rarispinus de Meijere	..	..	..	..	..	..	..	s	..	Indian Ocean, Indonesia
FIBULARIIDAE										
Echinocyamus elegans Mazetti	..	..	..	..	..	..	d	s	..	Red Sea, Gulf of Oman
LAGANIDAE										
Laganum fudsiyama africanum Mortensen	..	..	..	..	..	..	..	? s d	..	
ASTRICLYPEIDAE										
Echinodiscus auritus Leske	..	..	..	..	..	..	..	..	i	Tropical Indo-West Pacific
Echinodiscus bisperforatus Leske	..	..	..	..	..	i s	..	s	i	East Africa–East Indies
ECHINOLAMPADIDAE										
Echinolampas (*Palaeolampas*) *crassa* (Bell)	..	..	..	s	..	vd	s	..	..	
NEOLAMPADIDAE										
Tropholampas loveni (Studer)	..	..	d	..	..	d	..	..	..	
URECHINIDAE										
Urechinus naresianus A. Agassiz	..	..	vd	..	..	..	..	..	..	Greenland–Caribbean and Subantarctic
POURTALESIIDAE										
Pourtalesia alcocki Koehler	..	..	vd	..	..	..	..	..	..	Gulf of Oman
ASTEROSTOMATIDAE										
Eurypatagus parvituberculatus (H. L. Clark)	..	..	..	..	..	..	..	d	..	

DISTRIBUTION TABLE FOR OPHIUROIDEA (*cont.*)

	Luderitz Bay area	Lambert's Bay area	Cape Town area	False Bay	Cape Agulhas area	Mossel Bay–Knysna area	Port Elizabeth area	Durban area	Lourenço Marques area	Other localities
SCHIZASTERIDAE										
Brisaster capensis (Studer)	..	..	d	..	..	..	..	..	..	
Schizaster lacunosus (Linnaeus)	..	..	..	..	..	..	..	s d	..	Maldives, South China and Japan, N. Australia
BRISSIDAE										
Brissopsis lyrifera capensis Mortensen	..	d	s d	s	s d	vd	d	..	..	West Africa to Gulf of Guinea
Spatagobrissus mirabilis H. L. Clark	..	..	..	s	..	..	..	..	..	
LOVENIIDAE										
Echinocardium capense Mortensen	..	..	d	..	? s*	..	s d	..	..	
Echinocardium cordatum (Pennant)	..	? d*	? d*	s	..	i s	s	..	..	Norway–Morocco, Caribbean area, Japan–New Zealand, California
Lovenia elongata Gray	..	..	..	..	..	..	..	..	i	Tropical Indo-West Pacific
SPATANGIDAE										
Spatangus capensis Döderlein	d	d	d	s	s	s d vd	..	..	..	

* The samples represented by these records of Bell's (1904) are not in the British Museum collections and in my view need confirmation.

Key to the Echinoidea

1 Regular echinoids with periproct and peristome at the north and south poles of the test respectively 2

– Irregular, the periproct posterior, or even on the oral surface, away from the apical system, the mouth usually towards the anterior end of the lower side but sometimes still central. 40

2 Both interambulacral and ambulacral series of plates forming continuous imbricating series over the peristome (Fig. 226) ; test very stout and rigid ; primary spines massive, when mature losing their skin (and so often becoming encrusted by epizoic organisms), each encircled by a ring of markedly different secondary spines. CIDARIDAE* 3

– Peristome not continuously plated, bearing only more or less isolated ambulacral plates embedded in the skin but these are usually reduced to a single pair in each radius, the buccal plates, only the flexible Echinothuriidae with numerous primary peristomial plates ; (Figs 227–229) test moderately stout to delicate ; primary spines variable in development, in most species not conspicuously different in shape from the secondary spines, though sometimes markedly larger, retaining their skin covering* 11

3 Aboral and ambital primary spines superficially appearing smooth or with fine continuous longitudinal ridges (Fig. 230) 4

– Primary spines rugose or sometimes spinose, covered with tubercles or larger projections usually arranged in longitudinal series but individually discrete on at least part of the shaft, not forming continuous ridges (Fig. 231) 9

4 The bare test densely and very finely tuberculated for the numerous secondary spines ; even the uppermost primary interambulacral tubercles hardly, if at all, crenulate (Fig. 232) 5

– Secondary tubercles moderately coarse and their spines more or less spaced ; 7
primary tubercles, at least the upper ones, distinctly crenulate (Fig. 233) . .

5 Genital plates sunken, except around the pore which is almost central ; test almost perfectly globular, only the apical system abruptly flattened
Stereocidaris excavata Mortensen, 1932 (p. 217)

– Genital plates not distinctly sunken, the pores placed distally ; test more or less evenly convex above 6

6 Primary spines hardly tapering, relatively broad at the tip, with about 15 longitudinal ridges . . . . . ***Stereocidaris squamosa*** Mortensen, 1928 (p. 218)

– Primary spines markedly tapering distally, about 20 ridges
Stereocidaris capensis Döderlein, 1901 (p. 217)

7 Collar of primary spines very long (10+ mm in adult specimens) and conspicuously marked with dark red or brown spots
Acanthocidaris maculicollis (de Meijere, 1904) (p. 214)

– Primary spines with a very short collar (*c.* 1 mm long), not spotted 8

8 Oral primary spines with fine ridges, superficially appearing almost smooth
Kionocidaris striata Mortensen, 1932 (p. 216)

– Oral primary spines flattened with the edges strongly serrated
Histocidaris sp. aff. ***H. elegans*** (A. Agassiz, 1879) (p. 215)

9 Apical system fringed by a ring of spatulate enlarged secondary spines and their tubercles, which stand out from the very fine miliary spinelets otherwise covering the genital and ocular plates . . ***Eucidaris metularia*** (Lamarck, 1816) (p. 215)

– Secondary tubercles of similar small size scattered all over the apical system . 10

* The small *Salenia phoinissa* Agassiz & Clark (only known at h.d. *c.* 10 mm) should run down to 11 but does have a considerable superficial resemblance to the Cidaridae, even to having some (secondary) plates on the peristome. It can be distinguished by the imperforate primary tubercles.

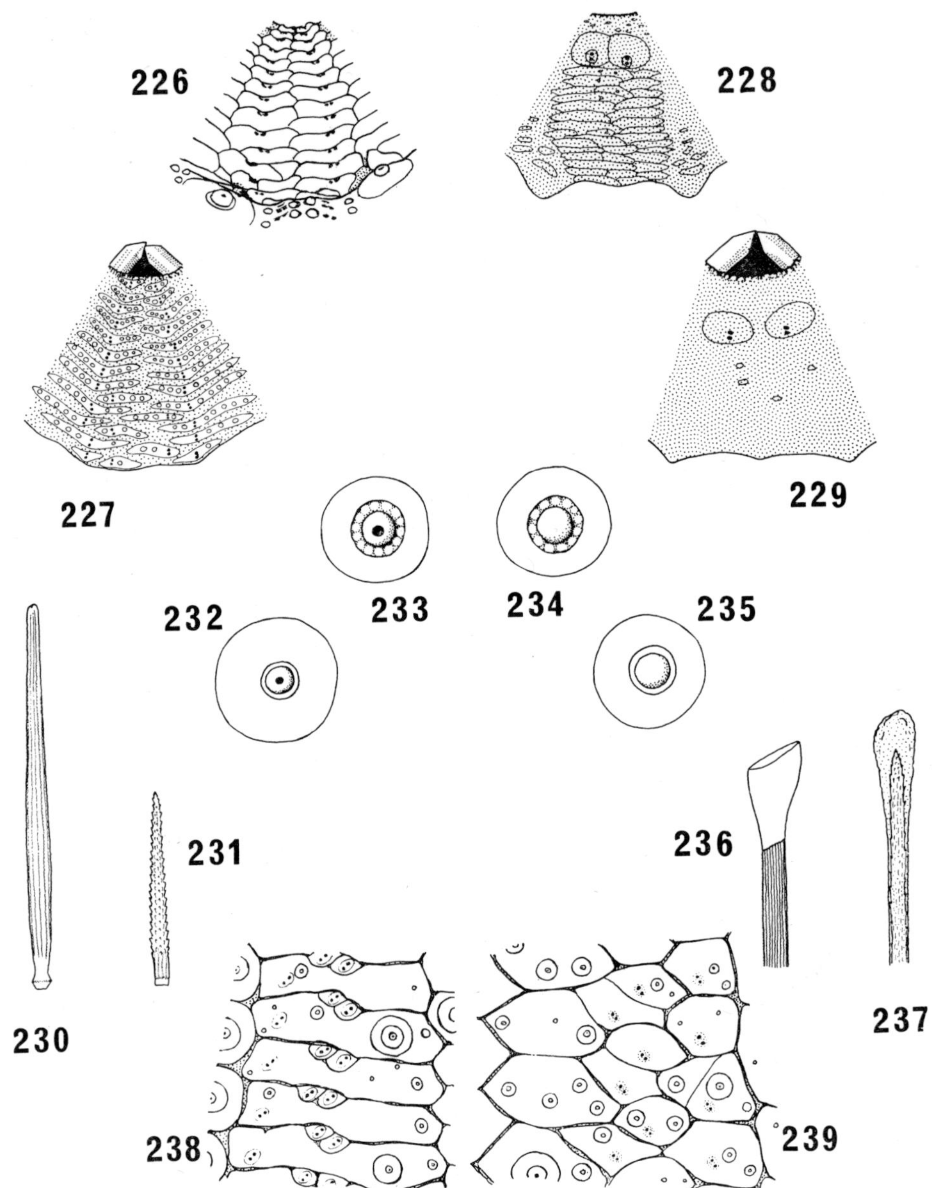

FIGS 226–239 Echinoidea. 226, 227, 228, 229 : parts of peristome of a cidarid, an echinothuriid, a diadematid and an echinid. 230, 231 : primary spines of *Stereocidaris squamosa* and *Prionocidaris pistallaris*. 232, 233, 234, 235 : primary interambulacral tubercles of *Stereocidaris squamosa, Chaetodiadema africanum, Salenia* sp. and *Coelopleurus maillardi*. 236, 237 : tip of primary spine from oral side of *Araeosoma paucispinum* and *Phormosoma bursarium*. 238, 239 : ambulacral plates from oral side of *Araeosoma paucispinum* and *Sperosoma biseriatum*.

10 Collar of primary spines marked with longitudinal stripes; no flange around the base of the shaft . . . . ***Prionocidaris pistillaris*** (Lamarck, 1816) (p. 216)

– No stripes on the collar of the primary spines, an expanded flange or partial disc around the base of the shaft, at least on the aboral side . ***Goniocidaris*** sp. (p. 215)

11 Ambulacral series of plates continuing over the peristome to the mouth (Fig. 227); test very flexible, the plates imbricating or partially separated by membrane. ECHINOTHURIIDAE 12

– Only the 10 buccal plates (two in each radius) present in the peristomial skin (Fig. 229), though secondary calcifications may also occur (Fig. 228); test usually rigid but occasionally (certain Diadematidae) somewhat flexible 16

12 Teeth tapering abruptly near the tip (Fig. 227); primary spines of oral side ending in a white 'hoof' (Fig. 236) 13

– Projecting parts of teeth narrow and constricted, tapering to attenuated tips; no terminal hooves on the oral primary spines (Fig. 237) 15

13 Pore-pairs in single series on the oral side ***Hygrosoma petersi*** (A. Agassiz, 1880) (p. 220)

– Pore-pairs on both oral and aboral sides in oblique or horizontal arcs forming three ventral series 14

14 Secondary ambulacral plates of oral side reduced to a small lozenge around the pore-pair and so aligned that the arcs are almost horizontal (Fig. 238) ***Araeosoma paucispinum*** H. L. Clark, 1924 (p. 219)

– Secondary ambulacral plates of oral side irregular, mostly bearing a tubercle as well as the pore-pair and not clearly distinct from the primary plates which are sub-divided, the arcs more or less oblique (Fig. 239) ***Sperosoma biseriatum*** Döderlein, 1901 (p. 221)

15 Primary aboral spines curved; peristome 30–32% of h.d. (Indian Ocean) ***Phormosoma bursarium*** A. Agassiz, 1881 (p. 221)

– Primary aboral spines straight; peristome up to 40%, rarely less than 35% of h.d. (Atlantic) . . . ***Phormosoma placenta africana*** Mortensen, 1934 (p. 221)

16 Tubercles of primary spines perforate (Fig. 233); teeth not keeled on the inside. DIADEMATACEA 17

– Tubercles imperforate (Fig. 234); teeth keeled. ECHINACEA 23

17 Tubercles not crenulate (Fig. 235); primary spines solid. PEDINIDAE ***Caenopedina capensis*** H. L. Clark, 1923 (p. 222)

– Tubercles crenulate; primary spines hollow or with a partially filled central core . 18

18 Test high, vertical diameter (v.d.) *c.* 75% of h.d.; size small, h.d. probably not > 30 mm; buccal plates large and joined to form a complete ring round the mouth. ASPIDODIADEMATIDAE ***Aspidodiadema africanum*** Mortensen, 1939 (p. 223)

– Test low, v.d. rarely > 55% of h.d., often much less; adult size up to 100 or even 180 mm h.d.; buccal plates small, each pair isolated. DIADEMATIDAE . 19

19 Test distinctly flexible, low, v.d. only *c.* 33% or less of h.d. 20

– Test not flexible, v.d. more than a third of h.d., usually about half 21

20 Adult size up to *c.* 180 mm h.d.; pore-pairs mostly contiguous with each other on oral side, the pore-zones wide adorally . ***Astropyga radiata*** (Leske, 1778) (p. 224)

– H.d. not known to exceed 50 mm; pore-pairs very small on oral side, well separated and inconspicuous . . ***Chaetodiadema africanum*** H. L. Clark, 1924 (p. 224)

21 Ambulacral spines of aboral side very fine and needle-like, with backwardly directed barbs near the tip (Fig. 244), contrasting sharply in thickness with the primary interambulacral spines . . . ***Echinothrix calamaris*** (Pallas, 1774) (p. 226)

– Aboral spines of ambulacra and interambulacra not markedly differentiated, the barbs on the ambulacral ones projecting sideways (Fig. 245) 22

22 Large tridentate pedicellariae with narrow blades only meeting at the tip, the ratio of valve length to the breadth at the middle of the blade *c.* 20 : 1 (Fig. 246), a red or orange ring around the anus in life ***Diadema setosum*** (Leske, 1778) (p. 226)

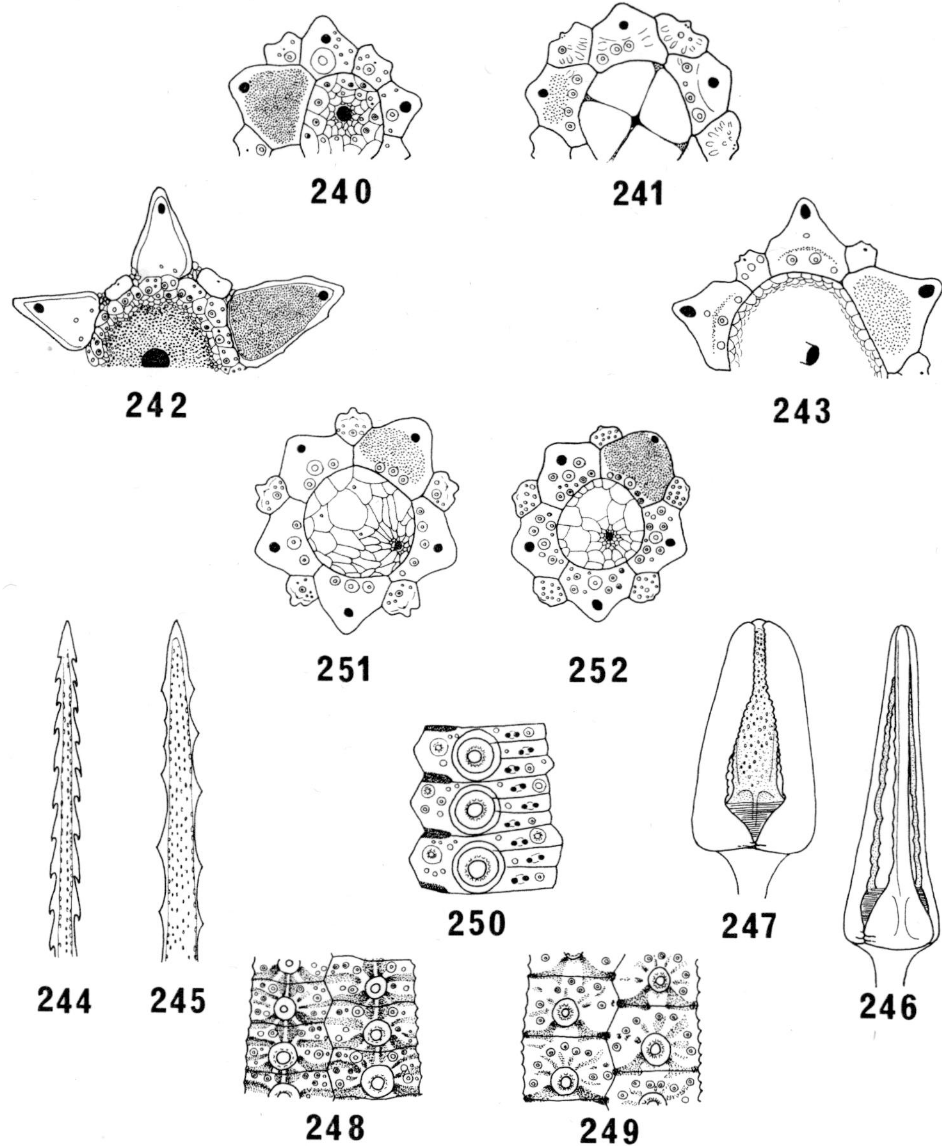

FIGS 240–252 Echinoidea. 240, 241 : apical systems of *Stomopneustes variolaris* and *Coelopleurus maillardi*. 242, 243 : parts of apical areas of *Astropyga radiata* and *Diadema savignyi*. 244, 245 : tips of aboral primary ambulacral spines of *Echinothrix calamaris* and *Diadema setosum*. 246, 247 : valves of tridentate pedicellariae of *Diadema setosum* and *D. savignyi*. 248, 249, 250 : parts of test of *Orechinus monolini* (interambulacral plates), *Temnopleurus reevesi* (IA plates) and *T. toreumaticus* (ambulacral plates). 251, 252 : apical systems of *Temnopleurus reevesi* and *T. toreumaticus*.

22 Tridentate pedicellariae with leaf- or spoon-shaped valves, length : breadth rarely much over 5 : 1 (Fig. 247) ; no anal red ring **_Diadema savignyi_** Michelin, 1845 (p. 225)

23 Ambulacral plates doubly compound, three to six trigeminate plates fusing so that up to 18 pore-pairs correspond to one primary tubercle, the limits of the individual plates not distinct. STOMECHINIDAE **_Stomopneustes variolaris_** (Lamarck, 1816) (p. 228)

– Ambulacral structure not so complex, most often simply trigeminate, if polyporous then with not more than five pore-pairs in each arc 24

24 Primary spines few, massive and very long, easily exceeding the h.d. ; size small, h.d. rarely as much as 50 mm 25

– Primary spines slender or stout but not markedly exceeding the h.d., usually relatively short ; adult size often exceeding 50 mm h.d. 27

25 Primary interambulacral tubercles crenulate (Fig. 234) ; apical system markedly asymmetrical, the periproct excentric, with an enlarged suranal plate among the periproctal plates. SALENIIDAE **_Salenia phoinissa_** A. Agassiz & H. L. Clark, 1908 (p. 227)

– Primary interambulacral tubercles not crenulate (Fig. 235) ; apical system not markedly excentric, the periproct covered by four similar triangular plates and resembling a four-spoked wheel (Fig. 241). ARBACIIDAE 26

26 Primary spines slightly flattened with only faint longitudinal ridges, almost cylindrical, coloured with alternate broad bands of purple and greenish-yellow ; bare aboral interambulacral areas purple or mauve **_Coelopleurus maillardi_** (Michelin, 1862) (p. 229)

– Primary spines markedly prismatic, with dull colour sometimes graduated but without alternating bands ; bare aboral interambulacral areas brown with white edges **_Coelopleurus interruptus_** Döderlein, 1910 (p. 229)

27 Tubercles more or less distinctly crenulate (except in the deep-water *Orechinus* and *Lamprechinus*) ; test with pores, pits or troughs at the angles of the sutures or else extensively sculptured (Figs 248–250). TEMNOPLEURIDAE . . . 28

– Tubercles smooth, not crenulate ; no angular pits, pores or sculpturing on the test 33

28 Test sculptured but no angular pores or pits (Fig. 248) ; spines and bare test white or very pale ; h.d. not known to exceed 20 mm 29

– Angular pores or pits present (Figs 249, 250) (though sometimes reduced in large specimens of *Salmacis*, h.d. approaching 100 mm) ; spines and test variously coloured ; size small to large 30

29 Apical system sculptured . . . **_Orechinus monolini_** (A. Agassiz, 1879) (p. 230)

– Apical system smooth . . . **_Lamprechinus nitidus_** Döderlein, 1905 (p. 230)

30 Rounded pits usually well developed at the angles of the plates 31

– Only small pores present at the angles 32

31 The whole apical system markedly excentric (Fig. 251) ; spines light khaki, not banded . . . . **_Temnopleurus reevesi_** (Gray, 1855) (p. 232)

– Anus nearly central and apical system roughly circular (Fig. 252) ; spines, at least on the oral side, banded dark olive or reddish and white **_Temnopleurus toreumaticus_** (Leske, 1778) (p. 233)

32 Primary ambulacral tubercles enlarged on every plate and forming a single vertical series ; apical system not markedly excentric ; h.d. up to *c.* 100 mm ; spines red, at least basally, otherwise usually banded red with yellow or green **_Salmacis bicolor_** L. Agassiz, 1846 (p. 232)

– Primary ambulacral spines reduced on alternate plates aborally, or offset to form a second partial series near the ambitus ; apical system excentric ; h.d. up to *c.* 50 mm ; spines banded green and white **_Salmaciella dussumieri erythracis_** (H. L. Clark, 1912) (p. 231)

33 Gill slits around edge of peristome cutting deeply into the lower edge of the test (Fig. 253). TOXOPNEUSTIDAE 34

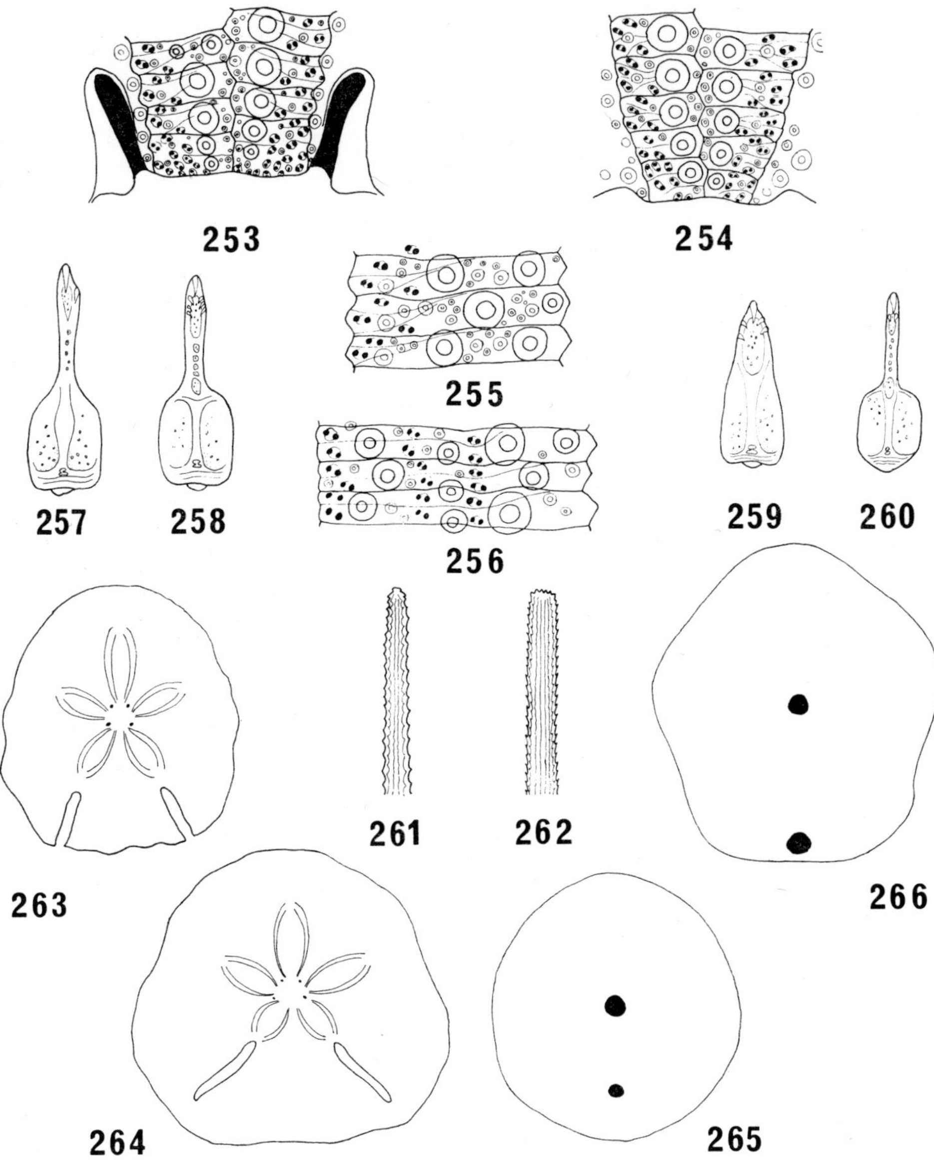

FIGS 253–266 Echinoidea. 253, 254: parts of test adjoining peristome of *Toxopneustes pileolus* and *Echinometra mathaei*. 255, 256: ambulacral plates of *Toxopneustes pileolus* and *Tripneustes gratilla*. 257, 258, 259, 260: valves of globiferous pedicellariae of *Echinometra mathaei*, *Polyechinus agulhensis*, *Parechinus angulosus* and *Echinus gilchristi*. 261, 262: tips of miliary spines of *Dermechinus horridus* and *Echinus gilchristi*. 263, 264, 265, 266: outlines of tests of *Echinodiscus auritus* and *E. bisperforatus* (dorsal view), *Laganum fudsiyama africanum* and *Clypeaster eurychorius* (ventral view).

33 Gill slits making only slight dents in the edge of the test (Fig. 254) . . . 35

34 Globiferous pedicellariae markedly enlarged, span when gaping up to 3 mm and clearly visible with the naked eye as spherical objects when alive or triangular ones when dead ; pore-pairs arranged in oblique arcs (Fig. 255)
Toxopneustes pileolus (Lamarck, 1816) (p. 234)

– Globiferous pedicellariae minute and individually inconspicuous ; pore-pairs arranged in nearly horizontal arcs, forming three vertical series (Fig. 256)
Tripneustes gratilla (Linnaeus, 1758) (p. 234)

35 Globiferous pedicellariae with an unpaired lateral tooth (Fig. 257) ; test either ovate or, if round, with the ambitus towards the top. ECHINOMETRIDAE . 36

– Globiferous pedicellariae more symmetrical, with one or more lateral teeth each side (Fig. 258) ; test round or pentagonal when viewed from above and the ambitus nearer the oral plane than the apical one. ECHINIDAE . . . 37

36 Test more or less ovate with the longest spines projecting sideways
Echinometra mathaei (de Blainville, 1825) (p. 239)

– Test circular when viewed from above, the longest spines projecting vertically from the flattened aboral side ***Echinostrephus molaris*** (de Blainville, 1825) (p. 239)

37 Ambulacral plates mostly quadrigeminate with four pore-pairs in each arc, sometimes three or five sporadically ***Polyechinus agulhensis*** (Döderlein, 1905) (p. 238)

– Ambulacra trigeminate throughout 38

38 Valves of globiferous pedicellariae approximately triangular in outline, the blade merging into the basal part (Fig. 259) ; primary spines fairly numerous
Parechinus angulosus (Leske, 1778) (p. 237)

– Globiferous pedicellariae with blades abruptly narrower than the base (Fig. 260) ; primary spines relatively few and widely spaced with a dense covering of secondary spines between 39

39 Peristome very small, usually smaller than apical system and only 10–14% of h.d. in adult specimens ; miliary spines very numerous with convexities projecting at right angles to the shaft (Fig. 261) ***Dermechinus horridus*** (A. Agassiz, 1879) (p. 235)

– Peristome of average size, larger than the apical system ; if projections occur on the miliary or secondary spines then they are directed distally (Fig. 262)
Echinus gilchristi Bell, 1904 (p. 236)

40 Lantern and teeth well developed and present throughout life ; mouth approximately central ; test usually markedly flattened, often discoidal (Figs 263–266). 41

– Lantern and teeth transitory in the young or absent throughout life ; mouth usually towards the anterior end of the oral side but sometimes approximately central ; test more or less ovate, the aboral side convex 46

41 Test with slots or lunules posteriorly (Figs 263, 264). ASTRICLYPEIDAE . 42

– Test entire (Figs 265, 266) 43

42 The two lunules open distally . . . ***Echinodiscus auritus*** Leske, 1778 (p. 243)

– Lunules enclosed within the circumference of the test (Fig. 264)
Echinodiscus bisperforatus Leske, 1778 (p. 243)

43 Size very small, length not known to exceed 10 mm ; test flattened ovate. FIBULARIIDAE . . . . ***Echinocyamus elegans*** Mazetti, 1893* (p. 242)

– Adult length 20 mm or more ; test discoidal or with a marginal horizontal flange and convex central part 44

44 Periproct separated from the posterior edge of the test by much more than its own width (Fig. 265). LAGANIDAE
Laganum fudsiyama africanum Mortensen, 1948 (p. 241)

– Periproct about its own width from the edge of the test (Fig. 266). CLYPEASTERIDAE 45

* *Echinocyamus elegans* can most easily be distinguished from the other very small irregular South African echinoid, *Trophalampas loveni*, by the fairly well-developed aboral pore-pairs forming petals. *T. loveni* is from very deep water.

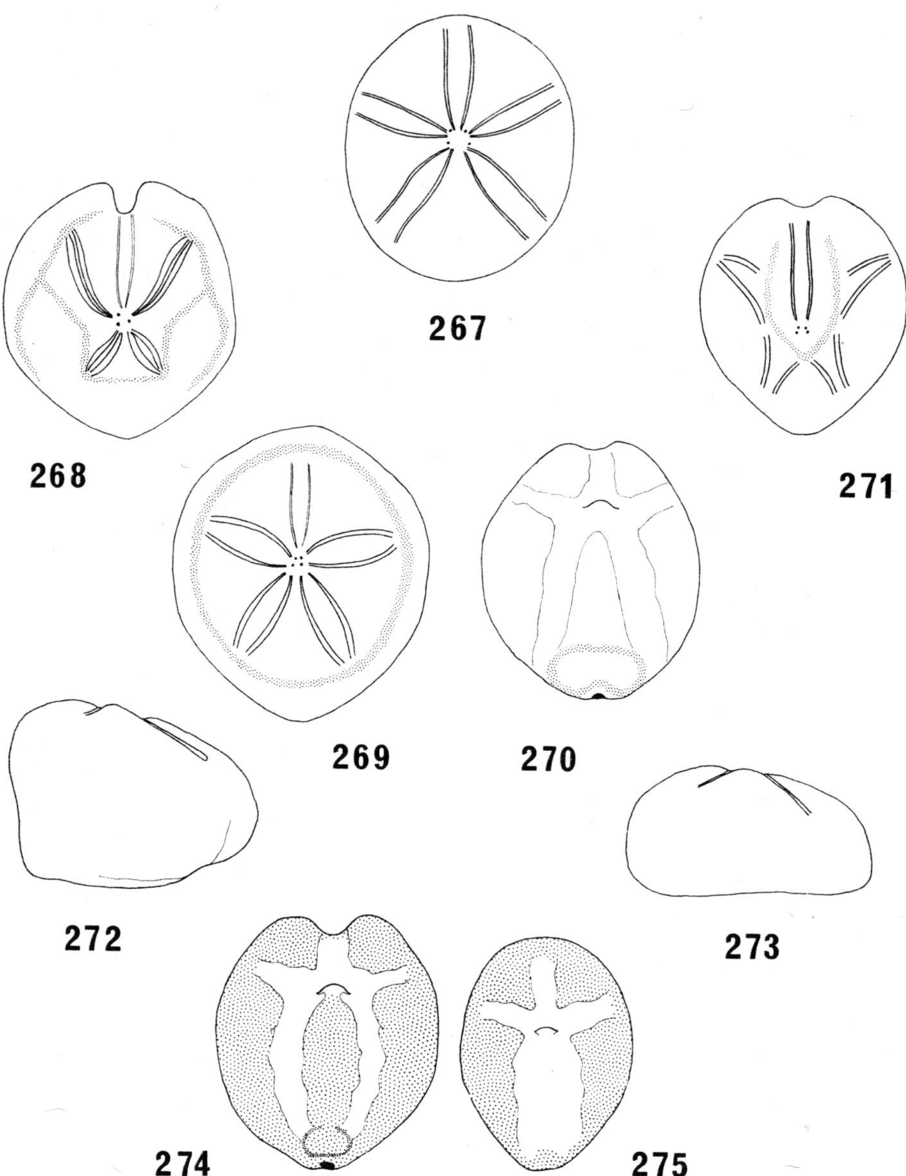

FIGS 267–275 Echinoidea : main features. 267, 268, 269 : *Echinolampas crassa, Brisaster capensis* and *Spatagobrissus mirabilis* (dorsal view). 270 : *Brissopsis lyrifera capensis* (ventral view). 271 : *Echinocardium cordatum* (dorsal view). 272, 273 : *Schizaster lacunosus* and *Brisaster capensis* (side view), 274, 275 : *Spatangus capensis* and *Eurypatagus parvituberculatus* (ventral view).

45 Petals open distally, especially the anterior one ; test distinctly convex centrally
Clypeaster eurychorius H. L. Clark, 1926 (p. 241)

– Petals closed distally ; test not much raised centrally
Clypeaster rarispinus de Meijere, 1902 (p. 241)

46 Peristome approximately in the centre of the oral surface, the interambulacrum posterior to it not markedly specialized ; fascioles (belts of crowded fine ciliated spinelets) absent or at most inconspicuous 47

– Peristome markedly anterior, rarely (Pourtalesiidae) sunk into a deep hollow in the anterior face of the test, otherwise the interambulacrum posterior to the peristome enlarged and more or less swollen to form a sternum ; at least one kind of fasciole well developed 49

47 Ambulacral pores forming open-ended petals aborally (Fig. 267). ECHINOLAMPADIDAE . . . . ***Echinolampas crassa*** (Bell, 1880) (p. 244)

– No petals on aboral side where the pores are rudimentary or absent . . . 48

48 Size very small, not known to exceed 8 mm length ; four genital pores visible in the male, the female with apical system sunk into a marsupium. NEOLAMPADIDAE
Tropholampas loveni (Studer, 1880) (p. 245)

– Adult length up to *c.* 50 mm ; often exceeding 20 mm ; only three genital pores and no marsupium. URECHINIDAE
Urechinus naresianus A. Agassiz, 1879 (p. 245)

49 Peristome sunk into a concavity in the abruptly truncated anterior end of the test, the posterior end constricted and produced into a short blunt 'tail' ; no conspicuous petals on the upper side. POURTALESIIDAE
Pourtalesia alcocki Koehler, 1914 (p. 246)

– Test more or less ovate ; peristome distinctly on the oral surface ; petals developed 50

50 A peripetalous fasciole present (Figs 268, 269) 51

– No peripetalous fasciole 54

51 Frontal ambulacrum deeply sunken ; posterior petals much shorter than the paired anterior ones (Fig. 268) ; a latero-anal fasciole linked to the peripetalous one but no separate subanal. SCHIZASTERIDAE 52

– Only a slight frontal notch, if any ; paired petals approximately the same length (Fig. 269) ; subanal fasciole present. BRISSIDAE 53

52 Two genital pores ; test very high posteriorly (Fig. 272) ; apical system distinctly posterior . . . . . ***Schizaster lacunosus*** (Linnaeus, 1758) (p. 248)

– Three genital pores (Fig. 268) ; test not unusually high, the highest point at the apical system which is near the centre (Fig. 273)
Brisaster capensis (Studer, 1880) (p. 248)

53 Peripetalous fasciole convex between anterior and posterior petals ; no frontal notch (Fig. 269) . . . . ***Spatagobrissus mirabilis*** H. L. Clark, 1923 (p. 250)

– Peripetalous fasciole concave laterally ; a slight frontal notch developed (Fig. 270)
Brissopsis lyrifera capensis Mortensen, 1907 (p. 249)

54 An inner fasciole present around the apical system and aboral part of the frontal ambulacrum (Fig. 271). LOVENIIDAE 55

– No inner fasciole 57

55 Conspicuous spaced long anterolateral aboral primary spines present, the areoles around their tubercles deeply sunken ; test markedly flattened, height *c.* 33% of length . . . . ***Lovenia elongata*** (Gray, 1845) (p. 252)

– Aboral spines superficially fairly uniform in length, none markedly elongated and their tubercles unspecialized ; test height rarely less than 50% of the length . 56

56 Frontal ambulacrum conspicuously sunken, with crowded pores in irregular series
Echinocardium cordatum (Pennant, 1777) (p. 251)

– Frontal ambulacrum not markedly sunken, pores in single series
Echinocardium capense Mortensen, 1907 (p. 251)

57 Subanal fasciole well developed (Fig. 274) ; labrum spine-covered ; frontal notch present ; test domed above. SPATANGIDAE
Spatangus capensis Döderlein, 1905 (p. 253)

– Subanal fasciole more or less markedly reduced (Fig. 275) ; labrum without primary spines ; no frontal notch ; test flat above. ASTEROSTOMATIDAE
Eurypatagus parvituberculatus H. L. Clark, 1924 (p. 247)

Subclass *PERISCHOECHINOIDEA* M'Coy

Order CIDAROIDA Claus

Echinoids with the test radially symmetrical, somewhat flattened spherical ; ambulacral plates simple, each with a single pore-pair forming a straight or sinuous vertical series ; interambulacral plates much larger and fewer, each with a single massive primary spine having a very conspicuous areole, encircled by a ring of scrobicular secondary spines, the mature primary spines often losing their skin covering and becoming vulnerable to encrustation by epizoic organisms, secondary spines very much smaller and conspicuously different in form, usually flattened and spatulate ; both ambulacral and interambulacral plates continuing, though imbricated, over the peristome to the mouth ; lantern present, the teeth not keeled.

Family **CIDARIDAE** Gray

See : Mortensen, 1928 : 61 ; Durham *et al.*, 1966 : U321.

A family of Cidaroida with the test rigid, strong and the interambulacra in two rows ; primary tubercles perforate, often crenulate but sometimes smooth.

Acanthocidaris maculicollis (de Meijere)

Porocidaris maculicollis de Meijere, 1903, *Tijdschr. ned. dierk. Vereen.* (2) **8** : 1.
Acanthocidaris maculicollis : Mortensen, 1928b : 329–333, pl. 43, figs 1, 2, pl. 44, fig. 1, pl. 54, figs 5, 6, pl. 83, figs 12–15 ; 1932 : 157–158, pl. 5, fig. 6, pl. 11, fig. 5.

H.d. up to 47 mm, v.d. 60–70%, apical system 37–50%, peristome 36–46% and longest primary spines 180–280% of h.d. ; up to nine interambulacral plates.

A species of Cidaridae with the test moderately high, markedly flattened above ; primary tubercles markedly crenulate (except the oral ones) ; primary spines tapering from the base, collar remarkably long, up to 30 mm on the ambital spines, shaft with *c.* 16 low longitudinal ridges, oral primaries more or less flattened with serrated edges ; apical system with all oculars insert in larger specimens, peristome slightly smaller than apical system. Colour : primary spines banded reddish, the collar with serially arranged dark red spots, especially on the upper side, naked test whitish, darker above and in the ambulacra.

Type locality. Lesser Sunda Is, Indonesia, 69–94 metres.

Southern African record. Off Natal ?

Eucidaris metularia (Lamarck)

Cidarites metularia Lamarck, 1816 : 56.
Eucidaris metularia : H. L. Clark, 1923 : 370 ; Mortensen, 1928b : 386–392, pl. 41, figs 1–8, pl. 73, fig. 6, pl. 86, figs 11–14 ; Kalk, 1958 : 206, 238 ; Macnae & Kalk, 1962 : 117, 118 ; B. I. Balinsky *In* : Macnae & Kalk, 1969 : 106, 130.

H.d. up to 25 mm, v.d. 56–66%, apical system 45–50%, peristome 50–60% and longest spines 70–120% of h.d. ; up to seven interambulacral plates.

A species of Cidaridae with the test moderately high, flattened above ; primary tubercles not crenulate ; primary spines relatively thick and short, cylindrical or slightly broader basally, tip crown-like, collar very short, shaft with low spinules in 12–18 longitudinal series, superficially appearing rugose ; apical system about half the h.d., oculars normally all exsert, even in the largest specimens, a circle of large secondary tubercles around the outer edge of the apical system bearing a fringe of spatulate spines much larger than the miliary spinelets scattered elsewhere ; peristome distinctly larger. Colour : primary spines brownish, secondaries darker, denuded test white.

Type locality. 'L'océan des Grandes Indes' (East Indies), 'les côtes de l'Ile de France' (Mauritius) [also 'Saint-Domingue' (St Domingo, West Indies), probably *E. tribuloides*].

Southern African records. 26/33/i ; 24/35/i.

Nature of bottom. Coral ; rock.

Goniocidaris sp.

Goniocidaris sp. Mortensen, 1928b : 154 (footnote).
? *Goniocidaris indica* Mortensen, 1939a : 6.

This is known only from a broken test and associated primary spine which Mortensen (1928) likened to the spines of *G. fimbriata* (de Meijere) from the East Indies, described as having long curving thorns on the shaft. However, in 1939 he thought it more likely to prove conspecific with his new species *G. indica* from off Tanzania, described as having primary spines nearly smooth (!) Presumably the spine has at least the partial disc or flange near its base characteristic of this group of species of *Goniocidaris*.

Southern African record. South of Cape St Blaize (34/22), 192 metres.

Nature of bottom. Sand, shells or rock.

Histocidaris sp. aff. *H. elegans* (A. Agassiz)

Histocidaris elegans : Mortensen, 1932 : 147–148. [See also Mortensen, 1928b : 72–77, pls.]

In Pacific material of *H. elegans* h.d. is up to 67 mm, v.d. 73–85%, apical system 41–49% (rarely less), peristome 28–39% and longest spines 180–220% of h.d. ; up to 13 interambulacral plates.

The specimen supposedly from South Africa which Mortensen determined as *Histocidaris elegans* in 1932 has h.d. 66 mm but v.d. only 38% of this, a discrepancy on which Mortensen surprisingly made no comment, though he noted that the primary spines are unusually thick and the number of interambulacral plates is unusually small at 9–10, so few being found at h.d. *c.* 40 mm in Pacific specimens of *H. elegans*, those of h.d. 60+ mm having 12–13.

Histocidaris elegans has the upper primary tubercles markedly crenulate ; the primary spines only slightly tapering, finely rugose, the oral ones with a double series of strong serrations ; the apical system nearly half h.d., rather bare with sparse secondary tubercles, the ocular plates all exsert ; the peristome distinctly smaller.

SOUTHERN AFRICAN RECORD. Presumed to be from South Africa.

Kionocidaris striata Mortensen

Kionocidaris striata Mortensen, 1932 : 165–168, figs 9–12, pl. 5, fig. 7, pl. 9, figs 4–6, pl. 11, figs 3, 9, pl. 12, figs 1, 2, pl. 13, figs 4, 6, 7.

H.d. 27 mm (in the two recorded specimens), v.d. *c.* 63%, apical system 45–48%, peristome 43–48% and longest spines 134–210% of h.d. ; up to seven interambulacral plates.

A species of Cidaridae with the test moderately high, flattened at least above ; upper primary tubercles distinctly crenulate ; primary spines varying in length but tapering slightly to a simple point, collar very short, shaft almost smooth but for *c.* 25 fine longitudinal striations, oral primaries similar but slightly flattened ; apical system with all oculars narrowly insert, sparsely tuberculated, similar in diameter to the peristome. Colour whitish, apical system and secondary spines partially olive-green.

TYPE LOCALITY. East of Durban (29/31), 126 metres.

Prionocidaris pistillaris (Lamarck)

Cidarites pistillaris Lamarck, 1816 : 55.
Cidaris sp. (part) Bell, 1904 : 168 (no. 92, part).
Prionocidaris baculosa : H. L. Clark, 1923 : 370 ; 1924 : 1 ; Macnae & Kalk, 1962 : 112, 117 ; B. I. Balinsky *In* : Macnae & Kalk, 1969 : 97, 101, 102, 106, 130, fig. 26b. [Non *Cidarites baculosa* Lamarck, 1816.]
Prionocidaris pistillaris : Mortensen, 1928b : 452–456, figs 139, 140 (1), pl. 49, pl. 50, pl. 51, fig. 1, pl. 73, fig. 18, pl. 86, figs 20, 21 ; 1932 : 170, pl. 9, figs 7–9, pl. 11, fig. 8 ; 1939b : 18–19.
Prionocidaris pistilaris (sic) Lopes, 1939 : 83, fourth pl., fig. 1.

H.d. up to 60 mm, v.d. 52–68%, apical system 37–46%, peristome 37–44% and longest spines 75–165% of h.d. ; up to 10 interambulacral plates.

A species of Cidaridae with the test moderately high, flattened above and below ; primary tubercles not crenulate ; primary spines usually only just exceeding the h.d. but rather variable in relative length, tapering, often more or less flattened, the tip often broadened, collar *c.* 4 mm long, the shaft beset with pointed tubercles in longitudinal series, oral primaries flattened, often with somewhat serrated edges ; apical system with all oculars broadly insert, closely covered with uniform small tubercles, similar in diameter to the peristome. Colour : denuded test mostly light greenish but pore zones reddish-brown and primary tubercles pink, collar of primary spines conspicuously *striped* with purple.

TYPE LOCALITY. Mauritius.

SOUTHERN AFRICAN RECORDS. 33/27/i ; 30/30/s ; 29/31/i, s ; 28/32/s ; 26/33/i ; intertidal–91 metres.

NATURE OF BOTTOM. Sand ; broken shells ; stones ; *Cymodocea* beds.

Genus *STEREOCIDARIS* Pomel

See : Mortensen, 1928b : 225.

A genus of Cidaridae with up to only about eight interambulacral plates in each series ; test densely covered with secondary tubercles, giving a shagreen-like appearance between the areoles of the primary spines ; primary tubercles with only traces of crenulation ; primary spines with fine longitudinal ridges, superficially appearing almost smooth, often somewhat flared at the tip.

Stereocidaris capensis Döderlein

Stereocidaris indica var. *capensis* Döderlein, 1901, *Zool. Anz.* **24** : 19.
Cidaris sp. (part) Bell, 1904 : 168 (no. 81, part).
Stereocidaris capensis : Döderlein, 1906 : 110–112, pl. 10, figs 3–6, pl. 12, fig. 2, pl. 36, fig. 4, pl. 37, fig. 1 ; Mortensen, 1928b : 269–271 (part), pl. 71, fig. 1.

H.d. up to 36 mm, v.d. 68–75%, apical system 44–51%, peristome 31–36% and longest spines 140–200% of h.d. ; up to six interambulacral plates.

A species of *Stereocidaris* with the test moderately high, the upper side convex ; primary spines with *c.* 20 distinct ridges, usually tapering more in their distal half than proximally ; apical system nearly half the h.d., oculars exsert, peristome distinctly smaller. Colour grey-white, scrobicular secondary spines encircling the primaries with darker tips.

TYPE LOCALITY. South from Knysna, 500 metres.

SOUTHERN AFRICAN RECORDS. 35/23/vd ; 33/28/d–vd ; 457–555 metres.

NATURE OF BOTTOM. Sand.

Stereocidaris excavata Mortensen

Stereocidaris sp. Mortensen, 1928b : 270–271, pl. 27, fig. 4.
Stereocidaris excavata Mortensen, 1932 : 151–154, fig. 5, pl. 2, figs 1, 2, pl. 3, figs 1–5, pl. 4, fig. 2, pl. 11, figs 1, 2 ; A. M. Clark, 1952 : 201.

H.d. up to 69 mm, v.d. 72–78%, apical system 43–53%, peristome 36–45%, longest spines 115–125% of h.d. ; up to eight interambulacral plates.

A species of *Stereocidaris* with the test relatively high, almost spherical except that the apical system is flattened or even sunken ; primary spines relatively slender, cylindrical, with *c.* 12 low ridges becoming higher distally to make a fluted tip ; apical system about half h.d., genital pores towards the middle of each plate, not distal as in the other species represented, oculars narrowly insert, peristome similar in size or smaller. Colour : denuded test creamy white, faintly green apically, primary spines whitish but collar brownish, scrobicular secondary spines greenish olive.

TYPE LOCALITY. Off East London, *c.* 120 metres.

SOUTHERN AFRICAN RECORDS. 36/21/d ; 33/27/d ; 120–177 metres.

NATURE OF BOTTOM. Sand.

Stereocidaris squamosa Mortensen

Cidaris sp. (part) Bell, 1904 : 168 (no. 81, part).

Stereocidaris squamosa Mortensen, 1928a, *Vidensk. Meddr dansk naturh. Foren.* **85** : 70 ; 1928b : 245–247, fig. 79, pl. 20, figs 4–6, pl. 70, fig. 7, pl. 80, figs 10–16 ; 1932 : 151 ; 1939b : 6–7, pl. 4, figs 2, 3.

H.d. up to 47 mm, v.d. 56–63%, apical system 46–49%, peristome 40–47% and longest spines 107–170% of h.d. ; up to six interambulacral plates.

A species of *Stereocidaris* with the test moderately high, slightly convex above ; primary spines cylindrical, only slightly tapering, shaft with *c.* 15 longitudinal series of low warts, superficially appearing smooth when well preserved but otherwise finely ridged ; apical system about half h.d., oculars exsert or narrowly insert, peristome slightly smaller. Colour cream to yellowish, scrobicular secondary spines olive green.

TYPE LOCALITY. Saya da Malha Bank, SE. from the Seychelles, 270 metres.

SOUTHERN AFRICAN RECORD. 29/31/374 metres.

Subclass *EUECHINOIDEA**

Superorder DIADEMATACEA Duncan

[Lepidocentroida and Aulodonta of Mortensen.]

Echinoids with the test radially symmetrical, periproct within the apical system, test either rigid or more or less flexible, the plates joined by sutures or membrane, sometimes bevelled at the edges ; ambulacral plates compound (in the taxa represented), consisting of three or more plates of which all but one may be more or less reduced and the pore-pairs often staggered in position so as to form arcs corresponding in number (but not position) with the compound plates ; interambulacral plates each with one or more primary spines ; tubercles perforate, either crenulate or smooth ; primary spines usually rather long and brittle, hollow or solid, sometimes the aboral or ambital ones exceeding the test diameter ; apical system showing a wide range of form ; peristome usually with only a single pair of ambulacral

* The remaining families of Echinoidea are grouped here in the superorders of the *Treatise* since the multiplicity of orders results too often, in this restricted context, in their coinciding with single families; using the higher taxa gives a better grouping.

(buccal) plates in each radius, sometimes also with secondary calcifications, except in Echinothuriids where series of ambulacral plates form two columns (distally somewhat irregular) in each radius ; lantern present, the foramen magnum outside the upper end of each tooth not bridged over by the epiphyses ; teeth grooved on the inside, not keeled.

Family **ECHINOTHURIIDAE** Wyville Thomson

See : Mortensen, 1935 : 80 ; Durham *et al.*, 1966 : U346.

A primarily deep-water family of Diadematacea with the test plates imbricating or separated by membranous spaces, especially the interambulacral coronal plates and the peristomial ambulacral ones, tending to collapse to a pancake under rough treatment but most species probably low hemispherical in life ; ambulacral plates compound, the smaller components very reduced, often limited to a lozenge around the pore-pair and offset from the abradial edge of the ambulacrum to form almost horizontal arcs of pores ; tubercles not crenulate, primary spines hollow, fragile, those of the oral side often with expanded clavate or hoof-like tips for walking on mud, some spines with poison sacs ; apical system highly modified and irregular in the adult, the periproctal plates very numerous, forming a cone and tending to encroach peripherally between the ocular and genital plates, the latter sometimes fragmented ; peristome with numerous ambulacral plates in series but somewhat irregular distally ; no globiferous pedicellariae.

Araeosoma paucispinum H. L. Clark

Araeosoma paucispinum H. L. Clark, 1924 : 4–5, pl. 2 ; Mortensen, 1935 : 251–254, fig. 139, pls 41, 42, pl. 79, figs 1–9.

H.d. up to 180 mm, apical system 17–20%, peristome 24–27% of h.d.

A species of Echinothuriidae with the test probably relatively high in life, rounded at the ambitus ; ambulacral plates on oral side with only inconspicuous primary tubercles, the pore-pairs in arcs on both oral and aboral sides, forming three vertical series throughout ; interambulacral plates on oral side each with a fairly conspicuous primary spine near the adradial edge forming a vertical series, though sometimes a few additional large spines may occur near the ambitus, aborally all tubercles and spines much reduced in size ; primary spines of oral side moderately long and curved, with a terminal white 'hoof' (usually broken off) ; peristome with numerous (up to 17) plates in each series and pores in double series distally ; teeth broad basally, tapering abruptly towards the tip ; large tridentate pedicellariae with the valves straight, somewhat compressed, contiguous for most of their length. Colour red or dark violet.

Type locality. ENE. of Durban, 380 metres.

Southern African records. 29/31/d ; 27/32/d ; 369–380 (? 280–454) metres.

Nature of bottom. Mud ; green mud.

Hygrosoma petersi (A. Agassiz)

Phormosoma petersii A. Agassiz, 1880, *Bull. Mus. comp. Zool. Harv.* **8** : 76.
? *Echinosoma petersii* : H. L. Clark, 1923 : 375 : 1924 : 5. [See under *Sperosoma biseriatum.*]
Hygrosoma petersii : Mortensen, 1927 : 284–285, figs 158.1, 161 ; 1935 : 202–208, figs 118, 119, pls 13–17, pl. 18, fig. 2, pl. 19, fig. 2, pl. 78, figs 1, 3–5, 24, 25.

H.d. up to 185 mm, apical system 13–17%, peristome 24–28% of h.d.

A species of Echinothuriidae with the test probably low hemispherical in life (*c.* 30% of h.d.) ; ambulacral plates on the oral side with variable tuberculation, their pore-pairs in single series (those of the primary plates sometimes reduced) whereas aborally they form two or three series, the smaller secondary plates being offset from the abradial edge of the ambulacrum ; both ambulacral and interambulacral plates of oral side towards the ambitus with multiple conspicuous primary tubercles, some of their areoles more or less confluent laterally, no membranous spaces between the plates ; primary spines of oral side with a terminal white hoof ; peristome with few plates, up to only six in each series, teeth tapering abruptly near the tip ; large tridentate pedicellariae with medially constricted slightly curved valves. Colour dark violet.

TYPE LOCALITY. SE. of U.S.A.

SOUTHERN AFRICAN RECORDS. 34/17/d, vd ; 200–1380 metres. [Also ? 29/31/d, vd, which is queried by Mortensen as unlikely to refer to this Atlantic species ; at least one locality off Natal is duplicated by the British Museum specimen which proved to be referable to *Sperosoma biseriatum.*]

NATURE OF BOTTOM. Mud.

Genus ***PHORMOSOMA*** Wyville Thomson

See : Mortensen, 1935 : 123.

A genus of Echinothuriidae with the test relatively well calcified above, often retaining its slightly arched shape in preserved material, the ambitus distinctly angular, the oral side flat and appearing markedly different from the aboral side owing to its numerous large sunken areoles round the primary tubercles ; tube-feet rather reduced on the oral side, the small pore-pairs forming such shallow arcs as to make a single vertical series, aborally the much larger pores aligned in very oblique arcs of three ; primary spines of oral side with clavate, skin-sheathed tips, not a hoof ; teeth very compressed and prolonged, tapering fairly evenly ; tridentate pedicellariae mostly with fairly broad rounded valves.

The differences between the two species of *Phormosoma* found off southern Africa, according to Mortensen, appear to be almost entirely of degree, such as the coarseness of the marginal fringe of spines, or else are only inferred, as with the shape of the aboral primary spines (slightly curved only in *P. bursarium*) since none of the specimens which Mortensen referred to *P. placenta africana* had any of these spines complete. Also both *P. bursarium* and *P. placenta africana* are distinguished from *P. placenta placenta* by having coarser aboral primary tubercles so that the difference between them in this respect may well be negligible. The maximum size difference is also negligible since H. L. Clark has recorded a specimen of *P. placenta* (i.e. *africana*) from the Atlantic side of the Cape with h.d. 120 mm ; *P. placenta* is not otherwise known to exceed 90 mm h.d. Although Mortensen does not comment on it, there may be a significant difference in the relative size of the peristome between the Atlantic and Indian Ocean specimens but this needs to be tested further, together with the shape of the aboral spines from better preserved specimens. In this event, it seems impracticable to provide diagnoses for *P. placenta africana* and *P. bursarium*.

Phormosoma bursarium A. Agassiz

Phormosoma bursarium A. Agassiz, 1881 : 99–101, pl. 10b ; Mortensen, 1935 : 135–142, figs 83–85, pl. 2, fig. 20, pl. 3, figs 1, 2, pl. 74, figs 11–15.
Phormosoma indicum Döderlein, 1906 : 130–133, fig. 20, pl. 15, figs 1, 2, pl. 38, figs 2, 3 ; H. L. Clark, 1924 : 3–4.

H.d. up to 128 mm, apical system 20–25%, peristome 30–32% of h.d.

TYPE LOCALITY. Philippines, 466 metres ('Challenger' st. 200).

SOUTHERN AFRICAN RECORDS. 29/31/d, vd ; 27/32/vd ; 421–840 metres.

NATURE OF BOTTOM. Mud.

Phormosoma placenta africana Mortensen

Phormosoma sp. aff. *P. bursarium* A. Agassiz : Bell, 1904 : 169 (ref. no. 139).
Phormosoma placenta : H. L. Clark, 1923 : 374–375 ; 1924 : 4.
Phormosoma placenta var. *africana* Mortensen, 1935 : 132–133 (*africanum* p. 134) ; pl. 2, fig. 10.

H.d. up to 120 mm, apical system 22–28% (usually *c.* 23%), peristome 32–40% (usually *c.* 36%) of h.d.

TYPE LOCALITY. Not given, presumably off the west coast of South Africa.

SOUTHERN AFRICAN RECORDS. 32/16/vd ; 33/16/vd ; 658–1850 metres.

NATURE OF BOTTOM. Green mud ; mud ; globigerina ooze.

Sperosoma biseriatum Döderlein

Sperosoma biseriatum Döderlein, 1901, *Zool. Anz.* **23** : 20 ; 1906 : 150–153, fig. 27, pl. 19, fig. 1, pl. 40, fig. 1 ; Mortensen, 1935 : 191–193, figs 113, 114b, pl. 11, pl. 76, fig. 13.
Phormosoma sp. aff. *P.* [now *Tromikosoma*] *tenue* A. Agassiz : Bell, 1904 : 169 (ref. no. 33).
? *Echinosoma petersii* : H. L. Clark, 1923 : 375 ; 1924 : 5 (Durban specimen). [Non *P. petersii* A. Agassiz, 1880.]

H.d. up to 183 mm, apical system 13–15%, peristome 18–25% of h.d.

A species of Echinothuriidae with the test probably rounded at the ambitus in life ; primary ambulacral plates on oral side fragmented, forming an irregular mosaic, the secondary plates large enough to bear tubercles as well as the pore-pairs, which are in irregular oblique arcs of three forming three series, aboral pore-pairs also irregular and forming more than one series but somewhat smaller; primary tubercles present on both sides, mostly irregular but in some cases forming series, on the interambulacral plates of the oral side not noticeably restricted to a single series along the adradial edge; no large membranous spaces between the plates; complete primary spines of oral side unknown but almost certainly with terminal hooves; peristome with fairly numerous plates, *c.* 8–11 in each series at h.d. > 100 mm; teeth tapering abruptly near the tip; large tridentate pedicellariae with very broad valves having undulating edges (not found in Bell's specimen), smaller ones slightly constricted above the base, the blade fairly narrow with straight almost parallel edges.

TYPE LOCALITY. Off Kenya, 1019 metres.

SOUTHERN AFRICAN RECORD. 29/31/vd; 804 metres. Mortensen gave no locality data for his South African specimen. He suspected that H. L. Clark's records of *Echinosoma* (i.e. *Hygrosoma*) *petersi* from Natal were incorrectly identified and since one of them coincides with Bell's record of a specimen in the British Museum collection which has proved to be referable to *Sperosoma*, I think it very likely that his suspicions were correct. This would extend the depth range to 350 metres.

Family **PEDINIDAE** Pomel

See: Mortensen, 1940: 62; Durham *et al.*, 1966: U357.

A family of deep-water Diadematacea with the test rigid, fragile, variable in shape (low in the species represented); ambulacral plates trigeminate (in the species represented) but without much reduction of the smaller components, interambulacral plates each with a single very large primary tubercle and spine; tubercles not crenulate; primary spines solid, long and tapering, easily exceeding the h.d. in length; apical system with genital plates normally forming a continuous ring around the well-defined periproct and all the oculars exsert (i.e. dicyclic), no periproctal cone; peristome with five pairs of buccal plates, those of each pair contiguous but separated from the other pairs, numerous small secondary platelets also present.

Caenopedina capensis H. L. Clark

Coenopedina (sic) *capensis* H. L. Clark, 1923: 375–378, fig. 4, pl. 21, figs 1, 2.
Caenopedina capensis: Mortensen, 1940: 112–113, fig. 62a, d, pl. 69, figs 1, 2.

H.d. 16 mm, v.d. 44%, apical system *c.* 53%, peristome *c.* 47% of h.d., longest spines easily exceeding h.d.; up to 10 interambulacral plates.

A species of Pedinidae with the test relatively low, ambulacral plates with pore-pairs in very slight arcs of three forming an almost straight vertical series; primary tubercles very conspicuous, primary spines with 25–30 longitudinal ridges, secondary spines similar but much smaller; globiferous pedicellariae not found in the type material. Colour of naked test off-white except around the periproct where it is 'rich bright purple in abrupt contrast', primary aboral spines more or less markedly purple but basally pink or red.

TYPE LOCALITY. WSW. of Cape Point (probably 37/17), 1200–1650 metres.

NATURE OF BOTTOM. Green mud.

Family **ASPIDODIADEMATIDAE** Duncan

See: Mortensen, 1940 : 6; Durham *et al.*, 1966 : U352.

A family of Diadematacea with the test rigid or almost so, fragile, high, almost spherical; ambulacral plates trigeminate (in the species represented), interambulacral plates each with a single large tubercle and spine up to four times h.d. in length, the tubercles strongly crenulate, the spines curving down distally, very easily broken, the lower ones expanded at the tip; apical system consisting of a ring of the five genital and ocular plates alternating (monocyclic) and similar in size and shape, the periproct with relatively small plates isolated in the membrane; peristome with the five pairs of buccal plates large and covering most of it.

Aspidodiadema africanum Mortensen

Aspidodiadema nicobaricum: H. L. Clark, 1923 : 371–372. [Non *A. nicobaricum* Döderlein, 1901.]

Aspidodiadema africanum Mortensen, 1939b, *Vidensk. Meddr dansk naturh. Foren.* **103** : 548; 1940 : 49–51, pl. 2, figs 11, 12, pl. 63, figs 1–4.

H.d. up to 17 mm, v.d. *c.* 75%, apical system 50%, peristome *c.* 44% of h.d. and up to 11 interambulacral plates in the holotype, h.d. 16 mm (no complete spines known).

A species of Aspidodiadematidae with the test subspherical; ambulacral plates with pore-pairs in an almost straight vertical series; primary tubercles oval in shape; pedicellariae including stout and slender tridentate ones, the latter with long straight valves only meeting terminally, no globiferous pedicellariae. Colour of (dried) test light purplish, tubercles and spines white.

TYPE LOCALITY. SE. of East London, 730–820 metres.

Family **DIADEMATIDAE** Gray

See: Mortensen, 1940 : 158; Durham *et al.*, 1966 : U350.

A family of Diadematacea, mainly from shallow water, with the test usually rigid but sometimes rather flexible when the plates are bevelled and imbricated (*Astropyga* and *Chaetodiadema*), shape more or less flattened; ambulacral plates trigeminate, the secondary ones hardly reduced, pore-pairs mostly in arcs (except in *Chaetodiadema*), interambulacral plates near the ambitus often with additional large tubercles but aborally with more or less extensive bare tracts, usually in the shape of an inverted V; tubercles usually distinctly crenulate; spines hollow or with a partially filled cavity, very slender, fragile and often very long; apical system usually monocyclic with all the oculars insert, sometimes (*Diadema*) two of them exsert, periproct more or less naked, usually conical and inflatable ; peristome with only the ten small, usually isolated buccal plates, besides some secondary platelets in many cases.

Astropyga radiata (Leske)

Cidaris radiata Leske, 1778, *Additamenta ad J. T. Klein : Naturalem dispositionem Echinodermatum.* Lipsiae. pp. 52–53, pl. 44, fig. 1.

Astropyga radiata : H. L. Clark, 1923 : 373 ; 1925 : 46–47 ; Mortensen, 1940 : 187–196, figs 111–115, pls (numerous) ; Kalk, 1958 : 238 ; Macnae & Kalk, 1962 : 108 ; 1969 : 130. [Non *Astropyga radiata* : Bell, 1904, which is *Chaetodiadema africana*.]

H.d. up to 180 mm, v.d. 29–35%, apical system 19–23%, peristome 19–27% of h.d. ; up to 44 interambulacral plates.

A large species of Diadematidae with the test slightly flexible (especially when wet), very low, the upper side sunken adapically and in the interambulacra ; ambulacral plates with pore-pairs in arcs of three, aborally almost horizontal but orally oblique or almost vertical, though the pore-zones are widened adorally, interambulacral plates with multiple series of large tubercles, up to nine on each plate at the ambitus but aborally with a conspicuous inverted V-shaped naked area separating the adradial series of tubercles from the triangular patch of interradial tubercles which extends above the ambitus ; spines with the central cavity more or less filled with meshwork, their length not exceeding half the h.d. ; apical system with oculars broadly insert and genital plates elongated into the interradii, especially in large specimens, h.d. > 120 mm, periproct conical, often inflated into a sphere in life ; peristome superficially appearing naked, though with numerous secondary plates besides the small widely separated pairs of buccal plates, those of each pair partly contiguous. Colour of adults reddish brown, in life with iridescent bright blue spots in the bare parts of the aboral interambulacra.

Type locality. Unknown.

Southern African records. 31/29/– [probably s] ; 26/33/i.

Nature of bottom. Rock.

Chaetodiadema africanum H. L. Clark

Astropyga radiata : Bell, 1904 : 168–169. (Non *Cidaris radiata* Leske, 1778.]

Chaetodiadema africanum H. L. Clark, 1924 : 2–3, pl. 1 ; 1925 : 49–50 ; Mortensen, 1940 : 232–235, figs 132a, b, 133, 134c, pl. 33, figs 3–6, pl. 72, figs 9–13, 22, 23.

H.d. up to 45 mm, v.d. 34–37% (but 'almost exactly one-fourth' in the holotype), apical system 34–37%, peristome 22–30% of h.d. ; up to 18 interambulacral plates.

A species of Diadematidae with the test slightly flexible and very low, flat above or slightly sunken adapically ; ambulacral plates with the pore-pairs small and aligned in a single vertical series on both sides, only at the ambitus of larger specimens tending to form oblique arcs, interambulacral plates each with two or three large tubercles at the ambitus but adapically and adorally the size of these gradually reduced, a V-shaped bare area usually distinct in each interambulacrum aborally ;

spines with the cavity more or less filled with meshwork, in length probably not exceeding half the h.d. ; apical system with oculars broadly insert, genital plates with short interradial angles ; peristome with squarish buccal plates, those of each pair fairly close but widely separated from the other pairs. Colour (preserved) pinkish in larger specimens, smaller ones light brown, purplish or whitish with a green tinge.

TYPE LOCALITY. East from Durban, 330 metres.

SOUTHERN AFRICAN RECORDS. 29/31/s, d ; 84–330 metres.

NATURE OF BOTTOM. Mud, sand and shells.

Genus ***DIADEMA*** Gray

See : Mortensen, 1940 : 241 ; Opinion 206, 1954, the name validated.

A genus of Diadematidae with the test not flexible, flattish hemispherical and of moderate height ; ambulacral plates with fairly well-developed regular primary tubercles, though smaller than the interambulacral ones, pore-pairs in arcs of three, the pore-zones slightly widened adorally, interambulacral plates with only inconspicuous naked areas ; primary spines hollow, very long, especially the aboral ones which easily exceed the h.d. in specimens not exposed to wave action, the ambulacral primary spines not dissimilar from the interambulacral ones, though somewhat smaller ; apical system usually with two oculars exsert, the others insert, genital plates not markedly elongated, periproct conical and largely naked ; peristome appearing almost naked, the buccal plates very small, the pairs widely separated.

Diadema savignyi Michelin

Diadema savignyi Michelin, 1845, *Mag. Zool., Paris* (2) **1845**, Zoophytes : 15–16.
Diadema setosa (or *setosum*) : Stephenson, Stephenson & du Toit, 1937 : 367 ; Eyre & Stephenson, 1938 : 43 ; Stephenson, 1944 : 306 [lapsus]. [Non *Echinometra setosa* Leske, 1778.]*
Diadema savignyi : Mortensen, 1940 : 265–269, figs 136, 141b, 143 ; Stephenson, 1944 : 277, 306, 347 ; Day & Morgans, 1956 : 271, 308.

H.d. up to 67 mm (? to 85 mm), v.d. 45–59%, apical system 22·0–28·5 (exceptionally 34)%, peristome 40–51% of h.d. ; up to 17 interambulacral plates.

A species of *Diadema* with large tridentate pedicellariae having broad valves, length : median breadth *c.* 4·5 : 1. Colour of adults usually black, including the anal cone. In daylight, when alive, the iridophores make a blue ring around the apical system from which two blue lines extend down each interradius, the width and continuity of these vary with light intensity ; also a white spot is present aborally in each interradius, though this is rarely conspicuous.

TYPE LOCALITY. Mauritius.

SOUTHERN AFRICAN RECORDS. 31/29/i ; 29/31/i ; 26/33/i.

NATURE OF BOTTOM. Rock.

* From correspondence between T. A. Stephenson and D. D. John about 1947, it seems likely that the records of *Diadema* from Port St John and Natal refer to *D. savignyi*.

Diadema setosum (Leske)

Echinometra setosa Leske, 1778, *Additamenta ad J. T. Klein : Naturalem dispositionem Echinodermatum*. Lipsiae. pp. 36, figs 1, 2, pl. 46, fig. 1, pl. 51, figs 1, 2.
Centrechinus setosus : H. L. Clark, 1923 : 372–373.
Diadema setosum : Mortensen, 1940 : 256–264, figs 140, 141a, pls (numerous). [Non *D. setosum* : Stephenson *et al.*, 1937–1944 ; see above.]
Diadema setosa : Kalk, 1958 : 206, 218, 238 ; Macnae & Kalk, 1962 : 108 ; B. I. Balinsky *In* : Macnae & Kalk, 1969 : 97, 102, 104, 106, 130.

H.d. up to 103 mm, v.d. 47–60%, apical system 20–30%, peristome 40–50% of h.d. ; up to 23 interambulacral plates.

A species of *Diadema* with large tridentate pedicellariae having narrow valves, length : median breadth 15–20 : 1. Colour of adults usually blackish, anal cone with a red ring, live specimens with a conspicuous white spot in each interambulacrum aborally and an iridescent blue pattern usually broken up into isolated blue spots in daylight.

Type locality. Amboina, Moluccas.

Southern African records. 26/33/i.

Nature of bottom. Rock ; coral ; *Cymodocea* beds.

Echinothrix calamaris (Pallas)

Echinus calamaris Pallas, 1774, *Spicilegia Zoologica*. Berolini. Fasc. 10 : 31–32, pl. 2, figs 4–8.
Echinothrix calamaris : Peters, 1855 : 116 ; H. L. Clark, 1923 : 373 ; Mortensen, 1940 : 285–290, figs 147–149 ; Kalk, 1958 : 206, 238 ; Macnae & Kalk, 1962 : 108 ; B. I. Balinsky *In* : Macnae & Kalk, 1969 : 97, 104, 106, 130.

H.d. up to 130 mm, v.d. 38–49 (usually *c.* 45)%, apical system 20–33%, peristome 34–46% of h.d. ; up to 24 interambulacral plates.

A species of Diadematidae with the test not flexible, rather low, less than half as high as broad, the upper side sunken around the apical system, the ambulacra bulging aborally ; ambulacral plates with regular but very small primary tubercles, pore-pairs in oblique arcs of three, pore-zones not widened near the peristome, interambulacral plates with several large primary tubercles near the ambitus, a median aboral area usually distinctly bare but not bifurcating ; primary spines hollow, long, the interambulacral ones about equal to the h.d. and verticillate, with numerous successive fine transverse ridges along their length, giving a very rough texture when stroked downwards, primary ambulacral spines markedly finer, needle-like, with downward-pointing barbs ; apical system with oculars usually all exsert, genital plates not markedly elongated, though extending between the uppermost interambulacral plates, periproct usually inflated to a sphere in life, studded with small white platelets ; peristome with buccal plates not quite contiguous, the pairs widely separated. Colour : usually at least partly dark but spines more often conspicuously banded with white and often other colours in life.

Type locality. East Indies.

Southern African records. 29/31/– ; 26/33/i ; 24/35/i.

Nature of bottom. Rock ; coral, *Cymodocea* beds.

Superorder ECHINACEA Claus

Echinoids with the main structures almost entirely radially symmetrical, though occasionally the rigid test is horizontally elongated in one plane or another ; periproct within the apical system ; ambulacral plates compound, except in the Saleniidae the secondary plates more or less markedly reduced and integrated with the primary ones, pore-pairs usually in arcs ; primary tubercles imperforate, smooth or sometimes crenulate ; spines relatively large, usually distinctly tapering from a fairly stout base, solid, or sometimes with a small axial cavity ; peristome with plates restricted to a buccal pair in each radius, though secondary platelets may also be present ; lantern present, epiphyses either bridging the foramen magnum outside the upper end of each tooth or not, teeth keeled on the inside.

Family **SALENIIDAE** L. Agassiz

See : Mortensen, 1935 : 320 ; Durham *et al.*, 1966 : U377.

A family of deep-water Echinacea with the test small to moderate in size, subspherical, not sculptured ; ambulacral plates compound but only bigeminate (consisting of two plates) in the species represented, the pore-pairs in single, almost straight, series ; primary interambulacral tubercles large and distinctly crenulate ; primary spines very large, cidarid-like with a naked collar near the base and the shaft rough ; apical system markedly asymmetrical with a large suranal plate in the offset periproct, the system dicyclic with all the oculars exsert ; gill notches distinct but shallow ; lantern with epiphyses not bridging the foramina ; globiferous pedicellariae apparently not developed.

Salenia phoinissa Agassiz & Clark

Salenia pattersoni : Döderlein, 1906 : 179–180, pl. 2, fig. 2, pl. 15, fig. 4. [Non *S. Pattersoni* A. Agassiz, 1878.]

Salenia phoinissa A. Agassiz & H. L. Clark, 1908, *Mem. Mus. comp. Zool. Harv.* **34** : 54 ; Döderlein, 1910 : 256–257 ; Mortensen, 1935 : 377–379, fig. 196c, pl. 84, fig. 1.

H.d. 9·2 mm (only the holotype known), v.d. *c.* 65%, apical system *c.* 55%, peristome *c.* 40% (Döderlein gives 32% and Mortensen's measurement produces 43%), longest spines probably *c.* 400% of h.d. ; seven interambulacral plates.

A species of Saleniidae with the ambulacral plates very narrow, one primary tubercle to each two pore-pairs ; the huge primary spines slightly swollen at the base and then evenly tapering. Colour : primary spines greenish-white with about five large brownish-red patches on the upper (aboral) side, spaced by their own length, small spines white, test greenish-white with dark reddish-brown patterning, especially on the apical system, oculars partly dark violet, making a stellate pattern.

TYPE LOCALITY. Agulhas Bank, 35/21, 102 metres.

Family **STOMECHINIDAE** Pomel

See : Mortensen, 1935 : 492 (Stomopneustidae) ; Durham *et al.*, 1966 : U403.

A family of Echinacea (the only recent member a large species from shallow water) with the test hemispherical, not sculptured ; ambulacral plates doubly compound, consisting of three to six trigeminate plates further fused and sharing a single large primary tubercle, interambulacral plates with enlarged secondary tubercles and spines, similar to the primary ones ; tubercles not crenulate ; spines strong, cylindrical and pointed, without a basal collar ; apical system relatively small, periproct with numerous small plates ; peristome larger ; gill notches fairly deep and wide ; lantern with foramina not bridged by the epiphyses ; globiferous pedicellariae developed as well as other kinds, but rare, armed with one or two teeth each side of the tip.

Stomopneustes variolaris (Lamarck)

Echinus variolaris Lamarck, 1816 : 47.

Stomopneustes variolaris : H. L. Clark, 1923 : 378 ; Mortensen, 1935 : 507–512, figs 301, 302, pl. 71, figs 3–5, pl. 72, figs 1, 2, pl. 89, figs 16–26 ; Stephenson, Stephenson & du Toit, 1937 : 367 ; Eyre & Stephenson, 1938 : 37, 38, pl. 7, fig. 1 (part) ; Stephenson, 1944 : 277, 306, 347 ; Kalk, 1958 : 198, 238 ; B. I. Balinsky *In* : Macnae & Kalk, 1969 : 99, 102, 104, 106, 130 ; Day, 1974 : 94.

H.d. up to 100 mm, v.d. usually 50–60%, apical system 14–18%, peristome 24–32%, longest spines 45–95% of h.d. ; up to 20 interambulacral plates.

A species of Stomechinidae with a moderately high test ; ambulacral plates with limits very indistinct in larger specimens, pore-zones broad, petaloid on the oral side and narrowing close to the peristome, the pore-pairs scattered irregularly, primary tubercles rather irregular, similar in size to those on the interambulacral plates ; apical system with usually two oculars insert and the other three exsert ; tridentate pedicellariae with narrow leaf-shaped valves. Colour in life black or dark purple, bare test greyish.

Type locality. South Seas ('les mers australes').

Southern African records. 31/29/i ; 29/31/i ; 28/32/i ; 26/33/i ; 23/35/i.

Nature of bottom. Rock ; coral.

Family **ARBACIIDAE** Gray

See : Mortensen, 1935 : 529 ; Durham *et al.*, 1966 : U409.

A family of Echinacea with the test small to moderate in size (h.d. up to 50 mm), low hemispherical in the species represented, not sculptured ; ambulacral plates trigeminate with large primary tubercles, pore-pairs either in arcs or straight series ; primary interambulacral tubercles smooth, often smaller than the ambulacral ones and usually reduced aborally ; primary spines large and smooth ; apical system with ocular plates usually all exsert, the periproct very distinctive having four (sometimes five) triangular valves resembling a quartered cake ; peristome often relatively large, with numerous small secondary calcifications ; gill notches shallow but with a superficial continuation on the test ; lantern with epiphyses not bridging the foramina ; no globiferous pedicellariae.

Genus *COELOPLEURUS* L. Agassiz

See : Mortensen, 1935 : 605.

A genus of Arbaciidae with the primary ambulacral tubercles in a regular series, larger than the interambulacral ones which are restricted to the oral side leaving a noticeable bare area aborally, pores small on the oral side, larger aborally ; primary spines easily exceeding the h.d. in length ; peristome with the pairs of buccal plates more or less separated (at least in *C. maillardi*).

Coelopleurus interruptus Döderlein

Coelopleurus floridanus : Döderlein, 1906 : 181–182, figs 34, 35, pl. 45, fig. 2. [Non *C. floridanus* A. Agassiz, 1872.]

Coelopleurus floridanus var. *interrupta* Döderlein, 1910 : 257–258.

Coelopleurus interruptus : H. L. Clark, 1923 : 379–381, pl. 21, fig. 3 ; Mortensen, 1935 : 626, pl. 88, fig. 30. [Non *C. interruptus* : H. L. Clark, 1925 : 74, which is *C. maillardi* Michelin.]

H.d. up to 43 mm. In the holotype at h.d. 18 mm, v.d. 64%, apical system 33%, peristome 47% of h.d. and up to seven interambulacral plates.

A species of *Coelopleurus* of which the details of the test structure are undescribed ; the massive primary spines more or less prismatic, distinctly tapering and curved ; pedicellariae including large tridentate ones as well as ophicephalous. Colour dull, primary spines greenish near the base but becoming dull purplish-red, sometimes distally with red spots on greenish, bare interambulacral areas predominantly brown with transverse violet bands, each area bordered laterally by white.

TYPE LOCALITY. Agulhas Bank, 102 metres.

SOUTHERN AFRICAN RECORDS. 35/21/d ; 33/25/s ; 55–102 metres.

Coelopleurus maillardi (Michelin)

Keraiophorus maillardi Michelin *In* : Maillard, 1862, *Notes sur l'Ile de la Réunion*, Paris, Annex A : 2, pl. 14.

Diadema saxatile : Bell, 1904 : 168. [Non *Echinus saxatilis* Linnaeus, 1758, i.e. *Diadema setosum* or *savignyi*.]

Coelopleurus interruptus : H. L. Clark, 1925 : 74. [Non *C. floridanus* var. *interrupta* Döderlein, 1910.]

Coelopleurus maillardi : Mortensen, 1935 : 627–631, pl. 67, fig. 3, pl. 68, fig. 4, pl. 69, figs 1–3, pl. 88, figs 22, 23.

H.d. up to 40 mm, rarely more than 20 mm, v.d. 55–62%, apical system 29–41%, peristome 44–55% of h.d. when that is 10–20 mm, 9–10 interambulacral plates at h.d. 20 mm.

A species of *Coelopleurus* with aboral primary ambulacral tubercles very large, the pore-pairs curving around them in shallow arcs of three ; primary ambulacral spines only faintly ridged, slightly flattened cylindrical, hardly tapering or curving distally, no large tridentate pedicellariae observed, all the larger ones being ophicephalous. Colour vivid, primary ambulacral spines banded alternately greenish-yellow and purple, bare interambulacral areas purple or mauve and edged with red or orange.

TYPE LOCALITY. Réunion (Bourbon), Mascarene Is.

SOUTHERN AFRICAN RECORDS. 32/28/d (140 metres) ; probably 29/31/– ; also Mozambique, no details.

Family **TEMNOPLEURIDAE** A. Agassiz

See : Mortensen, 1943a : 39 ; Durham *et al.*, 1966 : U418.

A family of Echinacea with the test of various shapes, usually more or less conspicuously sculptured with pits or abrupt depressions in or between the plates ; ambulacral plates compound, trigeminate, with the secondary plates reduced in breadth, pore-pairs in one vertical series or in arcs so as to make two or three series ; primary interambulacral tubercles usually crenulate but sometimes indistinctly so or smooth ; primary spines usually fairly short and smooth ; apical system of various forms, periproct often with an enlarged suranal plate when the anus is eccentric, small specimens, h.d. < 20 mm, often with only a few relatively large triangular periproctal plates, sometimes when these number four resembling the periproct of arbaciids ; peristome usually naked outside the buccal plates ; gill notches shallow ; lantern with the foramina bridged over ; globiferous pedicellariae present, also ophicephalous, triphyllous and usually tridentate, the valves of globiferous ones with or without lateral teeth, which may be single, unpaired or multiple, besides the large end tooth.

Lamprechinus nitidus Döderlein

Lamprechinus nitidus Döderlein, 1905, *Zool. Anz.* **28** : 622 ; 1906 : 190–191, pl. 23, figs 1, 2, pl. 35, fig. 11, pl. 46, fig. 6 ; H. L. Clark, 1923 : 383 ; Mortensen, 1943 : 335–336, fig. 194.

H.d. up to 16 mm, v.d. *c.* 50%, apical system 43–45%, peristome 39–41% of h.d. ; up to 12 interambulacral plates.

A species of Temnopleuridae with the test low hemispherical, moderately sculptured ; ambulacral plates with pore-pairs in very slight arcs of three forming an almost straight vertical series, pores relatively small, ambulacral and interambulacral plates each with a primary tubercle ; tubercles smooth ; longest spines about one-third of h.d. ; apical system smooth, oculars all exsert, periproct with three large plates in the holotype, anus eccentric ; globiferous pedicellariae with an unpaired lateral tooth. Colour white.

TYPE LOCALITY. South from Knysna (35/23), 500 metres.

Orechinus monolini (A. Agassiz)

Trigonocidaris monolini A. Agassiz, 1879, *Proc. Am. Acad. Arts Sci.* **14** : 203 ; 1881 : 111–112, pl. 6a, figs 8–10.

Orechinus monolini : Döderlein, 1906 : 196–198, pl. 25, fig. 1, pl. 35, fig. 6, pl. 46, fig. 5 ; H. L. Clark, 1923 : 383 ; Mortensen, 1943a : 330–333, figs 40b, 190–193, pl. 18, figs 23–29, pl. 49, figs 1–10.

H.d. up to 14 mm, v.d. 50–66%, apical system 35–44%, peristome 39–50% of h.d. ; up to 12 interambulacral plates.

A species of Temnopleuridae with the test of moderate height, hemispherical, more or less elaborately sculptured, especially aborally, often with radiating ridges and depressions around the interambulacral primary tubercles ; ambulacral plates with pore-pairs in very slight arcs of three forming an almost straight vertical series, pores relatively small, both ambulacral and interambulacral plates each with a primary tubercle ; tubercles slightly crenulate, at least on ambital plates ; spines fairly stout, the longest about a third of h.d. ; apical system markedly sculptured, all oculars exsert, periproct covered by four or five large plates, anus slightly eccentric ; globiferous pedicellariae with an unpaired lateral tooth. Colour (in preserved specimens) white.

TYPE LOCALITY. Near the Kermadec Is, north from New Zealand, 950 metres.

SOUTHERN AFRICAN RECORD. 35/23, 500 metres.

Salmaciella dussumieri erythracis (H. L. Clark)

Salmacis erythracis H. L. Clark, 1912, *Mem. Mus. comp. Zool. Harv.* **34** : 315–316, pl. 111, figs 4–6.
Salmacis dussumieri : H. L. Clark, 1924 : 6.
Salmaciella erythracis : Mortensen, 1943a : 144–146, pl. 6, figs 9–13.
Salmaciella dussumieri erythracis : A. M. Clark & Rowe, 1971 : 148.

H.d. up to 48 mm, v.d. 41–48%, apical system 20–24% peristome 23–32% of h.d. ; up to 22 interambulacral plates.

A subspecies of Temnopleuridae with the test low hemispherical or sub-conical, angular pores usually distinct in the ambulacra on the oral side but indistinct aborally ; ambulacral plates with pore-pairs in oblique arcs of three aborally but at the ambitus and orally one pair on each plate tends to be offset so as to form a second vertical series with half as many pores as the main one, primary tubercles present only on every second plate, except adorally ; tubercles distinctly crenulate ; spines rather spaced ; apical system distinctly eccentric with one ocular near the markedly offset anus often more or less insert, though a suranal plate may not be distinct ; globiferous pedicellariae of one kind, with one lateral tooth each side. Colour (in preserved specimens) off-white between the spines and on the bases of the primaries, giving way on the distal two-thirds to indistinct bands of olive and dull brown, the aboral secondary spines tending to be reddish basally.

TYPE LOCALITY. Zanzibar.

SOUTHERN AFRICAN RECORDS. 29/31/s ; Mozambique (no details) ; 66 metres.

NATURE OF BOTTOM. Sand, shells.

Note : A Survey specimen from Durban Harbour ('in shallows') with h.d. 25 mm may belong either to this subspecies or to *Salmacis bicolor*, which I recorded from there in 1952. It has v.d. only *c*. 36% and aborally only alternate ambulacral plates have a primary tubercle, agreeing better with *erythracis* but the angular pits in the test aborally and the brighter colour with purple in place of brown on the primary spines and all the secondaries basally vermilion agree with *bicolor*. Unfortunately the periproct is gaping and inverted so the position of the anus is indeterminable. A better range of material from Natal is needed before the true relationship of the urchins from that area can be established. Mortensen noted under the heading of *Salmaciella dussumieri* that the specimen from off Natal which H. L. Clark (1924) referred to this species 'probably belongs to *Salmaciella erythracis*' (which Mortensen distinguished specifically from *S. dussumieri*) but he did not think it safe to identify it definitely as *erythracis*.

Salmacis bicolor L. Agassiz

Salmacis bicolor L. Agassiz *In* : Agassiz & Désor, 1846, *Annls Sci. nat.* (3) **6** : 359, pl. 15, fig. 4 ; H. L. Clark, 1923 : 382 ; 1924 : 5 ; Lopes, 1939 : 84 ; Mortensen, 1943a : 112–117, figs 67a, 68a, pl. 4, figs 1–8, pl. 5, figs 1–3, 10–12, pl. 6, figs 1–8, pl. 46, figs 1, 13, 16, 19, 20 ; A. M. Clark, 1952 : 201 ; Day & Morgans, 1956 : 308 ; Macnae & Kalk, 1962 : 108, 112 ; B. I. Balinsky *In* : Macnae & Kalk, 1969 : 97, 102, 130.

H.d. up to 70 mm, v.d. 59–69%, apical system 19–21%, peristome 25–34% of h.d. ; up to 34 interambulacral plates.

A species of Temnopleuridae with the test very variable in shape, with pores or small pits usually distinct at the angles of especially the aboral plates, horizontal sutures sometimes deepened ; ambulacral plates with pore-pairs more or less distinctly in arcs of three, a primary tubercle on every plate similar in size to the primary interambulacral tubercles ; tubercles distinctly crenulate ; spines short, making a dense covering ; apical system relatively small, all oculars exsert, anus not markedly eccentric and no distinct suranal plate ; globiferous pedicellariae of two kinds, the larger without lateral teeth and the smaller with an unpaired lateral tooth. Colour green, primary spines basally red, otherwise banded red or purple with green or yellowish, secondaries red, denuded test dull dark greenish.

Type locality. Bombay.

Southern African records. 29/31/i, s ; 26/33/i ; intertidal–68 metres.

Nature of bottom. *Cymodocea* beds ; sandy mud ; mud, sand.

Genus ***TEMNOPLEURUS*** L. Agassiz

See : Mortensen, 1943a : 72.

A genus of Temnopleuridae with the test of moderate or small size, h.d. up to nearly 60 mm in one species but usually much less, shape varying, aboral horizontal sutures sculptured with deep pits of variable extent ; ambulacral plates with pore-pairs either in slightly oblique arcs of three or in a single nearly straight series, each plate with a primary tubercle similar to the interambulacral ones ; primary tubercles distinctly crenulate ; spines long and slender, the longer ones usually about two-thirds h.d. but sometimes equal to h.d. ; globiferous pedicellariae mostly with one lateral tooth each side below the main tooth.

Temnopleurus reevesi (Gray)

Toreumatica reevesi Gray, 1855, *Proc. zool. Soc. Lond.* **1855** : 39.

Temnopleurus reevesi : Döderlein, 1906 : 200–201, pl. 25, figs 3–6, pl. 46, fig. 7 ; H. L. Clark, 1923 : 381 ; 1924 : 6.

Temnopleurus (*Toreumatica*) *reevesi* : Mortensen, 1943a : 92–99, figs 56, 57, 58, 59a, 60, pl. 3, figs 1, 2, 12, 16–20.

H.d. up to 35 mm, exceptionally 45 mm, v.d. 47–53%, apical system 23·5–33%, peristome 26–36% of h.d. ; up to 20 interambulacral plates.

A species of *Temnopleurus* with the test of moderate height, hemispherical ; ambulacral plates with the pore-pairs in an almost straight vertical series ; apical system with one ocular sometimes insert near the anus which is markedly eccentric, an enlarged suranal plate well developed. Colour : spines light greenish-brown, not banded, lighter on the oral side, test brownish above, sometimes greenish ; young specimens sometimes white all over.

TYPE LOCALITY. 'China.'

SOUTHERN AFRICAN RECORDS. 35/21/d ; 29/31/s ; 66–102 metres.

NATURE OF BOTTOM. Sand and shells.

Temnopleurus toreumaticus (Leske)

Cidaris toreumatica Leske, 1778, *Additamenta ad J. T. Klein* : *Naturalem dispositionem Echinodermatum*, Lipsiae, pp. 91–92 (155), pl. 10, figs D, E (of Klein).
Temnopleurus toreumatica : Junod, 1899 : 281 (footnote) ; H. L. Clark, 1923 : 382–383 ; Mortensen, 1943a : 76–84, figs 53b, 55b, pl. 1, figs 1–3, 6–12, pl. 2, figs 9–15, 19–21, pl. 45, figs 27, 28 ; Macnae & Kalk, 1962 : 108, 112 ; B. I. Balinsky *In* : Macnae & Kalk, 1969 : 97, 102, 130.

H.d. up to 58 mm, v.d. 53–75 (rarely less than 40)%, mean 57%, apical system 18–25%, peristome 26–34 (rarely more than 35)% of h.d. ; up to 25 interambulacral plates.

A species of *Temnopleurus* with the test hemispherical or sub-conical ; ambulacral plates with pore-pairs usually in distinct arcs of three ; apical system with all oculars exsert, anus near the centre of the periproct but a suranal plate usually distinguishable. Colour variable, spines often olive-green with narrow purplish bands, denuded test greyish-olive.

TYPE LOCALITY. Unknown.

SOUTHERN AFRICAN RECORDS. 26/32/s (no depth given) ; 26/33/i.

NATURE OF BOTTOM. Sand, *Cymodocea* beds.

Family **TOXOPNEUSTIDAE** Troschel

See : Mortensen, 1943a : 382 ; Durham *et al.*, 1966 : U426.

A family of Echinacea with the test low to high hemispherical, large in the species represented, not sculptured ; ambulacral plates trigeminate or polyporous, pore arcs oblique or almost horizontal, primary tubercles forming regular vertical series as do the interambulacral ones ; primary interambulacral tubercles smooth ; primary spines forming a relatively short dense coat ; apical system usually somewhat eccentric with two oculars broadly insert, the rest exsert, periproct with irregular small plates, no enlarged suranal ; peristome large with scattered small secondary platelets in addition to the buccal plates ; gill notches sharp and deep in the species represented; lantern with the foramina bridged over and also with extra processes from the epiphyses supporting the teeth; globiferous pedicellariae present and either very numerous or very large in the species represented, without any lateral teeth.

Toxopneustes pileolus (Lamarck)

Echinus pileolus Lamarck, 1816 : 45.
Toxopneustes pileolus: H. L. Clark, 1923 : 386; Mortensen, 1943a : 472–480, figs 240b, 293a, 294, 295a, 296, 297a, b, 298, pls (numerous); Kalk, 1958 : 206, 238; Macnae & Kalk, 1962 : 110; B. I. Balinsky *In*: Macnae & Kalk, 1969 : 99, 104, 106, 130.

H.d. up to 150 mm, v.d. usually 40–50%, apical system 12–16%, peristome usually 30–37% of h.d., up to 41 interambulacral plates.

A species of Toxopneustidae with the test low hemispherical or subconical; ambulacral plates trigeminate with the pore-pairs in more or less shallow arcs tending to form three close vertical series in large specimens; apical system eccentric, usually two oculars insert, anus distinctly offset from the centre of the periproct; globiferous pedicellariae exceptionally large, up to 3 mm in span when open and easily visible with the naked eye, appearing almost spherical in life but triangular in preserved specimens where the poison glands are shrunken. Colour variable, creamy pink at Inhaca, according to Balinsky.

TYPE LOCALITY. Mauritius.

SOUTHERN AFRICAN RECORD. 26/33/i.

NATURE OF BOTTOM. Rock or coral.

Tripneustes gratilla (Linnaeus)

Echinus gratilla Linnaeus, 1758, *Systema naturae*. Holmiae. Ed. 10 : 664.
Tripneustes gratilla: H. L. Clark, 1923 : 387; Stephenson, Stephenson & du Toit, 1937 : 367; Eyre & Stephenson, 1938 : 38, 43; Eyre, Broekhuysen & Crichton, 1938 : 96; Lopes, 1939 : 84, pl. 4, fig. 2; Mortensen, 1943a : 500–508, figs 306, 307, pls (numerous); Stephenson, 1944 : 277, 306, 347; Day & Morgans, 1956 : 308; Kalk, 1958 : 198, 238; Macnae & Kalk, 1962 : 108, 112; B. I. Balinsky *In*: Macnae & Kalk, 1969 : 97, 101, 102, 106, 130; Day, 1974 : 54, 94.

H.d. up to 145 mm, v.d. 51–63% (occasionally more), apical system 14·0–18·5%, peristome 23–33% of h.d.; up to 41 interambulacral plates.

A species of Toxopneustidae with the test very variable in shape, usually moderately high hemispherical; ambulacral plates trigeminate with the pore-pairs in almost horizontal arcs in larger specimens and more or less spaced so as to form three quite separate vertical series, the middle one rather irregular; apical system with two oculars usually insert, anus not markedly eccentric; globiferous pedicellariae not enlarged but extremely numerous. Colour: generally with radiating vertical bands of brown and black (formed by the pedicellariae) against which the white or sometimes green or brown spines show up in contrast. In life often shaded by, for example, leaves of *Cymodocea* held by the tube feet.

TYPE LOCALITY. 'Indian Ocean.'

SOUTHERN AFRICAN RECORDS. 32/28/i; 29/31/i, s; 26/32/i; 26/33/i; 23/35/s; intertidal–*c.* 5 metres at least.

NATURE OF BOTTOM. *Cymodocea* beds; sand and shells.

Family **ECHINIDAE** Gray

See: Mortensen, 1943b: 1; Durham *et al.*, 1966: U431.

A family of Echinacea with the test usually hemispherical, circular or pentagonal when viewed from above, not sculptured, size moderate to large, ambulacral plates compound, usually trigeminate, sometimes polyporous, pore-pairs usually in oblique arcs, in larger specimens tending to form three vertical series, primary tubercles usually developed on every plate, smooth; primary spines of moderate length, usually slender and tapering evenly; apical system usually dicyclic with all oculars exsert, not markedly eccentric, periproct with small irregular plates, a suranal plate rarely distinct; peristome with secondary platelets except in *Dermechinus*; gill notches shallow; lantern with the foramina bridged over; globiferous pedicellariae present, valves with one or more lateral teeth each side, tridentate, triphyllous and ophicephalous pedicellariae also found.

Dermechinus horridus (A. Agassiz)*

Echinus horridus A. Agassiz, 1879, *Proc. Amer. Acad.* 14: 203; H. L. Clark, 1923: 384; 1924: 7–8.
Sterechinus horridus: Döderlein, 1906: 220–222, pl. 28, figs 1, 2, pl. 35, figs 2, 3, pl. 47, figs 10, 11.
Dermechinus horridus: Mortensen, 1943b: 112–117, figs 46–49, 50a, pl. 19, figs 6–10, pl. 20, figs 1–3, pl. 56, figs 22, 23, 29–31; A. M. Clark, 1974: 479–480.

H.d. up to 110 mm, v.d. 74–140% (mean of 13 specimens with h.d. $>$ 70 mm = 98%), apical system *c.* 12–16%, peristome 10–14% of h.d.; up to 38 interambulacral plates.

A species of Echinidae with the test fragile, hemispherical in smaller specimens but in larger ones, h.d. $>$ 70 mm, remarkably high, often almost spherical, the height sometimes considerably exceeding the h.d.; ambulacral plates trigeminate, pore-pairs very small, in oblique arcs aborally but the upper pair of each arc offset on the ambital plates, tending to form a second vertical series, a primary tubercle on each plate, ambulacra more or less distinctly sunken on the oral side and interambulacra swollen; primary spines relatively long and slender but brittle, probably not more than a third of h.d., a dense coat of secondary spines; apical system very small, densely covered with small tubercles, oculars all widely exsert; peristome usually even smaller; most globiferous pedicellariae with two to four lateral teeth, sometimes one or five, on each side of the blade which is abruptly narrower than the basal part. Colour: spines bright orange-red, naked test red and white.

* Döderlein (1906) followed by Mortensen (1943b) distinguishes South African specimens from Chilean and Australian ones as variety (i.e. subspecies in this case) *africanus* on account of some minor differences in the pedicellariae.

TYPE LOCALITY. Southern Chile, 320 metres.

SOUTHERN AFRICAN RECORDS. 32/17/d ; 33/17/d ; 34/18/d ; 35/23/vd ; 380–567 metres.

NATURE OF BOTTOM. Mud ; mud and sand ; mud and stones ; rocky.

Echinus gilchristi Bell

Echinus gilchristi Bell, 1904 : 170–171 [except nos. 145 and 200, which are *Polyechinus agulhensis* Döderlein] ; Döderlein, 1906 : 213–216, pl. 26, figs 1–5, pl. 35, figs 10, 14, pl. 46, fig. 9 ; H. L. Clark, 1923 : 384–385 ; 1924 : 6–7 ; 1925 : 112 ; Mortensen, 1943b : 57–60, fig. 18, pl. 15, figs 7–12, pl. 16, figs 1–9, pl. 55, figs 14, 16, 20, 26, 27.
Echinus hirsuta Döderlein, 1905, *Zool. Anz.* **28** : 623.
Echinus gilchristi var. *hirsuta* Döderlein, 1906 : 213–215.
Echinus gilchristi hirsutus : H. L. Clark, 1924 : 7.

H.d. up to 84 mm, v.d. 51–70%, apical system 18–24%, peristome 28–36% of h.d. ; up to 24 interambulacral plates.

A species of Echinidae with the test variable in shape from low hemispherical to moderately high subconical or subspherical, almost circular viewed from above, rather fragile ; ambulacral plates trigeminate, pore-pairs in oblique, slightly irregular, arcs, aborally usually only every second or third plate with a primary tubercle and these are smaller than the similarly sparse interambulacral primaries, in specimens of the forma *hirsuta* the primary tubercles are more numerous and may occur on every plate but are not regular in size ; primary spines up to about one-third of h.d. in length, appearing sparse aborally, except in forma *hirsuta*, contrasting with the much smaller but dense coat of secondary spines ; apical system with all oculars exsert, plates with a scattering of secondary tubercles ; peristome larger ; globiferous pedicellariae with usually a single lateral tooth each side of the blade which is abruptly narrower than the basal part. Colour : primary spines white or light greenish, small secondary ones aborally often dull greenish, bare test light greenish.

TYPE LOCALITY. South of Cape Seal, east of Knysna, 146 metres. No holotype was selected by Bell from among his samples numbered 34, 70, 83, 85, 138, 145 and 200. No. 34 (two specimens) from off Cape St Blaize has been dried ; nos. 70 and 85 (three specimens bottled together and not individually distinguished) from 16 km approximately south of Cape Point and from the south-west of False Bay are currently labelled as 'types' in the British Museum collection, as is sample 138 (one specimen) from south of Cape Seal. Although this specimen was bottled by Bell with no. 145 (which H. L. Clark, 1925 : 132, recognized as *Paracentrotus* – now *Polyechinus – agulhensis* Döderlein) it bears a tied-on tag with the number and (hopefully) no confusion should have arisen when they were separated. No. 200 was also referred to *P. agulhensis* by H. L. Clark. No. 83 from 5–6 km approximately north-east of Cape Point Lighthouse, i.e. within False Bay, cannot now be found ; probably it included the denuded specimens measured by Bell. In selecting a lectotype to narrow the type locality it seems best to choose the single specimen no. 138 (registered number 1903.8.1.126, h.d. 40 mm, v.d. 23 mm) owing to the impossibility of correlating the three other specimens labelled as 'types', nos. 70 and 85, with their respective stations, although one is much larger, h.d. 72 mm. (In fact one of these three, probably from the other station, is of forma *hirsutus*.)

SOUTHERN AFRICAN RECORDS. 26/13/d; 31/16/d; 32/17/d; 33/18/d; 34/17/d; 34/18/s, d; 34/18/FB/s; 34/22/s, d; 34/23/d; 35/23/vd; 34/24/d; 34/25/s, d; 51–500 metres.

NATURE OF BOTTOM. Rock; rock and shell; sand and rocks; mud, sand; dark green mud and sand; green sand, mud; green mud, black sand; mud.

Parechinus angulosus (Leske)

Cidaris angulosa var. *minor* Leske, 1778, *Additamenta ad J. T. Klein: Naturalem dispositionem Echinodermatum.* Lipsiae. p. 30 (94), pl. 3, figs A, B (of Klein).

Echinus angulosus: A. Agassiz, 1881: 115; Meissner, 1893: 183–185; Bell, 1904: 169–170; Thomson, 1913: 190–198.

Echinus juv. Bell, 1904: 171.

Protocentrotus angulosus: Döderlein, 1906: 204–207, pl. 27, figs 6–8, pl. 35, fig. 16, pl. 47, fig. 6; Mortensen, 1909: 58–61, pl. 8, figs 7, 8, pl. 9, figs 8, 10, pl. 16, figs 3, 9; Döderlein, 1910: 258; Koehler, 1914: 248–249.

Paracentrotus angulosus: Koehler, 1908: 641.

Protocentrotus annulatus Mortensen, 1909: 61–64, pl. 3, figs 2, 3, pl. 8, figs 9–13, pl. 16, figs 2, 5, 6, 10, 18, 20.

Parechinus angulosus: H. L. Clark, 1923: 385–386; Stephenson, Stephenson & du Toit, 1937: 358, 359, 363, 381; Bright, 1937: 59, 63, 76, 86, 87; Stephenson, Stephenson & Bright, 1938: 10; Eyre & Stephenson, 1938: 41; Eyre, Broekhuysen & Crichton, 1938: 96, 110; Eyre, 1939: 298, 304, 305; Stephenson, Stephenson & Day, 1939: 357, 368; Mortensen, 1943b: 148–156, figs 64–68, pl. 18, figs 8–19, 22, pl. 58, figs 20, 21, 26–32; Stephenson, 1944: 306, 317, 334, 347; A. M. Clark, 1952: 201; Day, Millard & Harrison, 1952: 395, 396; Scott, Harrison & Macnae, 1952: 321; Macnae, 1957: 362, 363, 367; Day, 1959: 502, 544; Morgans, 1959: 399, 401, 415, 421, 424, 425, 426, 427, pl. 19 (part); 1962: 303, 311; Grindley & Kensley, 1966: 13; Day, Field & Penrith, 1970: 81; Penrith & Kensley, 1970: 201, 208, 234; Cram, 1971: 321–337 (embryology).

Parechinus angulosus var. *pallidus* H. L. Clark, 1923: 386; 1924: 9; 1925: 117; Day, Millard & Harrison, 1952: 396.

H.d. up to 60 mm, v.d. 51–60% (occasionally more), apical system 16–20%, peristome 29–39% of h.d.; up to 25 interambulacral plates. In two specimens of forma *pallidus* at h.d. *c.* 30 mm the apical system is *c.* 21 and 22·5% and the peristome *c.* 39·5 and 42% of h.d.

A species of Echinidae with the test moderately strong, hemispherical, often rounded pentagonal when viewed from above; ambulacral plates trigeminate with the pores moderately large, the pairs in low arcs forming three fairly distinct vertical series, equally spaced or one slightly separated from the other two, each plate with a regular primary tubercle; spines variable, up to about one-fifth of h.d. in length but sometimes relatively shorter and stouter, larger ones forming a fairly dense coat, graduating to intermingled smaller spines without abrupt change in magnitude; apical system with oculars either all exsert or two may be insert, in larger specimens fairly densely covered with secondary tubercles; peristome considerably larger; globiferous pedicellariae with triangular tapering valves, without shoulders on the basal part, one to three lateral teeth each side. Colour very variable, most often purple but also grey, green, red or pale, white in the forma *pallidus* which also tends to have a reddish patch aborally in each ambulacrum and interambulacrum or else be brownish-grey and green apart from the spines; forma *annulatus* with banded spines.

TYPE LOCALITY. Unknown.

SOUTHERN AFRICAN RECORDS. 26/15/i; 29/16/i; 31/18/i; 32/17/s, d; 32/18/i, s; 33/17/s; 33/18/i, s; 34/18/i, s; 34/18/FB/i, s; 34/21/i; 34/22/i, s; 34/23/i, s; 33/25/i, s; 33/27/i; 32/27/i; 29/31/i; intertidal–98 metres; forma *pallidus* to 180 metres.

NATURE OF BOTTOM. Rock; rock and shell; sand, shell, rock; sand, shell; sand, shell and *Phyllochaetopterus*; sand; muddy sand; khaki mud; green mud; in caves or under stones; in crevices in rock pools.

Polyechinus agulhensis (Döderlein)

Paracentrotus agulhensis Döderlein, 1905, *Zool. Anz.* **28**: 623; 1906: 207–210, fig. 38, pl. 27, figs 1–4, pl. 35, fig. 17, pl. 47, fig. 1; H. L. Clark, 1923: 388; 1924: 9; 1925: 132.
Paracentrotus grandis H. L. Clark, 1923: 388–390, pl. 22; 1924: 9.
Polyechinus agulhensis: Mortensen, 1943b: 118–123, figs 51–54, pl. 17, figs 4, 5, pl. 21, figs 1–8, pl. 57, figs 16, 21–28.

H.d. up to 80 mm, v.d. 51–72%, apical system 21·0–26·5%, peristome 21·0–30·5% of h.d.; up to 21 interambulacral plates.

A species of Echinidae with the test varying in shape from sub-spherical to low hemispherical but usually moderately high; ambulacral plates quadrigeminate with steep or oblique arcs of four pore-pairs (occasionally three or five) and a large primary tubercle on each plate, forming regular series; primary spines stout but easily broken, the longest probably about a quarter of h.d., somewhat spaced but not widely so; secondary spines inconspicuous; apical system with all oculars widely exsert, bearing few secondary tubercles and appearing rather bare; peristome variable in size in comparison with the apical system; globiferous pedicellariae mostly with two (one to three) lateral teeth each side of the blade which is abruptly narrower than the basal part, though that has rather sloping 'shoulders'. Colour whitish in preserved specimens.

TYPE LOCALITY. South from Knysna, 500 metres.

SOUTHERN AFRICAN RECORDS. 31/15/vd; 33/17/vd; 36/21/vd; 35/23/vd; 500–1700 (? 1830) metres. Although Mortensen gives a minimum depth of 200 mm, I cannot trace this record. H. L. Clark (1925) correctly reidentified one of Bell's specimens (previously named *Echinus gilchristi*) ref. no. 145, from 10·3 km approximately south of Cape St Blaize in only 65 metres, as *P. agulhensis* (as well as no. 200 from deeper water) but I suspect that there is some error in the locality here especially as the identical locality was given for Bell's no. 34 which is *E. gilchristi*.

NATURE OF BOTTOM. Green mud; grey mud; green sand; green sand, mud; green mud, black sand; globigerina ooze.

Family **ECHINOMETRIDAE** Gray

See : Mortensen, 1943b : 277 ; Durham *et al.*, 1966 : U433.

A family of Echinacea with the test round or oval when viewed from above, not sculptured, size small to large ; ambulacral plates compound, trigeminate or polyporous, pores in oblique arcs, a primary tubercle on each plate in the species represented, forming a regular series ; primary tubercles smooth ; primary spines usually fairly long, strong and stout, at least basally ; apical system usually with two oculars insert, periproct with irregular plates, no distinct suranal ; peristome with scattered secondary platelets ; gill notches shallow ; lantern with the foramina bridged over ; globiferous pedicellariae present, with only one unpaired lateral tooth besides the terminal one, tridentate, triphyllous and ophicephalous pedicellariae also found.

Echinometra mathaei (de Blainville)

Echinus mathaei de Blainville, 1826, *Dictionnaire des Sciences Naturelles*. Paris. **37** : 94.
Echinometra mathaei : H. L. Clark, 1923 : 390–391 ; Stephenson, Stephenson & du Toit, 1937 : 367 ; Eyre & Stephenson, 1938 : 37, 38, 43, pl. 7, fig. 1 (part) ; Mortensen, 1943b : 381–393, figs 185, 186, 187a, b, 188–194, pl. 42, figs 1–10, pl. 47, figs 1–4, pl. 65, figs 16–26 ; Stephenson, 1944 : 306, 347 ; Day & Morgans, 1956 : 308 ; Kalk, 1958 : 198, 206, 214, 238 ; B. I. Balinsky *In* : Macnae & Kalk, 1969 : 99, 102, 104, 106, 130.

H.d. (maximum) up to 76 mm, v.d. 52–65%, apical system 19·5–25·0%, peristome 37·5–53·0% of *mean* h.d. ; up to 22 interambulacral plates. [Mortensen's percentages are of maximum h.d.]

A species of Echinometridae of moderate size, with the test more or less markedly ovate in shape, elongated horizontally, rarely almost circular when viewed from above, the ambitus lower than half the height, upper surface domed ; ambulacral plates quadrigeminate with pore-pairs in oblique arcs of four, occasionally five ; primary spines stout and evenly tapering to a sharp tip, longer ones about half h.d. ; apical system elongate parallel to the test, oculars all exsert ; peristome much larger. Colour very variable, ranging from black to white, often purple, brown or green, the tips of the spines often white.

TYPE LOCALITY. Mauritius.

SOUTHERN AFRICAN RECORDS. 30/30/i ; 29/31/i ; 28/32/i ; 26/32/i ; 26/33/i.

NATURE OF BOTTOM. Rock or coral, usually in crevices or hollows.

Echinostrephus molaris (de Blainville)

Echinus molaris de Blainville, 1826, *Dictionnaire des Sciences Naturelles*. Paris. **37** : 88.
Echinostrephus molare : H. L. Clark, 1923 : 387–388 ; Stephenson, Stephenson & du Toit, 1937 : 367 ; Eyre & Stephenson, 1938 : 37, 43.
Echinostrephus molaris : Mortensen, 1943b : 311–316, figs 149, 150a, b, pl. 35, figs 1–10, pl. 58, figs 1, 2, 4, 9 ; Stephenson, 1944 : 277, 306, 347.

H.d. up to 28 mm, v.d. 53–59%, apical system 17–24%, peristome 39–50% of h.d. ; up to 18 interambulacral plates.

A species of Echinometridae of small size, the test circular or pentagonal when viewed from above but peculiarly top-heavy when viewed from the side, the ambitus being towards the upper side which is markedly flattened ; ambulacral plates trigeminate with pore-pairs in oblique arcs of three ; primary spines longest on the aboral side, making an erect tuft, the laterally projecting spines relatively short (to fit the tubular burrow) ; apical system almost circular, all oculars exsert ; peristome much larger. Colour dark green with spines purplish.

TYPE LOCALITY. Unknown.

SOUTHERN AFRICAN RECORD. 29/31/i. This deep-burrowing tropical Indo-West Pacific echinoid should occur at Inhaca. It is easily overlooked because of its habits.

NATURE OF BOTTOM. Rock or coral in tubular holes at least its own height.

Superorder GNATHOSTOMATA Zittel

Echinoids with the rigid test showing both radial and bilateral symmetry, the mouth and apical system approximately polar in position but the periproct posterior, usually on the lower side remote from the apical system ; shape sometimes ovate but usually rounded or pentagonal and more or less markedly flattened, so much so as to be termed discoidal ; ambulacral plates not compound, in the included taxa modified aborally to form five petals – one of them anterior – delimited by the pore-pairs, the ambital podia and others beyond the petals markedly reduced ; spines not bristling but mostly inclined posteriorly and very reduced in size though numerous, forming a dense plush-like coat ; primary tubercles usually perforate and crenulate, small ; lantern present in modified flattened stellate form, teeth keeled.

Family **CLYPEASTERIDAE** L. Agassiz

See : Mortensen, 1948b : 5 ; Durham *et al.*, 1966 : U462.

A family of Gnathostomata with the test reaching a moderate or large size, sometimes exceeding 100 mm diameter in the two species represented, in shape more or less markedly flattened, often discoidal but with the edge rounded, not sharp, in outline approximately circular, oval or rounded pentagonal, not perforated by large slots ; ambulacral plates forming the petals alternately large primary plates and much smaller demiplates, the latter limited to the area between the two spaced pores of each pair and not reaching the ambulacral mid-line ; genital pores five ; aboral miliary (smallest) spines ending in a simple unmodified tip.

Genus ***CLYPEASTER*** Lamarck

See : Mortensen, 1948b : 47.

Diagnosis as for the family ; both species represented with the periproct inframarginal and spaced from the posterior edge of the test by about its own breadth or less.

Clypeaster eurychorius H. L. Clark

? *Clypeaster humilis* : H. L. Clark, 1923 : 392. [Non *Echinanthus humilis* Leske, 1778.]
Clypeaster eurychorius H. L. Clark, 1924 : 10–11, pl. 3 ; Mortensen, 1948c : 5.
Clypeaster (*Stolonoclypus*) *eurychorius* : Mortensen, 1948b : 94–96, figs 54, 55, pl. 30, fig. 2, pl. 31, figs 2, 3, pl. 32, fig. 3, pl. 33, fig. 2.

Length up to 145 mm, breadth 88–98% of length.

A species of *Clypeaster* with the test abruptly swollen centrally to about the edge of the petaloid area, maximum height about a fifth to a quarter of the length, flattened towards the ambitus, outline rounded pentagonal ; petals broad and open distally. Colour greenish-brown or green (possibly only when moribund).

TYPE LOCALITY. ENE. from Durban, *c.* 320 metres.

SOUTHERN AFRICAN RECORDS. 29/31/d ; 166–384 metres.

NATURE OF BOTTOM. Mud ; green mud ; broken shell.

Clypeaster rarispinus de Meijere

Clypeaster rarispinus de Meijere, 1902, *Tijdschr. ned. dierk. Vereen.* **8** : 7 ; Mortensen, 1948c : 5–6.
Clypeaster audouini Fourtau, 1904, *Bull. Inst. Egypt* **4** : 418, pl. 1, figs 1–3 ; H. L. Clark, 1923 : 392 ; 1924 : 9–10.
Clypeaster (*Leptoclypus*) *rarispinus* : Mortensen, 1948b : 58–62, fig. 43, pl. 6, figs 1–15, pl. 40, figs 4, 5, 8, 9, pl. 64, figs 11, 12, 14–17, 20.

Length up to 108 mm, breadth 95–102% of length.

A species of *Clypeaster* with the test not markedly swollen centrally, height about a sixth of the length or less, the edge flattened or sometimes swollen, outline rounded pentagonal, sometimes with a distinct bulge opposite the periproct ; petals narrow and closed distally.

TYPE LOCALITY. Java Sea, 37–91 metres.

SOUTHERN AFRICAN RECORDS. 29/31/s (? and d), 53–71 (18–110) metres.

NATURE OF BOTTOM. Mud ; coarse sand and rock ; sand and shell.

Family **LAGANIDAE** A. Agassiz

See : Mortensen, 1948b : 238 ; Durham *et al.*, 1966 : U472.

A family of Gnathostomata with the test reaching a moderate length, commonly 40 mm or more, more or less markedly flattened, usually discoidal but the edge not very sharp, outline circular, oval or angular, not perforated by large slots ; ambulacral plates forming the petals all similar and extending for half the width of the petal ; genital pores four or five ; aboral miliary spines ending in a crown-like expansion.

Laganum fudsiyama africanum Mortensen

? *Laganum decagonale* : H. L. Clark, 1923 : 303. [Non *Scutella decagonalis* de Blainville, 1826.]
Laganum fudsiyama var. *africanum* Mortensen, 1948b : 342–343, pl. 55, fig. 10.

Length (of specimen illustrated by Mortensen) 35 mm (*L. fudsiyama fudsiyama* up to 92 mm), breadth probably *c.* 95–100% of length.

A species of Laganidae with the test low conical above, height centrally about a quarter of the length, outline slightly pentagonal or almost circular ; petals relatively small, narrow and more or less open distally ; genital pores five, madreporic pores sunk into a sinuous curved groove ; periproct distant from the posterior edge of the test by several times its own breadth. (These details are mainly from *L. fudsiyama fudsiyama*, from which the two tests from Natal apparently only differ in having relatively larger periprocts.)

TYPE LOCALITY. Natal, 220 metres.

SOUTHERN AFRICAN RECORDS. 29/31/s, d? From H. L. Clark's 1923 record.

Note : Koehler (1922, Echinoderma of the Indian Museum, p. 95) has described *Laganum joubini*, an elliptical species with the test distinctly longer than broad, as from 'Mozambique'. Without positive records from within the range now covered, this species is omitted here.

Family **FIBULARIIDAE** Gray

See : Mortensen, 1948b : 156 ; Durham *et al.*, 1966 : U469.

A family of Gnathostomata with the test small, rarely as much as 15 mm long, ovate or somewhat flattened, especially below, but not discoidal, more or less markedly elongated, rarely almost circular in outline ; petals not very well defined (correlated partly with the small size), open distally, the individual pores relatively large and not linked in pairs externally by a common depression, the outer ones often alternating in position ; genital pores four ; aboral miliary spines ending in a crown-like expansion.

Echinocyamus elegans Mazetti

Echinocyamus elegans Mazetti, 1894, *Memorie R. Accad. Sci. Lett. Modena* (2) **10** : 216–217, fig. (also *Atti Soc. Nat. Modena* **12** : 240, fig.) ; H. L. Clark, 1923 : 393–394 ; Mortensen, 1948b : 184–185, fig. 111, pl. 46, figs 29–31 ; 1948c : 7.

Length up to 10 mm, breadth 70 + %.

A species of Fibulariidae with the test elliptical (but not markedly elongated) or rounded pentagonal, the height probably a quarter to a third of the length, aboral side convex, orally somewhat sunken around the pentagonal peristome ; petals with seven or eight pairs of large pores each side ; periproct slightly nearer to the posterior edge than to the peristome.

TYPE LOCALITY. Northern Red Sea (Ghubbet Soghra), 40–100 metres.

SOUTHERN AFRICAN RECORDS. 33/27/d ; probably 30/31/s ; 99–146 (? 238) metres.

NATURE OF BOTTOM. Fine sand ; coral and rock.

Family **ASTRICLYPEIDAE** Stefanini

See : Durham *et al.*, 1966 : U489. [Mortensen, 1948b : 345 (Scutellidae – part).]

A family of Gnathostomata with the test reaching a moderate to large size, always markedly flattened, the edge thin and sharp, outline often broad three-lobed with the posterior side almost straight, perforated by lunules (slots), in the species represented limited to one in each posterior ambulacrum ; ambulacral plates forming well-defined petals, the consecutive plates similar ; genital pores four ; aboral miliary spines ending in a glandular bag.

Genus *ECHINODISCUS* Leske

See : Mortensen, 1948b : 398.

A genus of Astriclypeidae with the test very flat, broadest posteriorly, with two lunules which may or may not break through the posterior edge of the test ; petals short, usually distinctly closed distally ; periproct midway between the peristome and the posterior edge.

Echinodiscus auritus Leske

Echinodiscus auritus Leske, 1778, *Additamenta ad J. T. Klein : Naturalem dispositionem Echinodermatum*. Lipsiae. p. 138 (202) ; H. L. Clark, 1923 : 395 ; Mortensen, 1948b : 400–406, figs 240, 242c, pl. 56, figs 2, 3, pl. 57, fig. 7, pl. 71, figs 1–5, 10–15, 17, 22 ; Macnae & Kalk, 1962 : 115, 118 ; B. I. Balinsky *In* : Macnae & Kalk, 1969 : 99, 102, 106, 130, fig. 26a ; Day, 1974 : 94.

Length up to 183 mm, breadth 96–104%.

A species of *Echinodiscus* with the test about as broad as long ; the lunules open, like deep notches, one each side posteriorly.

Type locality. Unknown.

Southern African records. 26/33/i ; 23/35/s ; intertidal–5 metres.

Echinodiscus bisperforatus Leske

Echinodiscus bisperforatus Leske, 1778, *Additamenta ad J. T. Klein : Naturalem dispositionem Echinodermatum*. Lipsiae. p. 132 (196), pl. 2, figs A, B (of Klein) : H. L. Clark, 1923 : 394–395 ; 1925 : 170 ; Lopes, 1939 : 84, pl. 5, fig. 2 ; Mortensen, 1948b : 406–411, figs 241a, 242a, b, pl. 58, figs 2, 6–8, pl. 71, figs 6–9, 18 ; A. M. Clark, 1952 : 202 ; Day, Millard & Harrison, 1952 : 395, 396 ; Day & Morgans, 1956 : 308 ; Macnae & Kalk, 1962 : 115, 118 ; B. I. Balinsky *In* : Macnae & Kalk, 1969 : 99, 106, 130 ; Day, 1974 : 52, 94.

Length up to 97 mm, breadth 107–120%.

A species of *Echinodiscus* with the test markedly broader than long ; the lunules closed distally, wholly within the test.

Type locality. Unknown.

Southern African records. 34/22/s ; 34/23/i ; 29/31/s ; 26/33/i ; 23/35/i, s ; intertidal–20 metres.

Nature of bottom. Muddy sand ; sandy mud ; sand.

Superorder ATELOSTOMATA Zittel

Echinoids with the test rigid, often fragile, showing a marked degree of bilateral symmetry, the odd ambulacrum being anterior, the periproct posterior and interradial, distant from the still approximately polar apical system and the peristome usually more or less anterior on the lower surface, even on the frontal face of the test, but occasionally nearly central, shape approximately ovate but more or less flattened, especially below, without becoming discoidal; ambulacral plates not compound, the adapical ones of at least the paired series often modified to form petals delimited by their enlarged pore-pairs; spines slender, brittle and mostly inclined posteriorly, rarely uniformly small and erect, aboral primaries sometimes markedly elongated, others on the sternum behind the mouth usually specialized as shovels for forward locomotion; primary tubercles usually distinctly perforate and crenulate; apical system with four or fewer genital pores; peristome usually broad semilunar or kidney-shaped with the posterior side concave, formed by a slightly arched labrum or lip, peristome otherwise pentagonal or occasionally broad elliptical or round, without labrum; lantern and teeth undeveloped; fascioles of some kind (belts of densely placed fine ciliated spinelets) present in the more highly modified heart-urchins.

Family **ECHINOLAMPADIDAE** Gray

See: Mortensen, 1948a: 263; Durham *et al.*, 1966: U506.

A family of Atelostomata with the test often reaching a large size, exceeding 100 mm in length, ovate but flattened below, domed above; all the ambulacra aborally petaloid, the pores paired but often unequal, the petals neither sunken nor closed but fading out gradually towards the ambitus; primary spines short and similar, forming a dense uniform coat; apical system central or slightly anterior, genital pores four; periproct transverse oval, on the lower surface; peristome broad pentagonal in the species represented, the posterior side straight, not concave, central in position or nearly so; no fascioles.

Echinolampas (Palaeolampas) crassa (Bell)

Palaeolampas crassa Bell, 1880, *Proc. zool. Soc. Lond.* **1880**: 48–49, pl. 4; 1904: 172; Döderlein, 1906: 236–237, pl. 30, fig. 1.
Echinolampas crassa: H. L. Clark, 1923: 397; 1924: 11–12; 1925: 182–183; Day, Field & Penrith, 1970: 82.
Echinolampas (*Palaeolampas*) *crassa*: Mortensen, 1948a: 293–295, figs 256, 267e, 270d, 277a, pl. 3, figs 4–7, pl. 4, figs 11, 12, pls 9, 10, pl. 14, figs 8, 9; Cram, 1971: 339–352, 8 figs (embryology).

Length up to 124 mm, breadth 87–97%, height 40–60% of length.

A species of Echinolampadidae with the test moderately high, very thick; petals broad, straight, extending almost to the ambitus, widely open distally, the pores only slightly uneven in extent on the paired petals. Colour yellowish-green.

TYPE LOCALITY. '? India.'

SOUTHERN AFRICAN RECORDS. 34/18/FB/s; 35/23/vd; 33/27/s; 15–500 metres. However, the maximum depth is only for a dead test, no other record exceeds 100 metres.

NATURE OF BOTTOM. Sand, shell; sandy.

Family **NEOLAMPADIDAE** Lambert

See : Mortensen, 1948a : 330 ; Durham *et al.*, 1966 : U629.

A family of Atelostomata with the test small, length less than 20 mm, approximately ovate ; all the ambulacra aborally non-petaloid, not at all sunken, the pores single and more or less rudimentary ; primary spines short and similar ; apical system slightly anterior to centre, genital pores three or four ; periproct round, on the truncated posterior face of the test ; peristome round or transverse elliptical, slightly anterior to centre ; no fascioles.

Tropholampas loveni (Studer)

Catopygus loveni Studer, 1880 : 878, pl. 2, fig. 1.
Tropholampas loveni : H. L. Clark, 1923 : 395–397 ; Mortensen, 1948a : 342–343, figs 310b, 312c, 314a, 315a, 316b.

Length up to 8 mm, breadth 78–93%, height *c.* 70% of length.

A species of Neolampadidae with the test high subconical above, in outline oval or tapering to a blunt angle posteriorly ; in females the apical system deeply sunken to form a marsupium, genital pores four. Colour pale yellow, turning green on preservation.

Type locality. South of the Cape of Good Hope, 213 metres.

Southern African records. 34/17/d ; 34/18/d ; 34/24/d ; 110–350 metres.

Nature of bottom. Sand, shell and rocks ; green sand, black specks ; black sandy mud ; green sand and mud ; very fine black sand.

Family **URECHINIDAE** Duncan

See : Mortensen, 1950 : 103 ; Durham *et al.*, 1966 : U535.

A family of deep-sea Atelostomata with the test reaching a moderate or large size, particularly fragile, ovate but flattened below ; all the ambulacra aborally non-petaloid, not sunken, the pores single and more or less rudimentary ; primary spines short, fine and uniform ; apical system elongate, ill-defined, central or nearly so, genital pores three or four ; periproct round or transverse oval, on the oblique posterior face of the test but visible from below ; peristome round, central or slightly anterior ; a subanal fasciole sometimes distinct.

Urechinus naresianus A. Agassiz

Urechinus naresianus A. Agassiz, 1879, *Proc. Amer. Acad. Sci.* **14** : 207 ; 1881 : 146–148, pl. 29, figs 1–4, pl. 30, pl. 30a, figs 10–14 (? 1–9), pl. 39, figs 29, 30, pl. 40, figs 56–58 ; Bell, 1904 : 173 ; H. L. Clark, 1923 : 398 ; 1924 : 12 ; 1925 : 186–187 ; Mortensen, 1950 : 111–113, fig. 105.

Length up to *c.* 50 mm, breadth 85% and height 44% in one intact specimen.

A species of Urechinidae with the test moderately low, ovate, flattened below, tapering posteriorly to a blunt point ; primary spines aborally limited to one in the centre of most plates but orally the plates mostly have multiple spines ; apical system very elongate, genital pores three, the two posterior ones well spaced from the third which is to the left of the madreporite ; periproct facing obliquely downwards. Colour light purple.

TYPE LOCALITY. West of the Crozet Is, Southern Ocean (*c.* 46°S 48½°E), 2926 metres.

SOUTHERN AFRICAN RECORDS. 33/17/vd ; 34/17/vd ; 1648–1847 metres.

NATURE OF BOTTOM. Green mud ; globigerina ooze.

Unfortunately Bell gave no locality details of his samples ; no key to the reference numbers can be traced.

Family **POURTALESIIDAE** A. Agassiz

See : Mortensen, 1950 : 131 ; Durham *et al.*, 1966 : U537.

A family of deep-sea Atelostomata with the test reaching a moderate size, rarely large, particularly fragile, vase- or bottle-shaped, truncated anteriorly by a deep depression leading to the peristome, posteriorly narrowly prolonged but terminally more or less blunt ; paired ambulacra aborally non-petaloid, not sunken, the pores single, frontal ambulacrum sunken into a deep groove orally ; primary spines long and curved backwards in the species represented ; apical system compact with all four genital plates close together, markedly anterior, genital pores usually four, sometimes two or three ; periproct rounded, on the upper side of a subanal rostrum in the species represented ; peristome elliptical, placed vertically in the anterior groove ; subanal fasciole usually present, encircling the rostrum.

Pourtalesia alcocki Koehler

Pourtalesia carinata : Bell, 1904 : 172. [Non *P. carinata* A. Agassiz, 1879.]
Pourtalesia alcocki Koehler, 1914, *Echinides du Musée Indien à Calcutta.* 1. *Spatangidés.* Calcutta. pp. 8–17, pl. 1, figs 1–14, pl. 16, figs 1–15 ; H. L. Clark, 1923 : 399 ; 1925 : 191 ; Mortensen, 1950 : 146–147.

Length up to 46 mm, breadth 43–49%, height 40–49% of length.

A species of Pourtalesiidae with the test moderately low, narrow vase-shaped, broadly truncated anteriorly and the constricted posterior rostrum also truncated, lower side with a distinct median keel ; primary spines with fairly large tubercles, the anterolateral ones of the oral side arranged in linear series, otherwise irregular ; genital pores four, large. Colour light purple.

TYPE LOCALITY. Gulf of Oman, 2376 metres.

SOUTHERN AFRICAN RECORD. 34/18/vd ; 1645 metres.

Family **ASTEROSTOMATIDAE** Pictet

See : Mortensen, 1950 : 33 (Asterostomidae) and 181 (Palaeopneustidae) ; also 1951 : 564 ; Durham *et al.*, 1966 : U614 (Asterostomatidae).

The only species included here – *Eurypatagus parvituberculatus* (H. L. Clark) – is of uncertain position. Clark included it in *Maretia* (family Spatangidae) but

Mortensen, while noting its affinity with *Maretia*, erected a new genus for it which he referred to the Palaeopneustidae because of the reduction of the subanal fasciole and the distally open petals. However, he admits that the Palaeopneustidae is not well defined and probably an unnatural group. In the *Treatise* it is lumped with *Asterostoma* and other genera in the Asterostomatidae with the comment that this group is polyphyletic. For the sake of consistency, the *Treatise* nomenclature is followed here but the diagnosis now given is mainly applicable to *Eurypatagus*.

A family of Atelostomata with the test reaching a moderate or large size, flattened ovate without a frontal depression ; paired ambulacra aborally petaloid but not sunken, the pores paired, petals prolonged and open distally, frontal ambulacrum not petaloid, the pore-pairs very reduced ; primary spines moderately long, mostly curved backwards ; apical system somewhat anterior, genital pores four ; periproct round or elongate oval, on the posterior face of the test ; peristome broad pentagonal with a more or less well-developed posterior lip (labrum), anterior of centre but not markedly so ; fascioles very variable in occurrence and development, a reduced subanal usually present in the species represented.

Eurypatagus parvituberculatus (H. L. Clark)

Lovenia elongata : Bell, 1904 : 173 ; ? H. L. Clark, 1923 : 404. [Non *Spatangus elongatus* Gray, 1845.]
Maretia parvituberculata H. L. Clark, 1924 : 13–15, pl. 4 ; 1925 : 227.
Eurypatagus parvituberculatus : Mortensen, 1950 : 260–262, figs 190b, 191, pl. 17, figs 1, 5–7, pl. 18, figs 9, 11, pl. 19, figs 2, 11, 19 ; 1951, pl. 12, fig. 7.

Length up to 75 mm, breadth 75–81%, height 25–29% of length.

A species of Asterostomatidae with the test low, markedly flattened above and below, rounded anteriorly, the posterior end tapering to a truncated tip ; petals with pores paired but small, posterior petals much longer than the paired anterior ones ; primary spines long and numerous, giving a shaggy appearance, the tubercles arranged in horizontal linear series laterally ; peristome distinctly anterior of centre, with a posterior labrum ; a subanal fasciole may be present in small specimens but is usually reduced or lacking in adults. Colour aborally rose-purple, paler below, primary spines banded.

Type locality. Off Durban, 229 metres.

Southern African records. 29/31/d ; 155–229 metres. Possibly the fragment from 27 fathoms (49 metres) off the Umhloti River mouth is conspecific with Bell's specimens (originally named *Lovenia elongata*) from 100 fathoms at 22·5 km further south-east of the same river.

Nature of bottom. Mud, sand, shells.

Family **SCHIZASTERIDAE** Lambert

See : Mortensen, 1951 : 204 ; Durham *et al.*, 1966 : U569.

A family of Atelostomata with the test reaching a moderate or large size, usually markedly heart-shaped with a deep frontal depression ; paired ambulacra aborally markedly petaloid and sunken, the pores paired as in the more or less deeply sunken frontal ambulacrum, posterior pair of petals much shorter than the anterior ones; primary spines fairly uniform, mostly curved backwards; apical system compact, usually posterior to the centre, genital pores four, three or two; periproct oval, transversely or longitudinally, on the truncated posterior face of the test, vertical or just visible from below; peristome semilunar with a well-developed posterior labrum, markedly anterior; fascioles including a conspicuous peripetalous one from which usually extends posteriorly on each side a latero-anal fasciole looping under the periproct.

Brisaster capensis (Studer)

Schizaster capensis Studer, 1880 : 884, pl. 2, fig. 4.
Schizaster fragilis (part) A. Agassiz, 1881 : 201–202; Bell, 1904 : 175. [Non *Brissus fragilis* Düben & Koren, 1846.]
Brisaster fragilis : H. L. Clark, 1923 : 399–400; 1924 : 12; 1925 (part) : 206–207.
Brisaster capensis : Mortensen, 1951 : 286–288, pl. 25, figs 4–10.
? *Brissaster fragilis* : Morgans, 1962 : 325.

Length up to *c.* 50 mm, breadth 90–96%, height 43–51% of length.

A species of Schizasteridae with the test fairly low, broadly rounded, almost flat above; anterior petals straight distally, though curved proximally; apical system slightly posterior of centre, genital pores three; peripetalous fasciole rather angular.

TYPE LOCALITY. South of the Cape of Good Hope, 34°S, 213 metres.

SOUTHERN AFRICAN RECORDS. 33/17/d; 34/18/d; 35/18/d; 122–400 metres. Bell's record of 33/26/29 fathoms (53 metres) cannot be confirmed since the sample (reg. no. 1903.8.1.129) is missing; only a broken specimen without precise locality (presumably no. 1903.8.1.91) is present. Morgans' 1962 record at only 16 metres in False Bay also needs confirmation.

NATURE OF BOTTOM. Green sand and specks; green mud; khaki and black sand, gravel; rock.

Schizaster lacunosus (Linnaeus)

Echinus lacunosus Linnaeus, 1758, *Systema Naturae*. Holmiae. Ed. 10 : 665.
Schizaster edwardsi : H. L. Clark, 1923 : 400 (implied identity by H. L. C., 1924). [Non *Schizaster edwardsi* Cotteau, 1889.]
Schizaster lacunosus : H. L. Clark, 1924 : 12; Mortensen, 1951 : 300–303, figs 140a, b, 141a, b, pl. 21, figs 5–10, 14–18, pl. 54, figs 1–3, 5, 7–9, 12, 14–17.

Length up to 82 mm, breadth 89–96%, height 68–76% of length.

A species of Schizasteridae with the test heart-shaped, high, especially posteriorly; anterior petals curved outwards distally as well as inwards proximally; apical system markedly posterior, genital pores two; periproct on the posterior face of the test but visible from below; peripetalous fasciole without sharp angles.

TYPE LOCALITY. 'Oceano Indico.'

SOUTHERN AFRICAN RECORDS. 34/21/s ; 34/20/s ; 34/22/s ; 34/23/s ; 29/31/s, d ; 42–200 metres. Also shallower if H. L. Clark's 1923 record with depth only 12–14 fathoms (22–25 metres) from off Tugela River, Natal, like one of his 1924 samples, should be included.

NATURE OF BOTTOM. Sand, shell ; mud.

Family **BRISSIDAE** Gray

See : Mortensen, 1951 : 353 ; Durham *et al.*, 1966 : U582.

A family of Atelostomata with the test reaching a moderate or large size, ovate, more or less high, with or without a frontal depression ; paired ambulacra aborally markedly petaloid but not deeply sunken, the pores paired, frontal ambulacrum not petaloid, sometimes sunken, its pores reduced, anterior and posterior petals not markedly dissimilar in length ; enlarged primary spines on conspicuous tubercles sometimes present within the peripetalous fasciole ; apical system compact, central or slightly anterior, genital pores four or three, sometimes two ; periproct longitudinally oval, on the truncated posterior face of the test, usually vertical ; peristome semilunar with a well-developed labrum, more or less markedly anterior ; both peripetalous and entirely separate subanal fascioles present, the latter sometimes with an anal branch each side of the periproct.

Brissopsis lyrifera capensis Mortensen

Brissopsis lyrifera var. *capensis* Mortensen, 1907, *Dan. Ingolf Exped.* **4**(2) : 156–157, pl. 18, figs 3, 23, pl. 19, figs 2, 9 ; 1951 : 387.

Brissopsis lyrifera : A. Agassiz, 1881 (part) : 189–190 (sts 141, 142) ; Bell, 1904 : 175 ; Döderlein, 1906 : 256–258, pl. 34, figs 4–8, pl. 49, figs 1, 2 ; H. L. Clark, 1923 : 401 ; 1924 : 12–13 ; 1925 (part) : 213–214 ; A. M. Clark, 1952 : 202. [Non *Brissus lyrifer* Forbes, 1841 ; nec *Brissopsis lyrifera* : Day, Field & Penrith, 1970.]

Brissopsis lyrifera capensis : Chesher, 1968, *Stud. trop. Oceanogr. Miami* **7** : 90–96, figs 8, 18, 19, pl. 21.

Length up to *c.* 90 mm, breadth 86–95%, height 56–71% of length. (Chesher's range of height is 62–71% of length but the Struys Point specimen measured by H. L. Clark (1925) and the lectotype (see below) both have height only 56% of length.)

A subspecies of Brissidae with the test variable in shape, more or less high, broad oval in outline, truncated posteriorly and with a shallow frontal notch, the frontal ambulacrum slightly sunken, anterior and posterior paired petals about equal in length, the anterior ones aligned at *c.* 45° to the longitudinal axis ; apical system slightly anterior of centre, genital pores four ; peripetalous fasciole concave laterally between the two pairs of paired petals, subanal fasciole bilobed, kidney-shaped.

TYPE LOCALITY. South from False Bay, 34°41'S 18°36'E, 180 metres ('Challenger' st. 141, from which a lectotype was selected, B.M. reg. no. 90.2.18.80, l/br/ht = 43/41/24 mm, unfortunately rather lower than most specimens).

SOUTHERN AFRICAN RECORDS. 30/16/d ; 31/16/d ; 32/16/d ; 32/17/d ; 33/17/d ; 33/18/d ; 34/17/d ; 34/18/s, d ; 35/18/d ; 34/19/s ; 35/19/d ; 34/20/s, d ; 36/21/d ; 35/23/vd ; 34/25/d ; 79–500 metres.

NATURE OF BOTTOM. Mud; green mud; black sandy mud; dark, muddy sand; green sand; khaki sand; green sand, black specks; sand and gravel; yellow sand, clay, rock; rock.

Spatagobrissus mirabilis H. L. Clark

Brissopsis lyrifera (part): A. Agassiz, 1881: 189–190 (Simon's Bay specimen); Day, Field & Penrith, 1970: 82. [Non *Brissus lyrifer* Forbes, 1841.]
Brissopsis sp. Mortensen, 1907, *Dan. Ingolf Exped.* **4** (2): 157–158, pl. 18, figs 9, 13.
Spatagobrissus mirabilis H. L. Clark, 1923: 402–404, pl. 23; Mortensen, 1951: 492–494, figs 258–260, pl. 28, figs 10–12, pl. 60, fig. 8; A. M. Clark, 1974: 480.

Length up to 112 mm; breadth 82–89%, height 47–55% of length.

A species of Brissidae with the test moderate in height, broad oval in outline, convex at both ends without frontal notch; anterior and posterior petals about equal in length, the anterior ones aligned at *c.* 75° to the long axis; apical system central or slightly anterior, genital pores four; peripetalous fasciole convex all round, similar in shape to the outline of the test, at least in larger specimens, subanal fasciole shield-shaped. Colour in life white.

TYPE LOCALITY. Near Hermanus, between False Bay and Cape Agulhas.

SOUTHERN AFRICAN RECORDS. 34/18/FB/s; 34/19/s; 26(? 16)–32 metres.

NATURE OF BOTTOM. Fine sand.

Family **LOVENIIDAE** Lambert

See: Mortensen, 1951: 83; Durham *et al.*, 1966: U609.

A family of Atelostomata with the test reaching a moderate or large size, usually markedly heart-shaped with a more or less deep frontal depression; paired ambulacra aborally petaloid, hardly sunken, if at all, the petals often broadest proximally but sometimes with the pore series more nearly parallel throughout, distally open, frontal ambulacrum not petaloid but usually more or less sunken, pores variable in development, anterior and posterior paired petals not markedly dissimilar in length; enlarged primary spines sometimes very long and with modified tubercles; apical system compact, nearly central or somewhat anterior, genital pores four, rarely three; periproct longitudinally oval, on the truncated posterior face of the test, usually vertical; peristome semilunar with a well-developed labrum, more or less markedly anterior; fascioles including an inner one encircling the frontal ambulacrum aborally and the apical area, also a subanal one and occasionally a peripetalous.

Genus ***ECHINOCARDIUM*** Gray

See: Mortensen, 1951: 149.

A genus of Loveniidae with the test fairly high, rather rectangular in side view being flattened above and below, broadest at about the middle of the length or only slightly anteriorly, truncated posteriorly, flattened anteriorly or with a more or less well-developed frontal depression ; aboral spines dense and fairly uniform in length, the primary spine tubercles not very conspicuous on the naked test ; spines well developed all over the sternum ; periproct on the almost flat posterior face of the test ; subanal fasciole shield-shaped but varying in relative breadth.

Echinocardium capense Mortensen

Echinocardium flavescens : A. Agassiz, 1881 : 175 ; Bell, 1904 : 174. [Non *Spatangus flavescens* O. F. Müller, 1776.]

Echinocardium capense Mortensen, 1907, *Dan. Ingolf Exped.* **4** (2) : 137–139, fig. 22, pl. 2, figs 5, 6, 11, pl. 16, fig. 12, pl. 17, figs 5, 6, 9, 13, 16, 35, 39 ; H. L. Clark, 1923 : 405 ; 1924 : 15 ; 1925 : 232 ; Mortensen, 1951 : 160–161.

Length up to 26 mm, breadth 84–100%, height 54–58% of length.

A species of *Echinocardium* with the test of moderate height, flattened anteriorly, the frontal ambulacrum hardly at all notched at the ambitus ; paired petals not broadened proximally or tapering, the two on the same side quite separate, their adjacent pore-series straight and at an obtuse angle to each other ; internal fasciole short and oval, not extending forwards more than half of the distance between the apical system and the anterior end.

TYPE LOCALITY. Cape Infanta area to East London area, 57–146 (? 238) metres. [Bell's specimens labelled as 'cotypes' have been amalgamated and the type locality cannot easily be narrowed down.]

SOUTHERN AFRICAN RECORDS. 33/17/d ; 34/25/s ; 33/25/s ; 33/28/d ; 55–310 metres.

NATURE OF BOTTOM. Green mud ; sand.

Echinocardium cordatum (Pennant)

Echinus cordatus Pennant, 1777, *British Zoology*. London. **4** : 58 (69), pl. 34, fig. 75.

Echinocardium australe : Bell, 1904 : 174.

Echinocardium cordatum : H. L. Clark, 1923 : 405 ; 1925 : 232–233 ; Mortensen, 1951 : 152–157, fig. 79, pl. 18, figs 1–3, 6–8 ; A. M. Clark, 1952 : 202 ; Scott, Harrison & Macnae, 1952 : 321 ; Day, Millard & Harrison, 1952 : 395, 396 ; Macnae, 1957 : 369, 371 ; Day, Field & Penrith, 1970 : 82. [Non *E. cordatum* : Macnae & Kalk, 1962, nec B. I. Balinsky *In* : Macnae & Kalk, 1969 which are *Lovenia elongata*.]

Length up to *c.* 80 mm in British waters but possibly not exceeding *c.* 40 mm in South Africa, breadth 85–100%, height 55–70% of length.

A species of *Echinocardium* with the test more or less high, especially posteriorly and deeply notched anteriorly ; paired petals broad basally and tapering, the sides more or less concave, especially the posterior side of the anterior petals and the anterior side of the posterior petals which together tend to form a continuous curve ; internal fasciole relatively long and extending more than half-way to the front end from the apical system, crossing the frontal ambulacrum in an almost straight line, not curved. Colour brownish.

TYPE LOCALITY. Unknown (Britain).

SOUTHERN AFRICAN RECORDS. 34/18/FB/s; 34/19/i; 34/22/s; 34/23/i; 33/25/s; 34/25/s; intertidal–58 metres.

NATURE OF BOTTOM. Sand; muddy sand; sand and shell; broken shell.

Lovenia elongata (Gray)

Spatangus elongatus Gray *In*: Eyre, 1845, *Journals of expeditions of discovery into central Australia and overland from Adelaide to King George's Sound.* London. Vol. 1, Appendix: 436, pl. 6, fig. 2.

Lovenia elongata: Mortensen, 1951: 97–104, figs 41a, 42, 48a, 49–51, 52a, pl. 7, figs 1–10, pl. 8, fig. 1, pl. 12, fig. 5, pl. 47, figs 10–23. [Non *Lovenia elongata*: A. Agassiz, 1881 (Simon's Bay specimen, which is *L. triforis* Koehler); nec Bell, 1904, which are *Eurypatagus parvituberculatus* (H. L. Clark), the same being also possibly true of H. L. Clark, 1923; nec Day, Field & Penrith, 1970.]

Echinocardium cordatum: Macnae & Kalk, 1962: 115; B. I. Balinsky *In*: Macnae & Kalk, 1969: 99, 106, 130. [Non *Echinus cordatus* Pennant, 1777.]

Length up to *c.* 85 mm, breadth 75–86%, height 33–35% of length.

A species of Loveniidae with the test relatively low, broadest proximal to the apical system, the outline heart-shaped with a more or less deep frontal notch; paired petals broad basally and markedly tapering, curving laterally at their distal ends, the adjacent sides of the two petals each side tending to form a continuous curve; anterolateral primary spines very long, sometimes reaching the posterior end of the test, their areoles markedly sunken around the tubercles, the spines on the sternum reduced anteriorly; periproct sunken, the posterior end of the test distinctly hollowed; inner fasciole narrow, subanal fasciole kidney-shaped, bilobed. Colour red or brown, primary spines banded violet.

TYPE LOCALITY. Western Australia.

SOUTHERN AFRICAN RECORD. 26/33/i. The specimen from Simon's Bay in 18–36 metres named *L. elongata* by Agassiz (1881), and renamed as probably *L. gregalis* Alcock by H. L. Clark, I have discovered by removing the spines from the apical system has only three genital pores and so runs down to *L. triforis* Koehler, a species otherwise known from the Bay of Bengal to Japan. There are two 'Challenger' labels with the specimen, both of which have 'Simon's Bay, 10–20 fms' (not Sagami Bay, which might have been a possibility), but I still regard the record with considerable suspicion since nothing similar appears to have been taken since in this very well-worked area. Also *Lovenia* is essentially a tropical genus.

NATURE OF BOTTOM. Sand.

Family **SPATANGIDAE** Gray

See: Mortensen, 1951: 1; Durham *et al.*, 1966: U605.

A family of Atelostomata with the test reaching a moderate or large size, often slightly asymmetrical, usually heart-shaped with a more or less deep frontal depression; paired ambulacra aborally petaloid, not markedly sunken, the petals broadest at about the middle of their length and usually closed distally or at least approximating, pores paired, frontal ambulacrum not petaloid, its pores small and in regular single series, anterior and posterior petals not markedly dissimilar in length ; primary spines of various lengths, sometimes with conspicuous enlarged tubercles ; apical system compact, subcentral, genital pores four, sometimes three ; periproct longitudinal oval, on the posterior face of the test ; peristome semilunar with a well-developed labrum, more or less markedly anterior in position ; only a subanal fasciole developed.

Spatangus capensis Döderlein

Spatangus raschi : A. Agassiz, 1881 : 171 ; Bell, 1904 : 173. [Non *S. raschi* Lovén, 1869.]
Spatangus capensis Döderlein, 1905, *Zool. Anz.* **28** : 624 ; 1906 : 261–263, pl. 33, fig. 1, pl. 48, fig. 4 ; H. L. Clark, 1923 : 404 ; 1924 : 13 ; Mortensen, 1951 : 16, pl. 1, figs 1–3 ; A. M. Clark, 1952 : 202 ; Morgans, 1962 : 317, 322 ; Day, Field & Penrith, 1970 : 82.

Length up to 125 mm, breadth 90–96%, height 45–57% of length.

A species of Spatangidae with the test moderate in height, domed, the frontal ambulacrum and groove not very deep ; paired petals approximately equal in length ; enlarged primary aboral tubercles forming horizontal series in the interambulacra ; periproct just visible from below ; subanal fasciole not markedly bilobed on its upper side. Colour purple.

Type locality. Simon's Bay, 70 metres. (Although Döderlein described the species from four syntypes, two at this locality and two from off Cape Town in 106 metres, it is one of the former which he illustrated, making it the best candidate for lectotype.)

Southern African records. 28/14/d ; 29/14/d ; 29/15/d ; 30/15/d ; 30/16/d ; 31/17/d ; 32/17/d ; 33/17/d ; 33/18/d ; 34/18/FB/s ; 34/19/s ; 34/20/s ; 34/21/s ; 34/22/d ; 35/23/vd ; 34/23/d ; 33/27/– ; 37–500 metres.

Nature of bottom. Dark grey mud and sand ; light green sand ; sand and gravel ; sand ; mud and sand ; green mud, black sand ; sand and rocks ; rock.

REFERENCES

To avoid unnecessary enlargement of this list and to restrict it as far as possible to works dealing with southern Africa or adjacent areas, details of some mainly irrelevant papers, especially those giving first descriptions of old species from outside the area and preliminary papers, are given in the text and omitted here.

References marked with an asterisk deal exclusively or predominantly with the fauna of southern Africa south of the Tropic of Capricorn. Others, such as the 'Challenger' reports, dealing with only isolated records from within the area, are annotated accordingly.

AGASSIZ, A. 1881. Echinoidea. *Rep. scient. Results Voy. Challenger*. Zool. **3** : 1–321, 45 pls. [*Echinus angulosus, Spatangus raschi, Lovenia elongata, Echinocardium flavescens, Brissopsis lyrifer, Schizaster fragilis*.]

*BALINSKY, B. I. 1969. The Echinoderms. *In* : Macnae, W. & Kalk, M. *A Natural History of Inhaca Island, Moçambique*. Revised ed. Johannesburg. pp. 96–107.

*BALINSKY, J. B. 1957. The Ophiuroidea of Inhaca Island. *Ann. Natal Mus.* **14** : 1–33, 7 figs, 4 pls, map.

BELL, F. J. 1884. Echinodermata. *In* : Coppinger, R. W. *Report on the zoological collections made in the Indo-Pacific Ocean during the voyage of H.M.S. 'Alert'*, 1881–2. London. pp. 117–177 ; 509–512, pls 8–17, 45. [Northern Mozambique.]

—— 1888. Descriptions of four new species of ophiuroids. *Proc. zool. Soc. Lond.* **1888** : 281–284, 1 pl. [*Pectinura capensis*.]

*—— 1904. The Echinoderma found off the coast of South Africa. 1. Echinoidea. *Mar. Invest. S. Afr.* **3** : 167–175.

*—— 1905a. 2. Asteroidea. *Mar. Invest. S. Afr.* **3** : 241–253.

*—— 1905b. 2. Ophiuroidea. *Mar. Invest. S. Afr.* **3** : 255–260, 1 pl.

*—— 1905c. 4. Crinoidea. *Mar. Invest. S. Afr.* **4** : 139–142, 3 pls.

*BERRISFORD, C. D. 1969. Biology and zoogeography of the Vema Seamount : a report on the first biological collection made on the summit. *Trans. R. Soc. S. Afr.* **38** : 387–398, 1 fig.

BIANCONI, J. 1866. Specimina zoologica Mosambicana. *Mem. Accad. Sci. Ist. Bologna* (2) **5** : 537–541, 1 pl. [No locality details.]

*BRIGHT, K. M. F. 1938. The South African intertidal zone and its relation to ocean currents. 2. An area of the southern part of the West coast. 3. An area of the northern part of the West coast. *Trans. R. Soc. S. Afr.* **26** : 49–65, 1 fig., 3 pls ; 67–88, 2 figs, 3 pls.

BROWN, R. N. R. 1910. Echinoidea from the Kerimba Archipelago, Portuguese East Africa. *Proc. R. Phys. Soc. Edinb.* **18** (1) : 36–44.

CARPENTER, P. H. 1888. Crinoidea. 2. Comatulae. *Rep. scient. Results Voy. Challenger*. Zool. **26** (60) : 1–399, 70 pls. [*Actinometra parvicirra*.]

CHERBONNIER, G. 1962. Ophiures. *Résult. scient. Expéd. océanogr. Belge Eaux Côt. Afr. Atlant. Sud* **3** (8) : 1–24, 7 pls.

CLARK, A. H. 1911. The recent crinoids of the coasts of Africa. *Proc. U.S. natn. Mus.* **40** : 1–51.

—— 1915. Die Crinoiden der Antarktis. *Dt. Südpol. Exped.* **16**. Zool. 8 : 101–209, 9 pls. [*Cominia occidentalis*.]

—— 1931. A monograph of the existing crinoids. 1 (3). Superfamily Comasterida. *Bull. U.S. natn. Mus.* **82** (1) 3 : 1–816, 82 pls.

—— 1937. Crinoidea. *Scient. Rep. John Murray Exped.* **4** (4) : 88–108, 1 pl.

—— 1941. A monograph of the existing crinoids. 1 (4a). Superfamily Mariametrida (except the family Colobometridae). *Bull. U.S. natn. Mus.* **82** (1) 4a : 1–603, 61 pls.

—— 1947. A monograph of the existing crinoids. 1 (4b). Superfamily Mariametrida (concluded) and Superfamily Tropiometrida (except the families Thalassometridae and Charitometridae). *Bull. U.S. natn. Mus.* **82** (1) 4b : 1–473, 43 pls.

—— 1950. A monograph of the existing crinoids. 1 (4c). Superfamily Tropiometrida (the families Thalassometridae and Charitometridae). *Bull. U.S. natn. Mus.* **82** (1) 4c : 1–383, 32 pls.

*CLARK, A. H. 1952. A new species of the crinoid genus *Cyclometra* from South Africa. *Trans. R. Soc. S. Afr.* **33** (2) : 189–192, 1 fig., 2 pls.

—— & CLARK, A. M. 1967. A monograph of the existing crinoids. 1 (5). Suborders Oligophreata (concluded) and Macrophreata. *Bull. U.S. natn. Mus.* **82** (1) 5 : 1–860, 53 figs.

CLARK, A. M. 1951. On some echinoderms in the British Museum (Natural History). *Ann. Mag. nat. Hist.* (12) **4** : 1256–1268, 4 figs, 1 pl. [*Decametra durbanensis.*]

*—— 1952. Some echinoderms from South Africa. *Trans. R. Soc. S. Afr.* **33** (2) : 193–221, 3 figs, 1 pl.

—— 1953. Notes on asteroids in the British Museum (Natural History). 3. *Luidia.* 4. *Tosia* and *Pentagonaster. Bull. Br. Mus. nat. Hist.* (Zool.) **1** : 379–412, 15 figs, 8 pls. [*Luidia africana, L. savignyi.*]

—— 1956. A note on some species of the family Asterinidae (class Asteroidea). *Ann. Mag. nat. Hist.* (12) **9** : 374–383, 4 figs, 2 pls. [*Patiria bellula* and *P. granifera* synonymized.]

—— 1970. Notes on the family Amphiuridae. *Bull. Br. Mus. nat. Hist.* (Zool.) **19** : 1–81, 11 figs.

—— 1972. Some crinoids from the Indian Ocean. *Bull. Br. Mus. nat. Hist.* (Zool.) **24** : 73–156, 17 figs. [*Comanthus parvicirrus, C. wahlbergi,* Colobometrid, *Pentametrocrinus varians, Democrinus chuni.*]

*—— 1974. Notes on some echinoderms from southern Africa. *Bull. Br. Mus. nat. Hist.* (Zool.) **26** (6) : 421–487, 16 figs, 3 pls, 1 map.

—— & ROWE, F. W. E. 1971. *Monograph of shallow-water Indo-West Pacific Echinoderms.* London. vii + 238 pp., 100 figs, 31 pls.

*CLARK, H. L. 1923. The Echinoderm fauna of South Africa. *Ann. S. Afr. Mus.* **13** : 221–435, 4 figs, 6 pls.

*—— 1924. Echinoderms from the South African Fisheries and Marine Biological Survey. 1. Sea-urchins (Echinoidea). *Rep. Fish. mar. biol. Surv. Un. S. Afr.* no. 4. Spec. Rep. no. 1 : 1–33, 4 pls.

—— 1925. *A catalogue of the recent sea-urchins (Echinoidea) in the collection of the British Museum (Natural History).* London. xxviii + 250 pp., 12 pls.

*—— 1926. Echinoderms from the South African Fisheries and Marine Biological Survey. 2. Sea-stars (Asteroidea). *Rep. Fish. mar. biol. Surv. Un. S. Afr.* no. 4. Spec. Rep. no. 7 : 1–34, 7 pls.

—— 1939. Ophiuroidea. *Scient. Rep. John Murray Exped.* **4** (2) : 29–136, 62 figs.

*CRAM, D. L. 1971. Life history studies on South African echinoids. 1. *Parechinus angulosus* (Leske) (Echinidae, Parechininae). 2. *Echinolampas (Palaeolampas) crassa* (Bell) (Echinolampadidae). *Trans. R. Soc. S. Afr.* **39** (3) : 321–337, 6 figs; 339–352, 8 figs.

*DAY, J. H. 1959. The biology of Langebaan lagoon. *Trans. R. Soc. S. Afr.* **35** : 475–547, 7 figs.

*—— 1969. *A guide to marine life on South African shores.* Cape Town. 300 pp., figs, 8 pls.

*—— 1974. The ecology of Morrumbene Estuary, Moçambique. *Trans. R. Soc. S. Afr.* **41** : 43–97, 7 figs.

—— MILLARD, N. A. H. & HARRISON, A. D. 1952. The ecology of South African estuaries. 3. Knysna : a clear open estuary. *Trans. R. Soc. S. Afr.* **33** : 367–413, 4 figs, 1 pl.

*—— & MORGANS, J. F. C. 1956. The ecology of South African estuaries. 7. The biology of Durban Bay. *Ann. Natal Mus.* **13** : 259–312, 1 pl., map.

—— FIELD, J. G. & PENRITH, M. J. 1970. The benthic fauna and fishes of False Bay, South Africa. *Trans. R. Soc. S. Afr.* **39** (1) : 1–108, 1 fig.

DÖDERLEIN, L. 1906. Die Echinoiden der deutschen Tiefsee-Expedition. *Wiss. Ergebn. dt. Tiefsee-Exped. 'Valdivia'* **5** (2) : 61–290, 42 pls. [14 South African spp., summary on pp. 271–272.]

*—— 1908. *Asterina luderitziana,* eine neue Art aus Südwest-Afrika. *Jb. nassau Ver. Naturk.* **61** : 296–298, 1 pl.

*DÖDERLEIN, L. 1910. Asteroidea, Ophiuroidea, Echinoidea. *In*: Schultze, L. Forschungsreise im westlichen und zentralen Südafrika. 4 (1). *Denkschr. med-naturw. Ges. Jena* **16** : 245–258, 2 pls.

—— 1926. Über Asteriden aus dem Museum von Stockholm. *K. svensk. VetenskAkad. Handl.* (3) **2** (6) : 1–22, 4 pls. [*Astropecten antares.*]

—— 1928. Seesterne. *Dt. Südpol. Exped.* **19**. Zool. 11 : 291–301, 4 pls. [*Marthasterias glacialis africana, Cribrella ornata, Callopatiria bellula.*]

DURHAM, J. W. ET AL. 1966. Echinozoans. *In*: Moore, R. C. (Ed.). *Treatise on Invertebrate Paleontology*. Kansas. Part U. Echinodermata **3** : U108–U672, 445 figs.

*ENGEL, H. 1949. *Ophioteresis beauforti* nov. spec. *Bijdr. Dierk.* **28** : 140–143, 2 figs.

*EYRE, J. 1939. The South African intertidal zone and its relation to ocean currents. 7. An area in False Bay. *Ann. Natal Mus.* **9** : 283–306, 4 figs, 5 pls.

*——BROEKHUYSEN, G. J. & CRICHTON, M. I. 1938. The South African intertidal zone and its relation to ocean currents. 6. The East London district. *Ann. Natal Mus.* **9** : 83–111, 2 figs, 3 pls.

*—— & STEPHENSON, T. A. 1938. The South African intertidal zone and its relation to ocean currents. 5. A sub-tropical Indian Ocean shore. *Ann. Natal Mus.* **9** : 21–46, 3 figs, 3 pls.

FISHER, W. K. 1911. Asteroidea of the North Pacific. 1. Phanerozonia and Spinulosa. *Bull. U.S. natn. Mus.* **76** (1) : 1–419, 122 pls.

—— 1919. Starfishes of the Philippine Seas and adjacent waters. *Bull U.S. natn. Mus.* **100** (3) : 1–546, 156 pls.

—— 1928. Asteroidea of the North Pacific. 2. Forcipulata (part). *Bull. U.S. natn. Mus.* **76** (2) : 1–245, 81 pls.

—— 1930. Asteroidea of the North Pacific. 3. Forcipulata (concluded). *Bull. U.S. natn. Mus.* **76** (3) : 1–356, 93 pls.

—— 1940. Asteroidea. *Discovery Rep.* **20** : 69–306, 23 pls. [*Astropecten irregularis pontoporeus, Patiria bellula, P. granifera, Patiriella exigua, Henricia ornata, Marthasterias glacialis* forma *varispina.*]

*GISLÉN, T. 1938. Crinoids of South Africa. *K. svensk. VetenskAkad. Handl.* (3) **17** (2) : 1–22, 26 figs, 2 pls.

*GRINDLEY, J. R. 1963. A specimen of the asteroid *Acanthaster planci* from the Mozambique coast. *Durban. Mus. Novit.* **6** (20) : 265–268, 1 fig.

*—— & KENSLEY, B. F. 1966. Benthonic marine fauna obtained off the Orange River mouth by the diamond dredger Emerson-K. *Cimbebasia* no. 16 : 1–14.

HERTZ, M. 1927a. Ophiuroiden. *Dt. Südpol. Exped.* **19**. Zool. 11 (1) : 1–56, 9 pls. [*Ophiactis africana capensis, Amphiura incana, Amphipholis minor, Ophiothrix aristulata.*]

—— 1927b. Ophiuroiden. *Wiss. Ergebn. dt. Tiefsee-Exped. 'Valdivia'* **22** (3) : 59–122, 7 figs, 4 pls. [*Ophiura carnea, Gymnophiura novembris, Ophiomisidium pulchellum, Astrophiura permira, Ophiuroglypha capensis, Ophiochasma nitida.*]

*JANGOUX, M. 1973. Les Astéries de l'Ile d'Inhaca (Mozambique) (Echinodermata, Asteroidea). *Ann. Mus. R. Afr. cent.* **8°**. Zool. no. 208 : 1–50, 13 figs, 7 pls.

*JUNOD, H. A. 1899. [Footnote.] *Bull. Soc. vaud. Sci. nat.* **35** : 281.

*KALK, M. 1958. Ecological studies on the shores of Moçambique. 1. The fauna of intertidal rocks at Inhaca Island, Delagoa Bay. *Ann. Natal Mus.* **14** : 189–242, 8 figs, 2 pls.

—— 1959. A general ecological survey of some shores in northern Moçambique. *Rev. Biol. Lisb.* **2** : 1–24, 4 pls.

KOEHLER, R. 1908. Astéries, Ophiures et Echinides de l'Expédition antarctique nationale Ecossaise. *Trans. R. Soc. Edinb.* **46** (3) 22 : 529–649, 16 pls. [*Cribrella ornata, Patiria bellula, Asterina calcarata, Amphiura capensis, Ophiothrix fragilis* forma *pentaphyllum, Ophiothrix triglochis, Paracentrotus angulosus.*]

—— 1914. Asteroidea, Ophiuroidea et Echinoidea. *In* : Michaelsen, W. *Beitr. Kennt. Meeresfauna Westafr.* **1** (2) : 129–303, 12 pls. [*Patiria bellula, Asterina exigua, Cribrella ornata, Amphiura capensis, Amphipholis squamata, Ophiothrix fragilis, Protocentrotus angulosus.*]

*Koehler, R. 1915. Description d'une nouvelle espèce d'*Astrophiura*, l'*Astrophiura cavellae*. *Bull. Inst. océanogr. Monaco* no. 311 : 1–16.

—— 1922. Ophiurans of the Philippine Seas and adjacent waters. *Bull. U.S. natn. Mus.* **100** (3) : 1–486, 103 pls.

—— 1923. Sur quelques ophiures des côtes de l'Angola et du Cap. *Göteborgs K. Vetensk.-o. vitterhSamh. Handl.* (4) **25** (5) : 1–11, 1 pl.

Lamarck, J. B. P. A. de. 1816. *Histoire naturelle des animaux sans vertèbres*. Ed. 1. Paris. **2** : 522–568 (Stellerides) ; **3** : 1–59 (Echinides) ; 60–76 (Fistulides).

*Lopes, A. P. 1939. Fauna de Moçambique : Inhaca. *Document. trim. Moçambique* **20** : 71–88, figs.

Lyman, T. 1882. Ophiuroidea. *Rep. scient. Results Voy. Challenger*. Zool. **5** : 1–386, 46 pls. [*Amphiura capensis*, *A. dilatata*, *A. squamata*, *A. incana*, *Ophioglypha costata*, *O. irrorata*, *Ophiocoma scolopendrina*, *Ophiactis carnea*, *O. flexuosa*, *Ophiothrix aristulata*. *O. triglochis*, *Gorgonocephalus verrucosus*, *Ophioscolex dentatus*, *Ophiopeza aster*, *Ophiomusium pulchellum*, *Ophiothamnus remotus*, *Ophiomyxa vivipara*.]

*Macnae, W. 1957. The ecology of the plants and animals in the intertidal regions of the Zwartkops estuary near Port Elizabeth, South Africa. 2. *J. Ecol.* **45** : 361–387.

*—— & Kalk, M. 1962. The fauna and flora of sand flats at Inhaca Island, Moçambique. *J. Anim. Ecol.* **31** : 93–128, 5 figs.

*—— —— 1969. *A natural history of Inhaca Island, Moçambique*. Revised ed. Johannesburg. iii + 163 pp., 11 pls, maps, figs.

Madsen, F. J. 1971. West African ophiuroids. *Atlantide Rep.* no. 11 [1970] : 151–243, 49 figs.

Matsumoto, H. 1917. A monograph of Japanese Ophiuroidea. *J. Coll. Sci. Tokyo* **38** (2) : 1–408, 100 figs, 7 pls.

Meissner, M. 1892. Asteriden gesammelt von Herrn Stabsarzt Dr. Sander. *Arch. Naturgesch.* **58** (1) : 183–190, 1 fig. [*Linckia multiforis*, *Asterina penicillaris*.]

—— 1893. Sanders Seeigel. *Sb. Ges. naturf. Berlin* **1892** : 183–185. [*Echinus angulosus*.]

Michelin, M. H. 1862. Echinides et Stellérides. *In* : Maillard, L. *Notes sur l'Ile de la Réunion*. Annexe A. Paris. 7 pp., 3 pls.

*Morgans, J. F. C. 1959. The benthic ecology of False Bay. 1. The biology of infratidal rocks observed by diving, related to that of intertidal rocks. *Trans. R. Soc. S. Afr.* **35** : 387–442, 11 figs, 4 pls.

*—— 1962. The benthic ecology of False Bay. 2. Soft and rocky bottoms observed by diving and sampled by dredging, and the recognition of grounds. *Trans. R. Soc. S. Afr.* **36** : 288–334, 5 figs.

Mortensen, T. 1909. Echinoiden. *Dtsch. Südpol. Exped.* **11** : 1–114, 19 pls. [*Protocentrotus angulosus*, *P. annulatus*.]

*—— 1925. On some Echinoderms from South Africa. *Ann. Mag. nat. Hist.* (9) **16** : 146–154, 5 figs, 1 pl.

—— 1927. *Handbook of the Echinoderms of the British Isles*. London. ix + 471 pp., 269 figs.

—— 1928. *A monograph of the Echinoidea*. 1. Cidaroidea. Copenhagen. 551 pp., 173 figs, 88 pls.

—— 1932. New contributions to the knowledge of the cidarids. *K. dansk. Vidensk. Selsk. Skr.* (9) **4** (4) : 145–182, 18 figs, 13 pls.

*—— 1933a. Echinoderms of South Africa (Asteroidea and Ophiuroidea). *Vidensk. Meddr dansk naturh. Foren.* **93** : 215–400, 91 figs, 12 pls.

*—— 1933b. A new giant sea-star, *Mithrodia gigas* n. sp. from South Africa. *Ann. S. Afr. Mus.* **32** (1) : 1–4, 1 fig., 1 pl.

—— 1935. *A monograph of the Echinoidea*. 2. Bothriocidaroida, Melonechinoida, Lepidocentroida and Stirodonta. Copenhagen. 647 pp., 377 figs, 89 pls.

—— 1936. Echinoidea and Ophiuroidea. *'Discovery' Rep.* **12** : 199–348, 53 figs, 9 pls. [*Parechinus angulosus*, *Ophiothrix triglochis*, *O. fragilis*, *Amphiura incana*, *Ophiuroglypha tumida*, *Ophiocten amitinum* var. *simulans*, *Dictenophiura anoidea*.]

MORTENSEN, T. 1939a. Report on the Echinoidea of the Murray Expedition. 1. *Scient. Rep. John Murray Exped.* **6** (1) : 1–28, 10 figs, 6 pls.

—— 1939b. New Echinoidea (Aulodonta). *Vidensk. Meddr dansk naturh. Foren.* **103**: 547–550.

—— 1940. *A monograph of the Echinoidea.* 3 (1). Aulodonta. Copenhagen. 370 pp., 196 figs, 77 pls.

—— 1943a. *A monograph of the Echinoidea.* 3 (2). Camarodonta 1. Copenhagen. 553 pp., 321 figs, 56 pls.

—— 1943b. *A monograph of the Echinoidea.* 3 (3). Camarodonta 2. Copenhagen. 446 pp., 215 figs, 66 pls.

—— 1948a. *A monograph of the Echinoidea.* 4 (1). Holectypoida, Cassiduloida. Copenhagen. 371 pp., 326 figs, 14 pls.

—— 1948b. *A monograph of the Echinoidea.* 4 (2). Clypeastroida. Copenhagen. 471 pp., 258 figs, 72 pls.

—— 1948c. Report on the Echinoidea of the Murray Expedition. 2. *Scient. Rep. John Murray Exped.* **9** (1) : 1–15, 1 pl.

—— 1950. *A monograph of the Echinoidea.* 5 (1). Spatangoida 1. Copenhagen. 432 pp., 315 figs, 25 pls.

—— 1951. *A monograph of the Echinoidea.* 5 (2). Spatangoida 2. Copenhagen. 593 pp., 286 figs, 64 pls.

MÜLLER, J. & TROSCHEL, F. H. 1842. *System der Asteriden.* Braunschweig. xx+134 pp., 12 pls.

*PENRITH, M. L. & KENSLEY, B. F. 1970. The constitution of the intertidal fauna of rocky shores of South West Africa. 1. Lüderitzbucht. *Cimbebasia* **A1** : 192–239, 4 figs, 6 pls, 1 map.

PETERS, W. 1851. Übersicht der von ihm an der Kuste von Mossambique eingesammelten Ophiuren, unter denen sich zwei neue Gattungen befinden. *Mber. preuss. Akad. wiss. Berlin* **1851** : 463–466.

—— 1852. Übersicht der Seesterne (Asteridae) von Mossambique. *Mber. preuss. Akad. wiss. Berlin* **1852** : 177–178. [*Astropecten hemprichi, Luidia maculata.*]

—— 1853. Abhandlung über die an der Kuste von Moçambique beobachteten Seeigel und insbesondere über die Gruppe der Diademen, von welcher hier ein Auszug folgt. *Ber. Akad. wiss. Berlin* **1853** : 484–488.

*SCOTT, K. M. F., HARRISON, A. D. & MACNAE, W. 1952. The ecology of South African estuaries. 2. The Klein River estuary, Hermanus, Cape. *Trans. R. Soc. S. Afr.* **33** : 283–331, 2 figs, 2 pls.

SIMPSON, J. C. & BROWN, R. N. R. 1910. Asteroidea of Portuguese East Africa collected by Jas. C. Simpson. *Proc. R. Phys. Soc. Edinb.* **18** : 45–60.

SLADEN, W. P. 1889. Asteroidea. *Rep. scient. Results Voy. Challenger.* Zool. **30** : 1–935, 118 pls. [*Luidia africana, Astropecten pontoporæus, Psilaster acuminatus, Parachaster pedicifer, Pseudarchaster tessellatus, Calliaster baccatus, Patiria bellula, Asterina exigua, Cribrella ornata, Stichaster felipes, Asterias* (*Stolasterias*) *africana.*]

SPENCER, W. K. & WRIGHT, C. W. 1966. Asterozoans. *In* : Moore, R. C. (Ed.). *Treatise on Invertebrate Paleontology.* Kansas. Part U. Echinodermata **3** : U4–107, 89 figs.

*STEPHENSON, T. A. 1944. The constitution of the intertidal fauna and flora of South Africa. *Ann. Natal Mus.* **10** : 261–358, 13 figs, 3 pls.

—— STEPHENSON, A. & BRIGHT, K. M. F. 1938. The South African intertidal zone and its relation to ocean currents. 4. The Port Elizabeth district. *Ann. Natal Mus.* **9** : 1–19, 1 fig., 4 pls.

—— —— & DAY, J. H. 1939. The South African intertidal zone and its relation to ocean currents. 8. Lambert's Bay and the West coast. *Ann. Natal Mus.* **9** : 345–380, 7 figs, 7 pls.

—— —— & DU TOIT, C. A. 1937. The South African intertidal zone and its relation to ocean currents. 1. A temperate Indian Ocean shore. *Trans. R. Soc. S. Afr.* **24** : 341–382, 8 figs, 4 pls.

STUDER, T. 1880. Uebersicht über die während der Reise S.M.C. Corvette 'Gazelle' um die Erde 1874–76 gessammelten Echinoiden. *Mber. K. preuss. Akad. Wiss.* **1880** : 861–885, 2 pls.

—— 1884. Verzeichniss der während der Reise S.M.S. 'Gazelle' um die Erde 1874–76 gessammelten Asteriden und Euryaliden. *Phys. Abh. K. Akad. Wiss. Berl.* **1883, 1884** (3) : 1–64, 5 pls.

THOMAS, L. P. 1975. The ophiacanthid genus *Amphilimna* (Ophiuroidea, Echinodermata). *Proc. biol. Soc. Wash.* **88** : 127-140, 1 fig.

THOMSON, J. S. 1913. Observations on the coloration of *Echinus angulosus* A. Agass. *Ann. Mag. nat. Hist.* (8) **12** : 190–199.

At a late stage in this work, too late for inclusion in the preliminary paper (Clark, 1974), type material of two little-known but relevant species of Ophiodermatidae was obtained on loan. This allowed assessment of their validity and affinities, as well as illustration to show certain details, which falls outside the scope of the main part of the text.

We are indebted to Dr D. Kuhlmann of the Museum für Naturkunde, Berlin and Dr H. Fechter of the Zoologische Sammlung des Bayerischen Staates, Münich, for the loan of the three syntypes of the first species, *Ophiochasma nitida* Hertz, also to Dr P. A. Andersson of the Naturhistoriska Riksmuseet, Stockholm, of a specimen which can be presumed to be the holotype of the second, *Ophioderma wahlbergi*.

Ophiochasma nitida Hertz

Fig. 276a, b

See p. 183.

Affinities. Hertz compared this species with *Ophiochasma stellatum* Ljungman and *Ophiarachnella nitens* (Koehler). In 1930 (*Vidensk. Meddr dansk naturh. Foren.* **89**) Koehler himself synonymized *nitens* with *O. stellatum*, which is easily distinguished from *O. nitida* by the smaller number of arm spines (five or six) and the even larger radial shields, as noted by Hertz, but the absence of pores between the first and second ventral arm plates I think may be correlated with the relatively small size of the type specimens. [The shape of the dorsal arm plates provides another distinction, *O. nitida* having the lateral angle about midway in the length, whereas in *O. stellatum* the angles are markedly prolonged, very acute and distinctly distal in position.]

In 1971 (Clark & Rowe : 124) I expressed doubt about the generic distinction between *Ophiochasma* and *Ophiarachnella*, which is based only on the relatively larger size of the radial shields in the former, a character very variable in species of *Ophiarachnella* such as *O. gorgonia* (Müller & Troschel). Indeed, I find some similarities between certain specimens in the British Museum (Natural History) collections named *Ophiarachnella marmorata* (Lyman) – a provisional synonym of *O. gorgonia* – though the granulated supplementary oral shields should provide a distinction for *O. nitida*. Both have about 10 arm spines and the size of the radial shields may be relatively similar. In 1909 (*Bull. Mus. comp. Zool. Harv.* **52**) H. L. Clark recorded *Ophiarachnella gorgonia* from many Indo-West Pacific localities including Natal and Mozambique but in 1923 he omitted the species from the South African fauna. The morphological limits of *O. gorgonia* also badly need further investigation, which should involve *Ophiochasma nitida*. The figures given here, together with Hertz's photographs, should facilitate comparison, while further South African material should reveal the extent of variation.

In comparison with *Ophiarachnella capensis*, apart from the much larger radial shields, narrower shape of the oral shields and probable granulation of the supplementary oral shields, *Ophiochasma nitida* differs in having the disc plates smooth

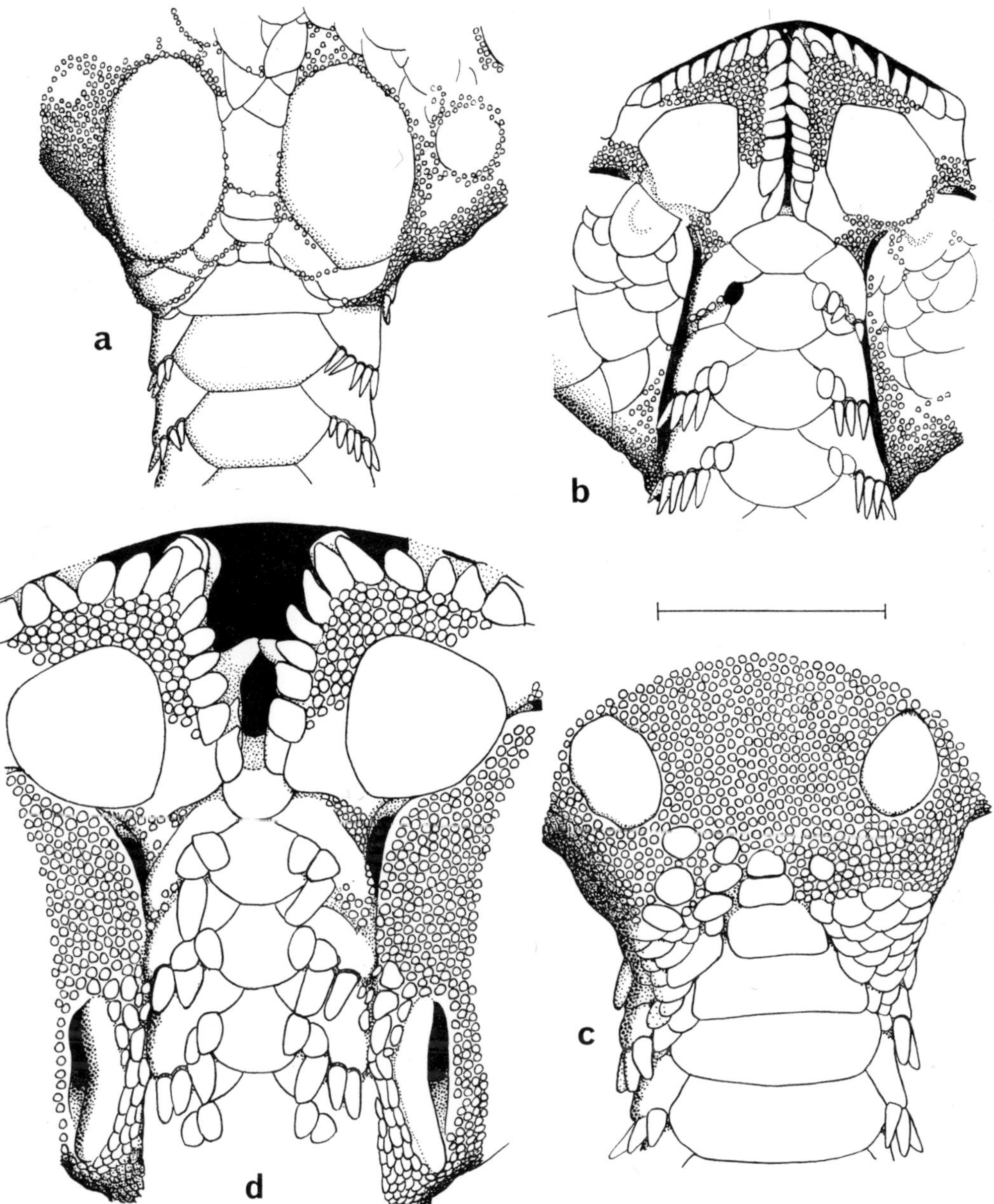

Fig. 276. a, b. Syntype of *Ophiochasma nitida* Hertz. Münich Museum no. 1623/1936, 'Valdivia' st. 105, d.d. 10·5 mm. a, dorsal and b. ventral partial views. c, d. Similar views of the presumptive holotype (or lectotype) of *Ophioderma wahlbergi* Müller & Troschel. Stockholm Museum no. 705, Port Natal (?), Wahlberg, d.d. 27 mm. [The scale represents 3 mm for a and b or 2 mm for c and d.]

under the granules, whereas when the granules of *Ophiarachnella capensis* are removed they leave behind a perforation in the surface of the plate.

Ophioderma wahlbergi Müller & Troschel

Fig. 276c, d

See p. 183.

SYNONYMY. This species was described by Müller & Troschel from material in the Stockholm Museum collected at Port Natal by Wahlberg. The only Wahlberg specimen in that museum has the original label completely faded and its labelling as *Ophioderma wahlbergi* dates from examination by Koehler. It agrees well in brown colour and size with the original description, having d.d. 27 mm and arm length 76–86 mm, giving a mean a.l./d.d. of 3·0/1 and a total diameter of 6–8 inches (Zoll). There are some bare disc scales, though nearly all are peripheral near the arm bases, most of the dorsal arm plates are undivided (e.g. all but seven of the 69 on one arm) and there are seven short blunt arm spines on most proximal segments, though a few have eight. I think that the presumption that this specimen can be regarded as the holotype is strong, especially as Wahlberg's sampling of marine species was normally sparse, so that only two out of his 10 ophiuroid samples include more than one specimen. However, as there must remain a small element of doubt, Dr Andersson thinks this specimen should be designated as lectotype of *Ophioderma wahlbergi*.

Although none of the four South African specimens (locality unfortunately unknown) named as *O. wahlbergi* by Bell have bare patches on the disc apart from the radial shields and their colour is black above rather than brown I believe they were correctly named. They show considerable variation in the extent of granulation covering the radial shields, the largest (d.d. 32 mm) having all the shields with a bare patch 1·8–3·5 mm long, whereas the smallest (d.d. 23 mm) has seven of the shields completely hidden and only very small patches of the other three exposed. H. L. Clark noted similar variation in his South African Ophiodermas, though it was his smallest specimen (d.d. 17 mm) which had a bare patch on all the shields. Döderlein's 10 (syn)types of *Ophioderma leonis* from Luderitz Bay (of which one at least has d.d. as much as 21 mm) evidently all have fully granule-covered discs, while Mortensen's two from Gordon Bay in False Bay, with d.d. 20 and 24 mm, have only a single shield between them partially exposed. The two Survey specimens (d.d. 11 mm and 15 mm, the former four-rayed) both have completely granule-covered discs. Although Döderlein describes the disc granulation of *O. leonis* as 'fein', his fig. 1a shows what I would call moderately coarse granulation, as in the Wahlberg and other specimens and Döderlein's tally of eight oral papillae each side of the jaw would include the second oral tentacle scale and so does not provide a distinction. Without other characters to support the deficient disc granulation in the lectotype of *O. wahlbergi* and the likely instability of this character itself, I consider that *O. leonis* should be regarded as a synonym.

As for the 'Challenger' *Ophioderma*, supposedly from Simon's Bay which Lyman named '? *O. tongana* Lütken', Mortensen was misled into thinking this conspecific with *O. leonis*. It has d.d./a.l. 13/65 = 1/5 and the arms conspicuously banded, ventrally as well as dorsally, while the disc has groups of fine dark dots (formed by single granules) on a light background. On this account I have renamed it *Ophioderma appressum* (Say), known from the western tropical Atlantic and think that its locality was more likely to have been Bahia, Brazil, where that species was collected by the 'Challenger'. Although Koehler (1914) reported *O. appressum* from Angola, Madsen (1951) doubts the correctness of this. Mortensen's figure of a jaw of the 'Challenger' specimen (drawn by C. C. A. Munro) shows eight oral papillae each side, as well as a second oral tentacle scale, rather than the usual five or six of *O. wahlbergi*, agreeing again with *O. appressum*.

INDEX

Names of families and subfamilies are printed in SMALL CAPITALS, new names in **bold** type, valid names in roman type and synonyms or combinations not valid in the present context in *italic* type. Page numbers for the principal references are given in **bold** type, for figures in *italic* type and the rest in roman.